CW01020992

Springer Finance

Springer
Berlin
Heidelberg
New York
Hong Kong
London
Milan
Paris
Tokyo

Springer Finance

Springer Finance is a new programme of books aimed at students, academics and practitioners working on increasingly technical approaches to the analysis of financial markets. It aims to cover a variety of topics, not only mathematical finance but foreign exchanges, term structure, risk management, portfolio theory, equity derivatives, and financial economics.

Credit Risk Valuation:
Risk-Neutral Valuation:
Pricing and Hedging of Finance Derivatives
Bingham, N. H. and Kiesel, R.
ISBN 1-85233-001-5 (1998)

Visual Explorations in Finance
with Self-Organizing Maps
Deboeck, G. and Kohonen, T. (Editors)
ISBN 3-540-76266-3 (1998)

Mathematical Models of Financial Derivatives
Kwok, Y.-K.
ISBN 3-981-3083-25-5 (1998)

Mathematics of Financial Markets
Elliott, R. J. and Kopp, P. E.
ISBN 0-387-98533-0 (1999)

Efficient Methods
for Valuing Interest Rate Derivatives
A. Pelsser
ISBN 1-85233-304-9 (2000)

Methods, Models and Applications
Ammann, M
ISBN 3-540-67805-0 (2001)

Credit Risk: Modelling, Valuation and Hedging
Bielecki, T. R. and Rutkowski, M.
ISBN 3-540-67593-0 (2001)

Mathematical Finance – Bachelier Congress 2000
– Selected Papers from the First World Congress
of the Bachelier Finance Society,
held in Paris, June 29–July 1, 2000
Geman, H., Madan, D. S., Pliska R. and Vorst, T.
(Editors)
ISBN 3-540-67781-X (2001)

Exponential Functionals of Brownian Motion
and Related Processes
M. Yor
ISBN 3-540-65943-9 (2001)

Financial Markets Theory:
Equilibrium, Efficiency and Information
Barucci, E
ISBN 3-85233-469-X (2003)

Financial Markets in Continuous Time
Dana, R.-A. and Jeanblanc, M.
ISBN 3-540-41722-9 (2003)

Weak, Convergence of Financial Markets
Prigent, J.-L.
ISBN 3-540-4233-8 (2003)

Incomplete Information and Heterogenous Beliefs
in Continuous-time Finance
Ziegler, A
ISBN 3-540-00344-4 (2003)

Stochastic Calculus Models for Finance:
Volume 1: The Binominal Assett Pricing Model
Shreve, S. E.
ISBN 3-540-40101-6 (2004)

Irrational Exuberance Reconsidered:
The Cross Section of Stock Returns
Külpmann, M
ISBN 3-540-14007-7 (2004)

Credit Risk Pricing Models: Theory and Practice
Schmid, B.
ISBN 3-540-40466-X (2004)

Bernd Schmid

Credit Risk Pricing Models

Theory and Practice

Second Edition
with 101 Figures
and 65 Tables

Dr. Bernd Schmid
Director
risklab germany GmbH
Nymphenburger Strasse 112–116
80636 Munich, Germany
risklab@gmx.de

Mathematics Subject Classification (2000):
35Q80, 60G15, 60G35, 60G44, 60H05, 60J10, 60J27, 60J35, 60J60, 60J65, 60J75, 62P05, 91B28, 91B30, 91B70, 91B84

Originally pulished with the title "Pricing Credit Linked Financial Instruments"
as volume 516 in the series:
Lecture Notes in Economics and Mathematical Systems,
ISBN 3-540-43195-0

ISBN 3-540-40466-X Springer-Verlag Berlin Heidelberg New York

Bibliographic information published by Die Deutsche Bibliothek
Die Deutsche Bibliothek lists this publication in the Deutsche Nationalbibliografie;
detailed bibliographic data available in the internet at *http.//dnb.ddb.de*

Springer-Verlag is a part of Springer Science+Business Media
springeronline.com

Cover design: design & production, Heidelberg

SPIN 10940907 42/3130 – 5 4 3 2 1 0 – Printed on acid-free paper

"Es mag sein,

dass wir durch das Wissen

anderer gelehrter werden.

Weiser werden wir nur durch uns selbst."

- Michel Eyquem de Montaigne -

Preface

This new edition is a greatly extended and updated version of my earlier monograph "Pricing Credit Linked Financial Instruments" (Schmid 2002). Whereas the first edition concentrated on the research which I had done in the context of my PhD thesis, this second edition covers all important credit risk models and gives a general overview of the subject. I put a lot of effort in explaining credit risk factors and show the latest results in default probability and recovery rate modeling. There is a special emphasis on correlation issues as well. The broad range of financial instruments I consider covers not only defaultable bonds, defaultable swaps and single counterparty credit derivatives but is further extended by multi counterparty instruments like index swaps, basket default swaps and collateralized debt obligations.

I am grateful to Springer-Verlag for the great support in the realization of this project and want to thank the readers of the first edition for their overwhelming feedback.

Last but not least I want to thank Uli Göser for ongoing patience, encouragement, and support, my family and especially my sister Wendy for being there at all times.

Stuttgart, November 2003 *Bernd Schmid*

Contents

1. Introduction

"Jede Wirtschaft beruht auf dem Kreditsystem, das heißt auf der irrtümlichen Annahme, der andere werde gepumptes Geld zurückzahlen."
- Kurt Tucholsky -

"Securities yielding high interest are like thin twigs, very weak from a capital-safety point of view if taken singly, but most surprisingly strong if taken as a bundle, and tied together with the largest possible number of differing external influences."
- British Investment Registry & Stock Exchange, 1904 -

1.1 Motivation

Although lending money is one of the oldest banking activities at all, credit evaluation and pricing is still not fully understood. There are many difficulties in assessing the impact of credit risk on prices in the bond and loan market. Two key problems are the data limitations and the model validation.

In general, credit risk is the risk of reductions in market value due to changes in the credit quality of a debtor such as an issuer of a corporate bond. It can be measured as the component of a debt instrument's yield that reflects the expected value of the risk of a possible default or downgrade. This so called credit risk premium is usually expressed in basis points. More precisely, according to the Dictionary of Financial Risk Management (Gastineau 1996), credit risk is

1. Exposure to loss as a result of default on a swap debt, or other counterparty instrument.
2. Exposure to loss as a result of a decline in market value stemming from a credit downgrade of an issuer or counterparty. Such credit risk may be reduced by credit screening before a transaction is effected or by instrument provisions that attempt to offset the effect of a default or require increased payments in the event of a credit downgrade.

3. A component of return variability resulting from the possibility of an event of default.
4. A change in the market's perception of the probability of an event of default, which affected the spread between two rates or reference indexes.

Although credit risk and default risk are quite often used interchangeably, in a more rigorous view, default risk is understood to be the risk that a debtor will be unable or unwilling to make timely payments of interest or principal. These definitions force several questions on us:

- What does it make important to consider these types of risk ?
- Why have credit risk modelling and credit risk management issues received renewed attention only recently ?

The last few years have seen dramatic developments in the credit markets with the declining of the traditional loan markets and the development of new markets. Corporate defaults have increased tremendously but haven't stopped investors from investing in risky sectors such as high-yield markets. In addition, banks have come up with new products to manage credit risks such as credit derivatives and asset backed securities. At the same time regulators have started changing their view on the credit markets and discussing their capital rules. More than a little this discussion has been driven by academics and practitioners who both have been developing new models for credit risk measurement and management that satisfy regulatory rules on the one hand and the needs for internal credit risk models on the other hand.

Regulatory Issues. There are growing regulatory pressures on the credit markets. Regulators wish to ensure that firms have enough capital to cover the risks that they run, so that, if they fail, there are sufficient funds to meet creditors' claims. Therefore, regulators set up capital rules which define the amount of capital a firm must have in order to enter a given position (the so called minimum capital requirements which are calculated based on a standardized approach). Roughly speaking, the amount necessary to put up against a possible loss depends on how risky the entered position is. As the valuation of credit risk still poses significant problems, the question raises, when, if at all, should regulators recognize banks' internal models for credit risk ? More than a decade has passed since the Basel Committee on Banking Supervision introduced its 1988 Capital Accord[1]. The business of banking, risk management practices, supervisory approaches, and financial markets each have undergone significant transformation since then. As a result, in June 1999 the Committee released a proposal to replace the 1988 Accord with a more risk-sensitive framework. The Committee presented even more

[1] For details on the 1988 Capital Accord see the webpage of the Bank for International Settlements (www.bis.org) and Ong (1999), chapter 1.

concrete proposals[2] in February 2001 and in April 2003. Based on the responses to the April 2003 consultative document the Committee is cosidering the need for further modifications to its proposals at the moment. The Committee aims to finalize the Basel II framework in the fourth quarter 2003. This version is supposed to be implemented by year-end 2006. A range of risk-sensitive options for addressing credit risk is contained in the new Accord. Depending on the specific bank's supervisory standards, it is allowed to choose out of at least three different approaches to credit risk measurement: The "standardized approach", where exposures to various types of counterparties will be assigned risk weights based on assessments by external credit assessment institutions. The "foundation internal ratings-based approach", where banks, meeting robust supervisory standards, will use their own assessments of default probabilities associated with their obligors. Finally an "advanced internal ratings-based approach", where banks, meeting more rigorous supervisory standards, will be allowed to estimate several risk factors internally. So banks should start improving their risk management capabilities today to be prepared for 2007 when they will be allowed to use the more risk-sensitive methodologies. At the same time academics must continuously develop further and better methods for estimating credit risk factors such as default probabilities.

Internal Credit Risk Models. Almost every day, new analytic tools to measure and manage credit risk are created. The most famous ones are Portfolio Manager™ of Moody's KMV[3] (see, e.g., Kealhofer (1998)), the Risk Metrics Group's CreditMetrics® and CreditManager™ (see, e.g., *CreditMetrics - Technical Document* (1997)), Credit Suisse Financial Products' CreditRisk+ (see, e.g., *CreditRisk+ A Credit Risk Management Framework* (1997)), and McKinsey & Company's Credit Portfolio View (see, e.g., Wilson (1997a), Wilson (1997b), Wilson (1997c), Wilson (1997d)). These models allow the user to measure and quantify credit risk at both, the portfolio and contributory level. Moody's KMV follows Merton's insight (see, e.g., Merton (1974)) and considers equity to be a call option on the value of a company's business, following the logic that a company defaults when its business value drops below its obligations. A borrower's default probability (i.e. the probability that a specific given credit's rating will change to default until the end of a specified time period) then depends on the amount by which assets exceed liabilities, and the volatility of these assets. CreditMetrics® is a

[2] For details on the new Capital Accord see the following publications of the Bank for International Settlements: *A New Capital Adequacy Framework* (1999), *Update on Work on a New Capital Adequacy Framework* (1999), *Best Practices for Credit Risk Disclosure* (2000), *Overview of the New Basel Capital Accord* (2001), *The New Basel Capital Accord* (2001), *The Standardised Approach to Credit Risk* (2001), *The Internal Ratings-Based Approach* (2001), *Overview of The New Basel Capital Accord* (2003).

[3] San Francisco based software company specialized in developing credit risk management software.

Merton-based model, too. It seeks to assess the returns on a portfolio of assets by analyzing the probabilistic behavior of the individual assets, coupled with their mutual correlations. It does this by using a matrix of transition probabilities (i.e. the probabilities that a specific given credit's rating will change to another specific rating until the end of a specified time period), calculating the expected change in market value for each possible rating's transition including default, and combining these individual value distributions via the correlations between the credits (as approximations of the correlations between relevant equities), to generate a loss distribution for the portfolio as a whole. CreditMetrics® has its roots in portfolio theory and is an attempt to mark credit to market. The model looks to the far more liquid bond market and the largely bond-driven credit derivatives market, where extensive data is available on ratings and price movements and instruments are actively traded. CreditRisk+ is based on insurance industry models of event risk. It does not make any estimates of how defaults are correlated. It rather considers the average default rates associated with each notch of a credit rating scheme (either a rating agency scheme or an internal score) and the volatilities of those rates. By doing so, it constructs a continuous, rather than a discrete, distribution of default risk probabilities. When mixed with the exposure profile of the instruments under consideration, it yields a loss distribution and associated risk capital estimates. CreditRisk+ is a modified version of the methodology the Credit Suisse Group has used to set loan loss provisions since December 1996. It has therefore evolved as a way to assess risk capital requirements in a data-poor environment where most assets are held to maturity and the only credit event that really counts is whether the lender gets repaid at maturity. In contrast to all other models it avoids the need for Monte Carlo simulation and therefore is much faster. The McKinsey model differs from the others in two additional important respects: First, it focuses more on the impact of macroeconomic variables on credit portfolios than other portfolio models do. Therefore it explicitly links credit default and credit migration behavior to the economic drivers. Second, the model is designed to be applied to all customer segments and product types, including liquid loans and bonds, illiquid middle market and small business portfolios as well as retail portfolios such as mortgages or credit cards.

To summarize, CreditMetrics® is a bottom-up model as each borrower's default is modeled individually with a microeconomic causal model of default. CreditRisk+ is a top-down model of sub-portfolio default rates, making no assumptions with regard to causality. Credit Portfolio View is a bottom-up model based on a macroeconomic causal model of sub-portfolio default rates. For a detailed overview and a comparison of these models see, e.g., Schmid (1997), Schmid (1998a), and Schmid (1998b). In addition, Gordy (1998) shows that, despite differences on the surface, CreditMetrics® and CreditRisk+ have similar underlying mathematical structures. Koyluoglu & Hickman (1998) examine the four credit risk portfolio models by placing them within a single general framework and demonstrating that they are only little different in theory and results, provided that the input parameters are somehow harmonized. Crouhy & Mark (1998) compare the models for a benchmark portfolio. It appears that the Credit Value at Risk numbers according to the various models fall in a narrow range, with a ratio of 1.5 between the highest and the lowest values. Keenan & Sobehart (2000) discuss how to validate credit risk models based on robust and easy to implement model performance measures. These measures analyze the cumulative accuracy to predict defaults and the level of uncertainty in the risk scores produced by the tested models. Lopez & Saidenberg (1998) use a panel data approach to evaluate credit risk models based on cross-sectional simulation.

For calculating the minimum capital requirements based on the new Capital Accord and for using internal credit risk models, financial institutions need mathematical models that are capable of describing the underlying credit risk factors through time, pricing financial instruments with regard to credit risk, and explaining how these instruments behave in a portfolio context.

1.2 Objectives, Structure, and Summary

During the last years we saw many theoretical developments in the field of credit risk research. Not surprisingly, most of this research concentrated on the pricing of corporate and sovereign defaultable bonds as the basic building blocks of credit risk pricing. But many of these models failed in describing real world phenomena such as credit spreads realistically. In chapter 6 we present a new hybrid term structure model which can be used for estimating default probabilities, pricing defaultable bonds and other securities subject to default risk. We show that it combines many of the strengths of previous models and avoids many of their weaknesses, and, most important, that it is capable of explaining market prices such as corporate or sovereign bond prices realistically. Our model can be used as a sophisticated basis for credit risk portfolio models that satisfy the rules of regulators and the internal needs of financial institutions.

In order to build a model for credit risk pricing it is essential to identify the credit risk components and the factors that determine credit risk. Therefore, we show in section 2 that default risk can mainly be characterized by default probabilities, i.e. the probabilities that an obligor defaults on its obligations, and recovery rates, i.e. the proportion of value still delivered in case of a default. The literature on modeling default probabilities evolves around three main approaches.

The historical method, discussed in section 2.3.1, is mainly applied by rating agencies to determine default probabilities by counting defaults that actually happened in the past. Sometimes not only default probabilities but also transition probabilities are of interest. Transition probabilities are the probabilities that obligors belonging to a specific rating category will change to another rating category within a specified time horizon. Transition matrices contain the information about all transition probabilities. E.g., rating agencies publish transition matrices on a regular basis. One problem of estimating transition matrices and transition probabilities is the scarcity of data. Therefore, we present the approach of Perraudin (2001) to estimate transition matrices only from default data. Sometimes there exist different transition matrices from different sources (e.g., different rating agencies). To combine information from different estimates of transition matrices to a new estimate, we show how a pseudo-Bayesian approach can be used. Finally, we discuss in depth, if transition matrices can be modeled as Markov chains.

The asset based method[4] as presented in section 2.3.3 relates default to the value of the underlying assets of a firm. All models in this framework are extensions of the work of Merton (1974), which has been the cornerstone of corporate debt pricing. Merton assumes that default occurs when the value of the firm's assets is less then the value of the debt at expiry. Extensions of this approach have been developed among others by Black & Cox (1976), Geske (1977), Ho & Singer (1982), Kim, Ramaswamy & Sundaresan (1992), Shimko, Tejima & Deventer (1993), Longstaff & Schwartz (1995b), Zhou (1997), and Vasicek (1997). We introduce not only Merton's classical approach but also so called first-passage default models that assume that a default can occur not only at maturity of the debt contract but at any point of time, and assume that bankruptcy occurs, if the firm value hits a specified (possibly stochastic) boundary or default point such as the current value of the firm's liabilities.

The intensity based method (or sometimes called reduced-form model) as introduced in section 2.3.4 relates default time to the stopping time of some exogenously specified hazard rate process. This approach has been applied among others by Artzner & Delbaen (1992), Madan & Unal (1994), Jarrow & Turnbull (1995), Jarrow, Lando & Turnbull (1997), Duffie & Singleton (1997), Lando (1998), and Schönbucher (2000). We give a lot of examples of specific intensity models and generalize the concept of default intensities to transition intensities. Finally, we discuss the generation of transition matrices from transition intensities as continuous time Markov chains.

In sections 2.3.1, 2.3.3, and 2.3.4 we review the three approaches, add some new interpretations, and summarize their advantages and disadvantages. Section 2.3.5 shows that one shouldn't only rely on theoretical models but always should consider the view and opinion of experts as well.

[4] The asset based method is sometimes called firm value method, Merton-based method, or structural approach.

In section 2.4 we give an overview of possible ways to model recovery rates. We show the dependence of recovery rates from variables such as the industry or the business cycle. We give some examples of specific recovery rate models and finally give a short introduction to Moody's model for predicting recovery rates called LossCalc^{TM}.

Asset based and intensity based methods can't only be applied for modeling default probabilities but also for pricing defaultable debt. In chapter 3 we show the two different concepts and give some examples of specific models.

Chapter 4 generalizes the discussion of pricing single defaultable bonds to the modeling of portfolios of correlated credits. We show how correlated defaults are treated in the asset based and in the intensity based framework. Finally, we give a short introduction to the copula function approach. The copula links marginal and joint distribution functions and separates the dependence between random variables and the marginal distributions. This greatly simplifies the estimation problem of a joint stochastic process for a portfolio with many credits. Instead of estimating all the distributional parameters simultaneously, we can estimate the marginal distributions separately from the joint distribution.

Credit derivatives are probably one of the most important types of new financial products introduced during the last decade. Traditionally, exposure to credit risk was managed by trading in the underlying asset itself. Now, credit derivatives have been developed for transferring, repackaging, replicating and hedging credit risk. They can change the credit risk profile of an underlying asset by isolating specific aspects of credit risk without selling the asset itself. In chapter 5 we explain these new products including single counterparty as well as multi counterparty products. Even more complicated products than pure credit derivatives are structured finance transactions (SPs), such as collateralized debt obligations (CDOs), collateralized bond obligations (CBOs), collateralized loan obligations (CLOs), collateralized mortgage obligations (CMOs) and other asset-backed securities (ABSs). The key idea behind these instruments is to pool assets and transfer specific aspects of their overall credit risk to new investors and/or guarantors. We give a short introduction to CDOs and show the so called BET approach for modeling CDOs.

A recent trend tries to combine the asset based and intensity based models to more powerful models, that are as flexible as intensity based models and explain the causality of default as well as asset based models. Examples are the models of Madan & Unal (1998) who assume that the stochastic hazard rate is a linear function of the default-free short rate and the logarithm of the value of the firm's assets, and Jarrow & Turnbull (1998) who choose the stochastic hazard rate to be a linear function of some index and

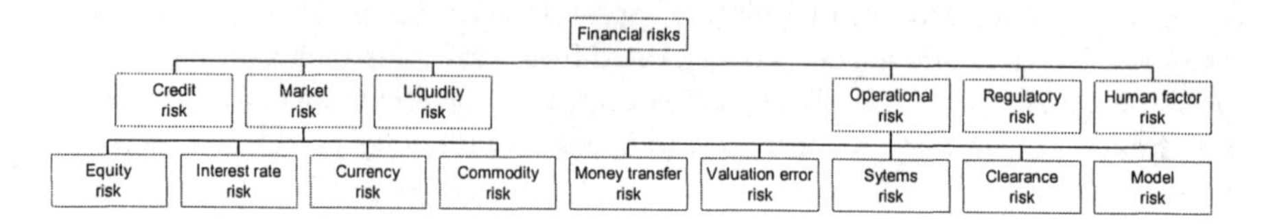

Fig. 1.1. Key risks of financial institutions.

the default-free short rate. Both models have the problem that their hazard rate processes can admit negative values with positive probability. Cathcart & El-Jahel (1998) use the asset based framework, but assume that default is triggered when a signalling process hits some threshold. Duffie & Lando (1997) model a default hazard rate that is based on an unobservable firm value process. Hence, they cover the problem of the uncertainty of the current level of the assets of the firm. The three-factor defaultable term structure model which we develop in section 6.2 is a completely new hybrid model. We directly model the short rate credit spread and assume that it depends on some uncertainty index, which describes the uncertainty of the obligor. The larger the value of the uncertainty index the worse the quality of the debtor is. In addition, we assume that the non-defaultable short rate process follows a mean reverting Hull-White process or a mean-reverting square root process with time-dependent mean reversion level. As such our model is an extension of the non-defaultable bond pricing models of Hull & White (1990) and Cox, Ingersoll & Ross (1985) to defaultable bond pricing. The non-defaultable short rate, the short rate credit spread and the uncertainty index are defined by a three-dimensional stochastic differential equation (SDE). We show that this SDE admits a unique weak solution by applying and generalizing results of Ikeda & Watanabe (1989a). This three-dimensional approach where we consider market and credit risk at the same time, serves as a basis for the application of advanced methods for credit risk management. In the past, financial institutions have disaggregated the various risks (see figure ??) generated by their businesses and treated each one separately. However, for reasons like the linkages between the markets, this approach needs to be replaced by an integrated risk management which allows comparison of risk levels across business and product units. In particular, as credit risk is one of the key risks, financial institutions need to be able to provide an accurate and consistent measurement of credit risk. Our hybrid model can serve as a basis for a stochastic approach to an integrated market and credit risk management.

By using no-arbitrage arguments we apply the model to the pricing of various securities subject to default risk: The counterparty to a contract may not be able or willing to make timely interest rate payments or repay its debt at maturity. This increases the risk of the investor which must be compensated by reducing the price of the security contract. Our model determines a fair value for such a defaultable security and compares its price to the value of an otherwise identical non-defaultable security. In section 6.3 we determine closed form pricing formulas for defaultable zero coupon bonds and various other types of fixed and floating rate debt such as defaultable floating rate notes and defaultable interest rate swaps. In addition, we show that the theoretical credit spreads generated by our model are consistent with the empirical findings of Sarig & Warga (1989) and Jones, Mason & Rosenfeld (1984). Especially, we demonstrate that the term structure of credit spreads implied by our model can be upward sloping, downward sloping, hump shaped or flat. And in contrast to many other models we are even able to generate short term credit spreads that are clearly different from zero.

In section 6.4 we develop closed form solutions for the pricing of various credit derivatives by pricing them relative to observed bond prices within our three-factor model framework. Although there are a lot of articles that have been written on the pricing of defaultable bonds and derivatives with embedded credit risk, there are only a few articles on the direct pricing of credit derivatives. Das (1995) basically shows that in an asset based framework credit options are the expected forward values of put options on defaultable bonds with a credit level adjusted exercise price. Longstaff & Schwartz (1995a) develop a pricing formula for credit spread options in a setting where the logarithm of the credit spread and the non-defaultable short rate follow Vasicek processes. Das & Tufano (1996) apply their model, which is an extension to stochastic recovery rates of the model of Jarrow et al. (1997), to the pricing of credit-sensitive notes. Das (1997) summarizes the pricing of credit derivatives in various credit risk models (e.g., the models of Jarrow et al. (1997), and Das & Tufano (1996)). All models are presented in a simplified discrete fashion. Duffie (1998a) uses simple no-arbitrage arguments to determine approximate prices for default swaps. Hull & White (2000) provide a methodology for valuing credit default swaps when the payoff is contingent on default by a single reference entity and there is no counterparty default risk. Schönbucher (2000) develops various pricing formulas for credit derivatives in the intensity based framework. Our work is different from all other articles in that we apply partial differential equation techniques to the pricing of credit derivatives.

In section 6.5 we construct a four dimensional lattice (for the dimensions time, non-defaultable short rate r, short rate credit spread s, and uncertainty index u) to be able to price defaultable contingent claims and credit derivatives that do not allow for closed form solutions, e.g., because of callability features or because they are American. In contrast to the trees proposed by Chen (1996), Amin (1995), or Boyle (1988) the branching as well as the probabilities do not change with a changing drift, which makes the lattice more efficient, especially under risk management purposes. The probabilities for each node in the four dimensional lattice are simply given by the product of the one dimensional processes. We close this section by giving an explicit numerical example for the pricing of credit spread options.

In section 6.6 we close the gap between our theoretical model and its possible applications in practice by demonstrating various methods how to calibrate the model to observed data and how to estimate the model parameters. This is an important add-on to other research in the credit risk field which is often only restricted to developing new models without applying them to the real world. Actually, the ultimate success or failure in implementing pricing formulas is directly related to the ability to collect the necessary information for determining good model parameter values. Therefore, we suggest two different ways how meaningful values for the parameters of the three stochastic processes r, s, and u can be found. The first one is the method of least squared minimization. Basically, we compare market prices and theoretical prices at one specific point in time and calculate the implied parameters by minimizing the sum of the squared deviations of the market from the theoretical prices. The second one is the Kalman filter method that estimates parameter values by looking at time series of market values of bonds. By applying a method developed by Nelson & Siegel (1987) we estimate daily zero curves from a time series of daily German, Italian, and Greek Government bond prices. The application of Kalman filtering methods in the estimation of term structure models using time-series data has been analyzed (among others) by Chen & Scott (1995), Geyer & Pichler (1996) and Babbs & Nowman (1999). Based on the parameter estimations we apply a lot of different in-sample and out-of-sample tests such as a model explanatory power test suggested by Titman & Torous (1989) and find that our model is able to explain observed market data such as Greek and Italian credit spreads to German Government bonds very well. Especially, we can produce more encouraging results than empirical studies of other credit risk models (see, e.g., Düllmann & Windfuhr (2000) for an empirical investigation of intensity based methods).

Based on our three-factor defaultable term structure model, in section 6.7 we develop a framework for the optimal allocation of assets out of a universe of sovereign bonds with different time to maturity and quality of the issuer. Our methodology can also be applied to other asset classes like corporate bonds. We estimate the model parameters by applying Kalman filtering methods as described in section 6.6. Based on these estimates we apply Monte Carlo simulation techniques to simulate the prices for a given set of bonds for a future time horizon. For each future time step and for each given portfolio composition these scenarios yield distributions of future cash flows and portfolio values. We show how the portfolio composition can be optimized by maximizing the expected final value or return of the portfolio under given constraints like a minimum cash flow per period to cover the liabilities of a company and a maximum tolerated risk. To visualize our methodology we present a case study for a portfolio consisting of German, Italian, and Greek sovereign bonds.

To summarize, this work contributes to the efforts of academics and practitioners to explain credit markets, price default related instruments such as defaultable fixed and floating rate debt, credit derivatives, and other securities with embedded credit risk, and develop a profound credit risk management. Models are developed to value instruments whose prices are default dependent within a consistent framework, to detect relative value, to mark to market positions, to risk manage positions and to price new structures which are not (yet) traded. We describe the whole process, from the specification of the stochastic processes to the estimation of the parameters and calibration to market data.

Finally, a brief note with respect to some of the terminology. In this work, risky refers to credit risk and not to market risk. Riskless means free of credit risk. Default free is a synonym to riskless or risk free. Default and bankruptcy are used as synonyms.

2. Modeling Credit Risk Factors

"While substantial progress has been made in solving various aspects of the credit risk management problem, the development of a consistent framework for managing various sources of credit risk in an integrated way has been slow."
- Scott Aguais and Dan Rosen, 2001 -

"Credit risk management is being transformed by the use of quantitative portfolio models. These models can depend on parameters that are difficult to quantify, and that change over time."
- Demchak (2000) -

2.1 Introduction

Usually investors must be willing to take risks for their investments. Therefore, they should be adequately compensated. But what is a fair premium for risk compensation ? To answer this question it is essential to determine the key sources of risk. As we are concerned with credit risk, this section is devoted to the identification of credit risk factors. We show the current practice of credit risk factor modeling and present these methodologies within a rigorous mathematical framework.

2.2 Definition and Elements of Credit Risk

Credit risk consists of two components, default risk and spread risk. Default risk is the risk that a debtor will be unable or unwilling to make timely payments of interest or principal, i.e. that a debtor defaults on its contractual payment obligations, either partly or wholly. The default time is defined as the date of announcement of failure to deliver. Even if a counterparty does not default, the investor is still exposed to credit risk: credit spread risk is the risk of reductions in market value due to changes in the credit quality of a debtor. The event of default has two underlying risk components, one associated with the timing of the event ("arrival risk") and the other with its magnitude ("magnitude risk"). Hence, for modeling credit risk on deal or counterparty level, we have to consider the following risk elements:

- Exposure at default: A random variable describing the exposure subject to be lost in case of a default. It consists of the borrower's outstandings and the commitments drawn by the obligor prior to default. In practice obligors tend to draw on commitments in times of financial distress.
- Transition probabilities: The probability that the quality of a debtor will improve or deteriorate. The process of changing the creditworthiness is called credit migration.
- Default probabilities: The probability that the debtor will default on its contractual obligations to repay its debt.
- Recovery rates: A random variable describing the proportion of value still delivered after default has happened. The default magnitude or loss given default is the proportion of value not delivered.

In addition, for modeling credit risk on portfolio level we have to consider joint default probabilities and joint transition probabilities as well.

2.3 Modeling Transition and Default Probabilities

The distributions of defaults and transitions play the central role in the modeling, measuring, hedging and managing of credit risk. They are an appropriate way of expressing arrival risk. Probably the oldest approach to estimating default and transition probabilities is the historical method that focuses on counting historical defaults and rating transitions and using average values as estimates. Because this method is very static, newer statistical approaches try to link these historical probabilities to external variables which can better explain the probability changes through time. Most of these econometric methods try to measure the probability that a debtor will be bankrupt in a certain period, given all information about the past default and transition behavior and current market conditions. Firm value or asset based methods implicitly model default or transition probabilities by assuming that default or rating changes are triggered, if the firm value hits some default or rating boundary. Intensity based methods treat default as an unexpected event whose likelihood is governed by a default-intensity process that is exogenously specified. Like in the historical method, under the other two approaches, the likelihood of default can be linked to observable external variables. Jarrow et al. (1997) make the distinction between implicit and explicit estimation of transition matrices, where implicit estimation refers to extracting transition and default information from market prices of defaultable zero-coupon bonds or credit derivatives. In sections 2.3.1, 2.3.3, and 2.3.4 we only consider explicit methods but sections 2.3.3, and 2.3.4 are also a basis for some implicit methods (see chapter 3).

2.3.1 The Historical Method

Ratings and Rating Agencies. Rating the quality and evaluating the creditworthiness of corporate, municipal, and sovereign debtors and providing transition and default[1] probabilities as well as recovery rates for creditors is the key business of rating agencies[2]. Basically, rating agencies inform investors about the investors' likelihood to receive the principal and interest payments as promised by the debtors. The growing number of rating agencies on the one hand and the increasing number of rated obligors on the other hand proves their increasing importance. The four biggest US agencies are Moody's Investors Service (Moody's), Standard & Poor's (S&P), Fitch IBCA and Duff & Phelps. Table 2.1 shows a list of selected rating agencies around the world. However, the actual number of rating agencies is very dynamic.

Ratings are costly: US$ 25,000 for issues up to US$ 500 million and $\frac{1}{2}$ basis point for issues greater than US$ 500 million. Treacy & Carey (2000) report a fee charged by S&P of 0.0325% of the face amount. By the way, according to Partnoy (2002) until the mid-70s, it was the investors, not issuers, who paid the fees to the rating agencies.

In rating debt, each agency uses its own system of letter grades. The interpretation of S&P's and Moody's letter ratings is summarized in table 2.2. The lower the grade, the greater the risk that the debtor will not be able to repay interest and/or principal. The rating agencies distinguish between issue and issuer credit ratings. For details on the exact definitions see appendix A.1.

In evaluating the creditworthiness of obligors rating agencies basically use the same methodologies than equity analysts do - although their focus may be on a longer time horizon. Even though the methods may differ slightly from agency to agency all of them focus on the following areas:

- Industry characteristics
- Financial characteristics such as financial policy, performance, profitability, stability, capital structure, leverage, debt coverage, cash flow protection, financial flexibility
- Accounting, controlling and risk management
- Business model: specific industry, markets, competitors, products and services, research and development
- Clients and suppliers
- Management (e.g., strategy, competence, experience) and organization

[1] According to Caouette, Altman & Narayanan (1998), page 194, for rating agencies "defaults are defined as bond issues that have missed a payment of interest, filed for bankruptcy, or announced a distressed-creditor restructuring".

[2] Rating agencies are providers of timely, objective credit analysis and information. Usually they operate without government mandate and are independent of any investment banking firm or similar organization.

Table 2.1. Selection of rating agencies.

Agency Name	Year founded
Canadian Bond Rating Service	1972
Capital Intelligence	1985
Credit Rating Services of India Ltd.	1988
Dominion Bond Rating Service	1976
Duff & Phelps	1932
Fitch IBCA	1913
Global Credit Rating Co.	1996
ICRA	1991
Interfax	1989
International Bank Credit Analysis	1979
Japan Bond Research Institute	1979
Japan Credit Rating Agency	1985
JCR-VIS Credit Rating Ltd.	1997
Korean Investors Services	1985
Malaysian Rating Corporation Berhad	1996
Mikuni & Co.	1975
Moody's Investors Service	1900
National Information & Credit Evaluation, Inc	1986
Nippon Investors Services	1985
Pakistan Credit Rating Agency	1994
Rating Agency Malaysia Berhard	1990
Shanghai Credit Information Services Co., Ltd.	1999
Shanghai Far East Credit Rating Co., Ltd.	1988
Standard's & Poor's	1941
Thai Rating and Information Services	1993

- Staff/team: qualifications, structure and key members of the team
- Production processes: quality management, information and production technology, efficiency
- Marketing and sales

After intense research, a rating analyst suggests a rating and must defend it before a rating committee. Obviously, the credit quality of an obligor can change over time. Therefore, after issuance and the assignment of the initial issuer or issuance rating, regularly (periodically and based on market events) each rating agency checks and - if necessary - adjusts its issued rating. A rating outlook assesses the potential direction of a long-term credit rating over the intermediate to longer term. In determining a rating outlook, consideration is given to any changes in the economic and/or fundamental business conditions. An outlook is not necessarily a precursor of a rating change and is published on a continuing basis.

- Positive means that a rating may be raised.
- Negative means that a rating may be lowered.

Table 2.2. Long-term senior debt rating symbols

Investment-grade ratings		
S&P	*Moody's*	*Interpretation*
AAA	Aaa	Highest quality, extremely strong
AA+	Aa1	High quality
AA	Aa2	
AA-	Aa3	
A+	A1	Strong payment capacity
A	A2	
A-	A3	
BBB+	Baa1	Adequate payment capacity
BBB	Baa2	
BBB-	Baa3	
Speculative-grade ratings		
S&P	*Moody's*	*Interpretation*
BB+	Ba1	Likely to fulfill obligations ongoing uncertainty
BB	Ba2	
BB-	Ba3	
B+	B1	High risk obligations
B	B2	
B-	B3	
CCC+	Caa1	Current vulnerability to default
CCC	Caa2	
CCC-	Caa3	
CC		
C	Ca	In bankruptcy or default, or other marked shortcoming
D		

Source: Caouette et al. (1998)

- Stable means that a rating is not likely to change.
- Developing means a rating may be raised, lowered, or affirmed.
- N.M. means not meaningful.

If there is a tendency observable, that may affect the rating of a debtor, the agency notifies the issuer and the market. In case of Moody's the debtor is set on the rating review list, in case of Standard & Poor's the obligor is set on the credit watch list. Credit watch highlights the potential direction of a short- or long-term rating. It focuses on identifiable events (such as mergers, recapitalizations, voter referendums, regulatory action, or anticipated operating developments) and short-term trends that cause ratings to be placed under special surveillance by the rating agency's analytical staff. Ratings appear on credit watch when such an event or a deviation from an expected trend occurs and additional information is necessary to evaluate the current rating. A listing does not mean a rating change is inevitable and rating changes may occur without the ratings having first appeared on credit watch.

Let us finally mention that local currency and foreign currency country risk considerations are a standard part of the rating agency's analysis for credit ratings on any issuer or issue. Currency of repayment is a key factor in this analysis. An obligor's capacity to repay foreign currency obligations may be lower than its capacity to repay obligations in its local currency due to the sovereign government's own relatively lower capacity to repay external versus domestic debt. These sovereign risk considerations are incorporated in the debt ratings assigned to specific issues. Foreign currency issuer ratings are also distinguished from local currency issuer ratings to identify those instances where sovereign risks make them different for the same issuer.

Estimation of Transition Matrices. As already mentioned, in the historical method, transition and default probabilities are associated with either external or bank-internal ratings. The default studies of the rating agencies summarize the default experience of issuers of public debt by comparing historical ratings of defaulting issuers with ratings of public issuers that did not default. They determine default rates based on all bonds of a given rating class, regardless of their age. The bond issuers are the basic units of account. The most important concepts are marginal and cumulative historical default and survival rates. This concept of assigning a default rate to a rating is called calibration.

Let there be a finite state space of ratings $R \in \{1, 2, ..., K\}$. These states represent the various credit classes, with state 1 being the highest and state K being bankruptcy. In case of S&P's long-term debt rating symbols $1 \triangleq AAA$, $2 \triangleq AA+$, ..., $22 \triangleq D$.

Definition 2.3.1.

1. *The historical marginal year t default (survival) rate* $d_R^{[Y_0,Y_1]}(t)$ $\left(s_R^{[Y_0,Y_1]}(t) = 1 - d_R^{[Y_0,Y_1]}(t)\right)$ *based on the time frame* $[Y_0, Y_1]$, $Y_0 \leq Y_1 - t$, *in the past, is the average issuer-weighted default (survival) rate for an* R*-rated issuer in its* $t-$*th year as experienced between the years* Y_0 *and* Y_1. *Formally,*

$$d_R^{[Y_0,Y_1]}(t) = \frac{\sum_{Y=Y_0}^{Y_1} M_R^Y(t)}{\sum_{Y=Y_0}^{Y_1} N_R^Y(t)},$$

   where $N_R^Y(t)$ *is the number of issuers with rating* R *at the start of year* Y *that have not defaulted until the beginning of year* $Y+t-1$ *and* $M_R^Y(t)$ *is the number of issuers rated* R *at the start of year* Y *and defaulted in the* $t-$*th year, i.e. during year* $Y+t-1$.
2. *The historical cumulative* $\tau-$*year default (survival) rate* $cd_R^{[Y_0,Y_1]}(\tau)$ $\left(cs_R^{[Y_0,Y_1]}(\tau) = 1 - cd_R^{[Y_0,Y_1]}(\tau)\right)$ *based on time frame* $[Y_0, Y_1]$, $Y_0 \leq Y_1 - \tau$, *is the probability that an* R*-rated issuer will default (not default) within* τ *years. Formally,*

$$cs_R^{[Y_0,Y_1]}(\tau) = \prod_{t=1}^{\tau}\left(1 - d_R^{[Y_0,Y_1]}(t)\right) \text{ and } cd_R^{[Y_0,Y_1]}(\tau) = 1 - cs_R^{[Y_0,Y_1]}(\tau).$$

3. *For each rating R we define the historical $\tau-$year transition rate to rating $\check{R}$, $tr_{R,\check{R}}^{[Y_0,Y_1]}(\tau)$, based on the time frame $[Y_0, Y_1]$, $Y_0 \leq Y_1 - \tau$, as the average rate for an R-rated issuer to be rated $\check{R}$ at the end of its $\tau-$th year as experienced between the years Y_0 and Y_1. Formally,*

$$tr_{R,\check{R}}^{[Y_0,Y_1]}(\tau) = \frac{\sum_{Y=Y_0}^{Y_1} M_{R,\check{R}}^{Y}(\tau)}{\sum_{Y=Y_0}^{Y_1} N_R^Y},$$

where N_R^Y is the number of issuers with rating R at the start of year Y and $M_{R,\check{R}}^{Y}(\tau)$ is the number of issuers rated R at the beginning of year Y and rated $\check{R}$ at the end of the $\tau-$th year, i.e. at the end of year $Y+\tau-1$. The historical $\tau-$year transition matrix based on the time frame $[Y_0, Y_1]$, $Y_0 \leq Y_1 - \tau$, is denoted by $Tr^{[Y_0,Y_1]}(\tau) = \left(tr_{R,\check{R}}^{[Y_0,Y_1]}(\tau)\right)_{R,\check{R}}$, and fully describes the probability distribution of ratings after τ years given the rating today. For simplicity we denote the $1-$year transition matrix $Tr^{[Y_0,Y_1]}(\tau)$ by $Tr^{[Y_0,Y_1]}$.

Remark 2.3.1.
The historical marginal year 1 default rate $d_R^{[Y_0,Y_1]}(1)$ based on the time frame $[Y_0, Y_1]$, $Y_0 \leq Y_1 - 1$, and the historical cumulative $1-$year default rate $cd_R^{[Y_0,Y_1]}(1)$ based on the same time frame are obviously the same. For simplification we call both the one-year default rate based on the time frame $[Y_0, Y_1]$, $Y_0 \leq Y_1 - 1$. If we are not interested in the specific time frame $[Y_0, Y_1]$ we just say one-year default rate. In additon, for simplicity we usually write the marginal and cumulative default, survival and transition probabilities and transition matrices without the superscript $[Y_0, Y_1]$.

As an example, table 2.3 shows the historical NR[3]-adjusted average one-year corporate transition and default rates for the S&P ratings *AAA*, *AA*,,...,*CCC* based on the time frame [1980, 2002] (see *Special Report: Ratings Performance 2002* (2003)). Such a table is called historical transition matrix. Each row corresponds to the initial rating, i.e. the rating at the beginning of the year, each column corresponds to the rating at the end of the year. Table 2.4 shows the historical one-year transition and default rates for the S&P ratings *AAA*, *AA*,,...,*CCC* based on the time frame [1980, 2002] where the class NR is considered explicitly. Of course, the average one-year transition and default rates change continuously and depend on the specific time frame. For comparison

[3] Entities whose ratings have been withdrawn (e.g., because their entire debt is paid off, there has been a calling of the debt, there has been a merger or acquisition etc.) change to NR. However, the details of individual transactions to NR are usually not known.

Table 2.3. S&P's historical NR-adjusted worldwide corporate average one-year transition and default rates (in %, based on the time frame [1980,2002]).

Initial Rating	Rating at year end							
	AAA	*AA*	*A*	*BBB*	*BB*	*B*	*CCC*	*D*
AAA	93.06	6.29	0.45	0.14	0.06	0.00	0.00	0.00
AA	0.59	90.99	7.59	0.61	0.06	0.11	0.02	0.01
A	0.05	2.11	91.43	5.63	0.47	0.19	0.04	0.05
BBB	0.03	0.23	4.44	88.98	4.70	0.95	0.28	0.39
BB	0.04	0.09	0.44	6.07	82.73	7.89	1.22	1.53
B	0.00	0.08	0.29	0.41	5.32	82.06	4.90	6.95
CCC	0.10	0.00	0.31	0.63	1.57	9.97	55.82	31.58

Source: Standard & Poor's (*Special Report: Ratings Performance 2002* 2003).

Table 2.4. S&P's historical worldwide corporate average one-year transition and default rates (in %, based on the time frame [1980,2002]).

	Rating at year end								
I.R.	*AAA*	*AA*	*A*	*BBB*	*BB*	*B*	*CCC*	*D*	*NR*
AAA	89.37	6.04	0.44	0.14	0.05	0.00	0.00	0.00	3.97
AA	0.57	87.76	7.30	0.59	0.06	0.11	0.02	0.01	3.58
A	0.05	2.01	87.62	5.37	0.45	0.18	0.04	0.05	4.22
BBB	0.03	0.21	4.15	84.44	4.39	0.89	0.26	0.37	5.26
BB	0.03	0.08	0.40	5.50	76.44	7.14	1.11	1.38	7.92
B	0.00	0.07	0.26	0.36	4.74	74.12	4.37	6.20	9.87
CCC	0.09	0.00	0.28	0.56	1.39	8.80	49.72	27.87	11.30

Source: Standard & Poor's (*Special Report: Ratings Performance 2002* 2003).

to tables 2.3 and 2.4, tables 2.5 and 2.6 show the average one-year transition and default rates based on the time frame $[1980, 2001]$. The universe of obligors of these matrices is mainly large corporate institutions (e.g., industrials, utilities, insurance companies, banks, other financial institutions, real estate companies) around the world. Ratings for sovereigns and municipals are not included. Table 2.7 shows the historical sovereign foreign currency one-year transition rates based on the time frame $[1975, 2002]$. The sovereign transition matrix is quite different from a corporate one. For example, the entries far away from the diagonal are zero. Therefore, extreme rating changes are more unlikely for sovereigns than they are for corporates. The probability mass is more concentrated on the diagonal, i.e. the probability of staying in the same rating class is very high.

Apart from the regularly published data of the rating agencies (e.g., Standard & Poor's yearly publications "Ratings Performance: Transition and Stability"

Table 2.5. S&P's historical NR-adjusted worldwide corporate average one-year transition and default rates (in %, based on the time frame [1980,2001]).

Initial Rating	Rating at year end							
	AAA	*AA*	*A*	*BBB*	*BB*	*B*	*CCC*	*D*
AAA	93.28	6.16	0.44	0.09	0.03	0.00	0.00	0.00
AA	0.62	91.64	7.04	0.53	0.06	0.09	0.02	0.01
A	0.06	2.20	91.77	5.26	0.44	0.18	0.04	0.05
BBB	0.04	0.25	4.64	89.45	4.35	0.75	0.25	0.27
BB	0.03	0.07	0.45	6.31	83.07	7.58	1.20	1.29
B	0.00	0.09	0.31	0.41	5.38	82.78	4.31	6.71
CCC	0.13	0.00	0.27	0.81	1.75	10.08	58.20	28.76

Source: Standard & Poor's (*Special Report: Ratings Performance 2001* 2002).

Table 2.6. S&P's historical worldwide corporate average one-year transition and default rates (in %, based on the time frame [1980,2001]).

	Rating at year end								
I.R.	*AAA*	*AA*	*A*	*BBB*	*BB*	*B*	*CCC*	*D*	*NR*
AAA	89.62	5.92	0.43	0.09	0.03	0.00	0.00	0.00	3.93
AA	0.60	88.29	6.78	0.51	0.05	0.09	0.02	0.01	3.66
A	0.06	2.10	87.79	5.04	0.43	0.17	0.04	0.05	4.33
BBB	0.03	0.23	4.36	84.43	4.15	0.73	0.23	0.26	5.58
BB	0.02	0.06	0.41	5.75	75.98	7.05	1.09	1.22	8.42
B	0.00	0.08	0.27	0.35	4.77	74.15	3.90	5.96	10.53
CCC	0.11	0.00	0.22	0.67	1.45	8.95	51.34	24.72	12.53

Source: Standard & Poor's (*Special Report: Ratings Performance 2001* 2002).

Table 2.7. S&P's historical sovereign foreign currency average one-year transition rates (in %, based on the time frame [1975,2002]).

Initial Rating	Rating at year end							
	AAA	*AA*	*A*	*BBB*	*BB*	*B*	*CCC*	*SD*
AAA	97.37	2.63	0.00	0.00	0.00	0.00	0.00	0.00
AA	2.79	95.53	0.56	0.00	0.56	0.56	0.00	0.00
A	0.00	2.70	95.50	1.80	0.00	0.00	0.00	0.00
BBB	0.00	0.00	6.50	87.80	4.07	1.63	0.00	0.00
BB	0.00	0.00	0.00	6.92	83.08	7.69	1.54	0.77
B	0.00	0.00	0.00	0.00	12.00	81.33	5.33	1.33
CCC	0.00	0.00	0.00	0.00	0.00	0.00	16.67	83.33
SD	0.00	0.00	0.00	0.00	0.00	20.00	10.00	70.00

Source: Standard & Poor's (*Special Report: Ratings Performance 2001* 2002).

and Moody's yearly special reports "Corporate Bond Defaults and Default Rates") there are a lot of academic empirical studies on historical default and transition probabilities for different financial instruments. The most detailed default statistics available are those for corporate bonds stratified by bond ratings. The statistics currently available for loans are less comprehensive. Summaries of these academic works are given, e.g., by Caouette et al. (1998), chapters 15 and 16, by Carty & Lieberman (1998), and by Carty (1998).

Desirable Properties of Transition Matrices. If we estimate transition matrices as described above they represent only a limited amount of observation with sampling errors and usually they don't show all the properties one might desire:

- The estimate of a transition matrix should have a huge data basis to ensure that the granularity of the estimate is small. This point is important when estimating rare events such as the transition from AAA to default. As Lando & Skodeberg (2002) state, the maximum-likelihood estimator for the one-year transition probability from AAA to default will be non-zero even if there have been no direct or indirect defaults (such as default through a sequence of downgrades) during the observation period.
- There should be no inconsistencies in rank order across credit ratings:
  - The higher the rating the smaller the default probability.
    The transition probabilities should decrease as the migration distance increases.
  - The transition probabilities to a given rating should be greater for adjacent ratings than they are for more distant rating categories.
- There should be a mean-reversion effect in credit ratings. The higher the rating the stronger this effect.

Considering tables 2.3 - 2.8 and 2.20 - 2.22 we can find easily a lot of inconsistencies, for example:

- The default probability of an A-rated counterparty in table 2.21 is zero which is a result of the scarcity of data for unlikely events. But A-rated counterparties should have a positive default probability.
- The default probability of a BB-rated counterparty in table 2.22 is higher than the default probability of a B-rated counterparty. But low-rated firms should always be more risky then high-rated firms.
- In table 2.3 the probability of a AA-rated counterparty migrating to B is greater than migrating to BB. But BB is more adjacent to AA than B.
- In table 2.20 the probability of a AAA-rated counterparty migrating to BB is higher than the probability of a AA-rated counterparty migrating to BB. But AA is more adjacent to BB than AAA.
- In table 2.22 the probability that a B-rated counterparty stays in rating category B is higher than the probability that a BB-rated counterparty stays in rating class BB. But the mean-reversion effect of better credit rating categories should be higher than for worse rating classes.

Table 2.8. S&P's historical NR-adjusted worldwide corporate average one-year transition and default rates (in %, based on the time frame [1980,2002]) under the assumption that D is an absorbing state.

Initial Rating	Rating at year end							
	AAA	*AA*	*A*	*BBB*	*BB*	*B*	*CCC*	*D*
AAA	93.06	6.29	0.45	0.14	0.06	0.00	0.00	0.00
AA	0.59	90.99	7.59	0.61	0.06	0.11	0.02	0.01
A	0.05	2.11	91.43	5.63	0.47	0.19	0.04	0.05
BBB	0.03	0.23	4.44	88.98	4.70	0.95	0.28	0.39
BB	0.04	0.09	0.44	6.07	82.73	7.89	1.22	1.53
B	0.00	0.08	0.29	0.41	5.32	82.06	4.90	6.95
CCC	0.10	0.00	0.31	0.63	1.57	9.97	55.82	31.58
D	0.00	0.00	0.00	0.00	0.00	0.00	0.00	100.00

Source: Standard & Poor's (*Special Report: Ratings Performance 2002* 2003).

If straightforward compilation of historical data does not provide transition matrices without inconsistencies we have to transform the observed transition matrices into matrices that meet the properties we desire. Therefore, we calculate the transition matrix which is closest to the observed one and satisfies some additional constraints. For example, we can determine a transition matrix that is closest to table 2.22 and shows a consistent rank order of default probabilities. What we mean by closest is explained in the following subsection.

Comparing Transition Matrices. If we assume that the default state D is an absorbing state we can extend the NR-adjusted transition matrix by an additional row of the following kind: The probability of migrating from D to any other rating category equals 0 and the probability of staying in D equals 1 (see figure 2.8 compared to figure 2.3). Note that the resulting matrix is quadratic. If we want to assess the statistical similarity between two different NR-adjusted $K \times K$ transition matrices Tr and $\widetilde{Tr}$ it is convenient to use a norm [4] $\|\cdot\|$ (as an absolute measure for an individual matrix) or a scalar metric[5] d (as a relative comparison of two matrices) which capture the characteristics of the given matrices.

[4] Let $m, n \in \mathbb{N}$. The function $\|\cdot\| : M(m,n) \to \mathbb{R}$ is called a matrix norm, if the following conditions are satisfied:

- $\|\cdot\| > 0$ for all $A \neq 0$, $A \in M(m,n)$,
- $\|\alpha A\| = |\alpha| \, \|A\|$ for all $\alpha \in \mathbb{R}$, $A \in M(m,n)$,
- $\|A + B\| \leq \|A\| + \|B\|$ for all $A, B \in M(m,n)$.

Here $M(m,n)$ denotes the vector space of all matrices with dimensions $m \times n$ and real valued entries.

[5] Let $m, n \in \mathbb{N}$ and $M(m,n)$ denote the vector space of all matrices with dimensions $m \times n$ and real valued entries. A function

- Israel, Rosenthal & Wei (2001) use the L^1 norm and the L^1 metric (average absolute difference):

$$\|Tr\|_{L^1} = \frac{1}{K^2} \sum_{R,\check{R}} \left| tr_{R,\check{R}} \right|,$$

$$d_{L^1}\left(Tr, \widetilde{Tr}\right) = \frac{1}{K^2} \sum_{R,\check{R}} \left| tr_{R,\check{R}} - \widetilde{tr}_{R,\check{R}} \right|.$$

- Bangia, Diebold, Kronimus, Schagen & Schuermann (26) use the L^2 norm and the L^2 metric (average root-mean-square difference):

$$\|Tr\|_{L^2} = \frac{1}{K^2} \sqrt{\sum_{R,\check{R}} tr_{R,\check{R}}^2},$$

$$d_{L^2}\left(Tr, \widetilde{Tr}\right) = \frac{1}{K^2} \sqrt{\sum_{R,\check{R}} \left(tr_{R,\check{R}} - \widetilde{tr}_{R,\check{R}}\right)^2}.$$

A number of authors use a generalized norm $\|\cdot\|^*$ (i.e. not all necessary norm conditions are satisfied) or a generalized metric d^*(i.e. not all necessary conditions of a metric are satisfied):

- Based on the work of Shorrocks (1978) Geweke, Marshall & Zarkin (1986) develop eigenvalue-based generalized norms for transition matrices:

$$\|Tr\|_P^* = \frac{1}{K-1} \left(K - trace\left(Tr\right)\right),$$

$$\|Tr\|_D^* = 1 - \left|\det\left(Tr\right)\right|,$$

$$\|Tr\|_E^* = \frac{1}{K-1} \left(K - \sum_{i=1}^{K} \left|\xi_i\left(Tr\right)\right|\right),$$

$$\|Tr\|_2^* = 1 - \left|\xi_2\left(Tr\right)\right|,$$

where $trace\,(Tr)$ is the sum of the diagonal elements of Tr, $\det(Tr)$ denotes the determinant of Tr, and $\xi_i\,(Tr)$ denotes the $i-$th eigenvalue when arranged in the sequence from the largest to the smallest eigenvalue.

$$d : M\,(m,n) \times M\,(m,n) \to \mathbb{R}, \quad (x,y) \mapsto d\,(x,y),$$

is called a metric for $M\,(m,n)$ if the following conditions are satisfied for all $x, y, z \in M\,(m,n)$:

- $d\,(x,y) \geq 0$,
- $d\,(x,y) = 0 \Longleftrightarrow x = y$,
- $d\,(x,y) = d\,(y,x)$,
- $d\,(x,z) \leq d\,(x,y) + d\,(y,z)$.

- Arvanitis, Gregory & Laurent (1999) compare the similarity of two transition matrices by computing a scalar ratio of matrix norms $\|\cdot\|$:

$$d^*_{AGL}\left(Tr,\widetilde{Tr}\right)=\frac{\left\|Tr\widetilde{Tr}-\widetilde{Tr}Tr\right\|}{\|Tr\|\cdot\left\|\widetilde{Tr}\right\|}.$$

  $d^*_{AGL}\left(Tr,\widetilde{Tr}\right)$ is bounded between 0 and 2. It equals 0 if Tr and $\widetilde{Tr}$ have the same eigenvectors and increases the more dissimilar the eigenvectors are. Arvanitis, Gregory and Laurent suggest that one-year transition matrices can be assumed to be similar if $d^*_{AGL}\left(Tr,\widetilde{Tr}\right)\leq 0.08$. However, they don't explain why 0.08 is sufficiently small and what value would be sufficiently large to reject similarity.
- One characteristic of transition matrices is that these matrices are diagonally dominant, i.e. most of the probability mass is concentrated on the diagonal. Hence, there is little overall migration. Therefore, Jafry & Schuermann (2003) suggest to subtract the identity matrix I from the matrix under consideration to get the so called mobility matrix which captures the dynamic part of the original matrix. Then they calculate the average singular value of the mobility matrix (which is the same as the sum of the square-roots of all eigenvalues of the mobility matrix) and define a generalized norm as the average singular values of the mobility matrices:

$$\|Tr\|^*_{JS}=\frac{1}{K}\sum_{i=1}^{K}\sqrt{\xi_i\left((Tr-I)^T(Tr-I)\right)}.$$

Estimation of Transition Matrices from Default Data. Sometimes it is very hard to derive meaningful empirical transition matrices because of the scarcity of the available data. E.g., if we want to determine an empirical transition matrix for emerging market sovereign bonds we face the problem that according to Perraudin (2001) until 1990, only five non-industrial sovereigns were rated. During the 1990s many Asian countries were rated for the first time and finally until the end of the 1990s many of the Eastern European and Latin American countries were rated for the first time. Hence, the rating history of such countries is very short. Therefore, Perraudin (2001) suggests to use one additional source of information, the data on default events for non-rated countries. How to combine the transition and default information to derive estimates of transition matrices is described in detail in Hu, Kiesel & Perraudin (2001). We follow their work and give a brief overview of the methodology.

Basically, the approach suggests to model sovereign defaults and sovereign ratings within a common maximum likelihood, ordered probit framework[6].

[6] For an introduction to ordered probit models see, e.g., Greene (2000), pp. 875 ff.

The quality of each obligor is assumed to be driven by some explanatory variables. To find out which variables are significant for the credit quality of a sovereign obligor one should consider some of the following empirical studies:

- Cantor & Packer (1996), Haque, Kumar & Mathieson (1996), Juttner & McCarthy (1998), and Monfort & Mulder (2000) examine the determinants of sovereign credit ratings.
- Edwards (1984) identifies the key drivers of sovereign defaults.
- Burton & Inoue (1985), Edwards (1986), Cantor & Packer (1996), Eichengreen & Mody (1998), Min (1998), and Kamin & Kleist (1999) examine the determinants of spreads of sovereign debt.

The factors used in these studies can be classified as follows:

- Liquidity variables such as debt-service-to-exports, interest-service ratio, liquidity gap ratio. These variables reflect a country's short-run financing problems.
- Solvency variables such as reserves-to-imports, export fluctuations, debt to GDP ratio. These variables reflect a country's medium to long-term ability to service its debt.
- Macroeconomic fundamentals such as inflation rate, real exchange rate, GDP growth rate, export growth rate. These variables reflect a country's long-run prospects and are used to assess the quality of a country's government and the economic dynamics within an economy.
- External shocks such as treasury rate, real oil price.

Let there be K rating categories $\{1, 2, ..., K\}$. The quality of each obligor is assumed to be driven by a latent variable Y consisting of an index of macroeconomic variables and a random error:

$$Y = \beta^T X + \varepsilon, \tag{2.1}$$

where X is the n-dimensional vector of explanatory variables, β is the n-dimensional parameter vector and $\varepsilon \sim N(0, 1)$. Y is unobservable. What we can observe is the rating of the obligor. Given the initial rating $R \in \{1, 2, ..., K-1\}$ of an obligor at the beginning of the year we assume that the obligor's rating $\check{R}$ at the end of the year equals

$$\begin{cases} K, & \text{if } Y \leq 0 = Z_{K-1}, \\ K-1, & \text{if } Z_{K-1} < Y \leq Z_{K-2}, \\ \vdots & \\ 2, & \text{if } Z_2 < Y \leq Z_1 \\ 1, & \text{if } Z_1 < Y, \end{cases} \tag{2.2}$$

where $Z = (Z_1, ..., Z_{K-1})$ is a vector of thresholds such that $0 = Z_{K-1} < ... < Z_1$. Z is unknown and must be estimated with β. Then the one-year transition probabilities $tr_{R,\check{R}}$ for all $\check{R} \in \{1, 2, ..., K\}$ are given by

$$\begin{cases} tr_{R,K} = \Phi\left(-\beta^T X\right), \\ tr_{R,K-1} = \Phi\left(Z_{K-2} - \beta^T X\right) - \Phi\left(-\beta^T X\right), \\ \vdots \\ tr_{R,2} = \Phi\left(Z_1 - \beta^T X\right) - \Phi\left(Z_2 - \beta^T X\right), \\ tr_{R,1} = 1 - \Phi\left(Z_{K-1} - \beta^T X\right), \end{cases} \tag{2.3}$$

where Φ is the operator of the standard normal distribution. The log-likelihood function can be obtained readily, and optimization can be done as usual[7]. For many sovereigns and years, ratings are not available. But at least we might know whether a sovereign has defaulted or not. Then we can still form a likelihood for the observation by including the conditional probability that default does or does not occur:

$$\begin{cases} tr_{R,K}(1) = \Phi\left(-\beta^T X\right), \\ \sum_{j=1}^{K-1} tr_{R,j}(1) = 1 - \Phi\left(-\beta^T X\right). \end{cases} \tag{2.4}$$

If ratings are observed we get likelihood entries of the kind shown in equation (2.3). If ratings are not observed we get entries of the kind shown in equation (2.4). Combining these likelihood entries allows us to consider more sovereigns and and more years than only using sovereigns and years where ratings are available. The default/non-default observations help to better estimate the $\beta's$ whereas the rating observations help to estimate the $\beta's$ and the thresholds $Z_{K-1}, ..., Z_1$. As soon as the estimates $\widehat{\beta}$ and the threshold estimates $\widehat{Z}_{K-1}, ..., \widehat{Z}_1$ have been obtained, the ratings for each obligor and each year can be determined by calculating $\widehat{\beta}^T X$ and identifying the range $\left(\widehat{Z}_j, \widehat{Z}_{j-1}\right]$ into which it falls. This allows to create a rating history for all sovereigns over the entire rating history. Given these histories, we can generate estimates of rating transition matrices as usual.

A Bayesian Approach to the Estimation of Transition Matrices. To combine information from different estimates of transition matrices a pseudo-Bayesian approach can be used. E.g., suppose we want to combine information from a Standard and Poor's transition matrix $Tr_{S\&P}$ with information from a transition matrix estimated with the ordered probit method Tr_{OP}. Then we use the Standard and Poor's matrix as prior and the second estimate as an update to get a new estimation Tr according to:

$$Tr = A \cdot Tr_{S\&P} + (I - A) \cdot Tr_{OP},$$

where $A = diag(a_1, ..., a_K)$ is a $K \times K-$ dimensional diagonal matrix with diagonal elements $a_1, ..., a_K \in [0,1]$. Since both matrices $Tr_{S\&P}$ and Tr_{OP}

[7] See, e.g., Greene (2000), pp. 820-823.

are estimates of the true transition matrix, this method corresponds to a pseudo-Bayes approach. There are at least two well-known approaches for finding appropriate values for the diagonal elements of A :

- A global approach based on a goodness of fit χ^2 statistic (see, e.g., Duffy & Santner (1989)).
- A local approach considering each row of the transition matrix individually - also based on goodness of fit statistics (see, e.g., Bishop, Fienberg & Holland (1975)). To explain the local approach we denote the number of observations with initial rating R by N_R and the number of transitions from rating R to rating $\check{R}$ by $M_{R,\check{R}}$. Then, we estimate the weighting factor for an individual row, a_R, by

$$a_R = \frac{K_R}{N_R + K_R},$$

where

$$K_R = \frac{N_R^2 - \sum_{\check{R}} M_{R,\check{R}}^2}{\sum \left(M_{R,\check{R}} - N_R \cdot (Tr_{S\&P})_{R,\check{R}} \right)^2}.$$

The better the Standard & Poor's transition matrix $Tr_{S\&P}$ fits the observations, the larger is the weight of this matrix.

Transition Matrices as Markov Chains in Discrete Time. So far we have shown how to calculate an average historical τ-year transition matrix. If we want to use such a matrix for credit risk modeling one possible assumption is that this matrix is valid for the next τ-year period and the rating process is modeled as a Markov chain[8] generated by this transition matrix. A big part of the theory of modeling transition matrices as Markov chains is adopted from classical survival analysis (for a comprehensive introduction see, e.g., Kalbfleisch & Prentice (1980)). Lancaster (1990) emphasizes applications to economics, especially to unemployment spells. Klein & Moeschberger (1997) apply the theory in biology and medicine. The most interesting applications to rating transition matrices are covered in the works of Jarrow et al. (1997), Skodeberg (1998), Lando (1999), Kavvathas (2000), Lando &

[8] A stochastic process $X = \{X_n, n \in \mathbb{N}\}$ in discrete time is a collection of random variables. A Markov chain (named in honor of Andrei Andreevich Markov) is a stochastic process with what is called the Markov property: the process consists of a sequence X_1, X_2, X_3,.... of random variables taking values in a "state space", the value of X_n being "the state of the system at time n". The (discrete-time) Markov property says that the conditional distribution of the "future"

$$X_{n+1},\ X_{n+2}, \ldots$$

given the "past", X_1, ... , X_n, depends on the past only through X_n. In other words, knowledge of the most recent past state of the system renders knowledge of less recent history irrelevant.

Skodeberg (2002), and Christensen & Lando (2002). Basically, in the context of transition matrices the Markov property means that only the rating at the beginning of the period under consideration and not the entire history is relevant for determining the transition probabilities.

The Time Homogeneous Case. Assume that the rating process is modeled as a time homogeneous[9] K-state Markov chain with ratings $R \in \{1, 2, ..., K\}$, where 1 is the best rating and K is default. Using S&P notation AAA corresponds to 1, AA to 2, ..., D to K. As a basic building block we consider the quadratic $K \times K$ dimensional $\tau-$year ($\tau \in \mathbb{R}^+$) transition matrix

$$Tr(\tau) = \begin{pmatrix} tr_{1,1}(\tau) & tr_{1,2}(\tau) & \ldots & tr_{1,K}(\tau) \\ tr_{2,1}(\tau) & tr_{2,2}(\tau) & \cdots & tr_{2,K}(\tau) \\ \vdots & \vdots & \ddots & \vdots \\ tr_{K,1}(\tau) & tr_{K,2}(\tau) & \cdots & tr_{K,K}(\tau) \end{pmatrix}.$$

Obviously, all entries must be non-negative and each row has to sum to one, i.e.

$$\sum_{R=1}^{K} tr_{R,\check{R}}(\tau) = 1, \; tr_{R,\check{R}}(\tau) \geq 0 \text{ for all } R, \check{R} \in \{1, 2, ..., K\}.$$

If default is an absorbing state $tr_{K,R}(\tau) = 0$ for all $R \in \{1, 2, ..., K-1\}$ and $tr_{K,K}(\tau) = 1$. The two assumptions, Markov property and time homogeneity, imply that we can get transition matrices for $n \cdot \tau$ years ($n \in \mathbb{N}$) by multiplying the transition matrix $Tr(\tau)$ n times with itself:

$$Tr(n \cdot \tau) = Tr(\tau)^n.$$

Given this structure, we can calculate the probability of default occurring after $n \cdot \tau$ years for a counterparty with initial rating $R \in \{1, 2, ..., K-1\}$ by

$$\begin{aligned} \sum_{\check{R} \neq K} tr_{R,\check{R}}(n \cdot \tau) &= 1 - tr_{R,K}(n \cdot \tau) \\ &= 1 - \left(Tr(\tau)^n\right)_{RK}. \end{aligned}$$

Example 2.3.1.

Based on table 2.8, which shows the historical NR-adjusted worldwide average one-year corporate transition and default rates for the S&P ratings AAA, AA,,...,CCC, we calculate the corresponding two-year transition matrix. As not all rows of S&P's matrix sum up to one we modify it slightly to satisfy this condition:

[9] Time homogeneity means that for each time period (of the same duration) the same transition matrix is used.

$$Tr(2) = (Tr(1))^2$$

$$= \begin{pmatrix} 93.06 & 6.29 & 0.45 & 0.14 & 0.06 & 0.00 & 0.00 & 0.00 \\ 0.59 & 91.00 & 7.60 & 0.61 & 0.06 & 0.11 & 0.02 & 0.01 \\ 0.05 & 2.12 & 91.44 & 5.64 & 0.47 & 0.19 & 0.04 & 0.05 \\ 0.03 & 0.23 & 4.44 & 88.98 & 4.70 & 0.95 & 0.28 & 0.39 \\ 0.04 & 0.09 & 0.44 & 6.07 & 82.72 & 7.89 & 1.22 & 1.53 \\ 0.00 & 0.08 & 0.29 & 0.41 & 5.32 & 82.05 & 4.90 & 6.95 \\ 0.10 & 0.00 & 0.31 & 0.63 & 1.57 & 9.97 & 55.83 & 31.59 \\ 0.00 & 0.00 & 0.00 & 0.00 & 0.00 & 0.00 & 0.00 & 100.00 \end{pmatrix}^2$$

$$= \begin{pmatrix} 86.64 & 11.59 & 1.32 & 0.32 & 0.12 & 0.01 & 0.00 & 0.00 \\ 1.09 & 83.00 & 13.90 & 1.53 & 0.18 & 0.22 & 0.04 & 0.04 \\ 0.11 & 3.88 & 84.02 & 10.22 & 1.10 & 0.43 & 0.09 & 0.15 \\ 0.06 & 0.51 & 8.05 & 79.72 & 8.15 & 2.03 & 0.51 & 0.97 \\ 0.07 & 0.19 & 1.07 & 10.49 & 69.16 & 13.18 & 2.09 & 3.75 \\ 0.01 & 0.15 & 0.57 & 1.07 & 8.86 & 68.24 & 6.82 & 14.28 \\ 0.15 & 0.02 & 0.52 & 1.07 & 2.74 & 13.88 & 31.68 & 49.94 \\ 0.00 & 0.00 & 0.00 & 0.00 & 0.00 & 0.00 & 0.00 & 100.00 \end{pmatrix}.$$

Transition Probabilities and the Business Cycle. Historical transition matrices provide average transition probabilities sampled over many different firms and years. As described in the previous subsection, in practice it is quite common to use such transition matrices, implicitly assuming that transition probabilities and especially default probabilities are constant over time. These historical transition matrices are considered as good approximations for the actual transition matrices if the volume of data is big enough. According to Jarrow et al. (1997) the assumption of time homogeneity is more reasonable for investment grade bonds, than it is for speculative grade bonds. Indeed, one should keep in mind that such assumptions neglect that economic conditions change continuously and that there is a strong relationship between the transition probabilities and the business cycle. But using historical transition and default rate estimates as approximations for future transition and default rates neglects default rate volatility, especially at low rating levels, and the correlation to the business cycle. But actual transition and default probabilities are very dynamic and can strongly vary depending on general economic conditions. Hence, the historical observations may not reflect the current credit environment. As an example for a very volatile rate consider the CCC one-year default rate curve of figure 2.1: it shows the recessions of the years 1982, at the end of the 80's/beginning 90's, 1995, and the beginning 2000's.

Duffie & Singleton (2003) compute four-quarter moving averages of the ratio of the total number of rating upgrades to the total number of downgrades, separately for investment and speculative-grade rating classes, and call it U/D ratio. They compare these ratios to the four-quarter moving average

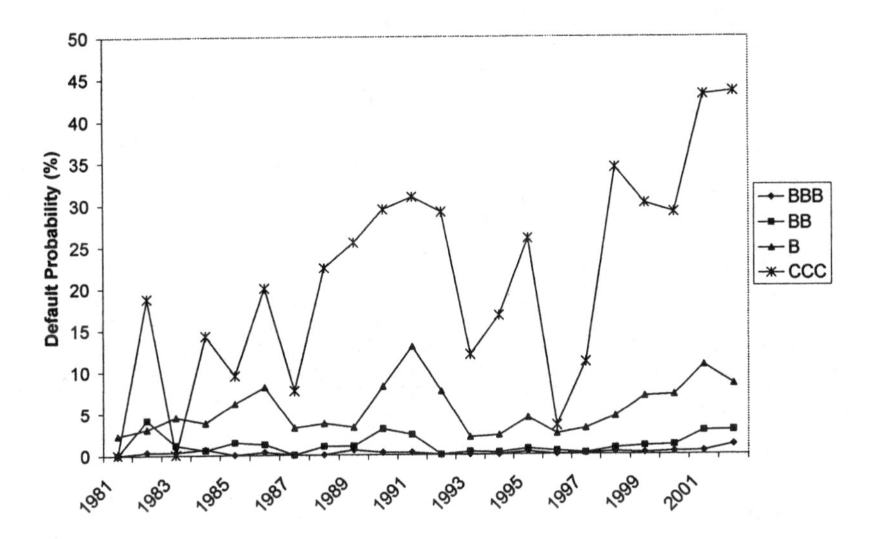

Fig. 2.1. S&P one-year default rates (%) by rating class for the years [1980, 2002]. Source: Standard & Poor's (*Special Report: Ratings Performance 2002* 2003).

of U.S. gross domestic product (GDP) growth rates over the sample period 1983 – 1997. They find a GDP - U/D ratio sample correlation of 0.198 for investment-grade issues and a sample correlation of 0.652 for speculative-grade issues. This result supports the previous finding that time homogeneity is more reasonable for investment grade bonds, than it is for speculative grade bonds.

Nickell, Perraudin & Varotto (1998) examine whether probabilities of moving between rating classes over one-year horizons vary across different stages of the business cycle. As data they use the universe of notional unsecured Moody's long-term corporate and sovereign bond ratings from December 1970 to December 1997 - a sample of 6534 obligor rating histories. First, they define three categories: peak, normal times and trough, depending on GDP values recorded in the sample period. Then they estimate the unconditional transition matrix (table 2.9) as well as three transition matrices (tables 2.10 - 2.12) that are conditional on the state of the economy.

Nickell et al. (1998) draw several conclusions:

- In business cycle peaks, low-rated obligors have much less rating volatility and are less prone to downgrades.
- The volatility of investment grade bonds falls sharply in business cycle peak years and rises in business cycle troughs.

Table 2.9. Unconditional rating transition matrix based on the time frame [1970,1997] and Moody's unsecured long-term corporate and sovereign bond ratings (entries in %).

Initial Rating	Rating at year end								
	Aaa	*Aa*	*A*	*Baa*	*Ba*	*B*	*Caa*	*Ca/C*	*D*
Aaa	90.4	8.7	0.8	–	0.0	–	–	–	–
Aa	1.1	89.5	8.9	0.4	0.1	0.0	–	–	–
A	0.1	2.3	92.1	5.0	0.5	0.1	0.0	–	0.0
Baa	0.0	0.2	5.4	89.1	4.4	0.6	0.1	–	0.1
Ba	0.0	0.0	0.5	5.4	85.7	6.7	0.2	0.0	1.3
B	0.0	0.1	0.2	0.7	6.8	83.0	1.9	0.5	6.9
Caa	–	–	–	0.9	2.5	8.0	66.6	3.7	18.4
Ca/C	–	–	–	–	0.9	5.6	15.0	57.9	20.6
D	–	–	–	–	–	–	–	–	100.0

Source: (Nickell et al. 1998).

Table 2.10. Conditional transition matrix based on the time frame [1970,1997] and Moody's unsecured Moody's long-term corporate and sovereign bond ratings, business cycle trough (entries in %).

Initial Rating	Rating at year end								
	Aaa	*Aa*	*A*	*Baa*	*Ba*	*B*	*Caa*	*Ca/C*	*D*
Aaa	89.6	10.0	0.4	–	–	–	–	–	–
Aa	0.9	88.3	10.7	0.1	0.0	–	–	–	–
A	0.1	2.7	91.1	5.6	0.4	0.0	–	–	0.0
Baa	0.0	0.3	6.6	86.8	5.6	0.4	0.2	–	0.1
Ba	–	0.1	0.5	5.9	83.1	8.4	0.3	0.0	1.7
B	–	0.1	0.2	0.8	6.6	79.6	2.2	1.0	9.4
Caa	–	–	–	0.9	1.9	9.3	63.0	1.9	23.1
Ca/C	–	–	–	–	–	5.9	5.9	64.7	23.5
D	–	–	–	–	–	–	–	–	100.0

Source: (Nickell et al. 1998).

- Default probabilities are very sensitive to the business cycle.

We basically did the same analysis as Nickell et al. (1998) but used S&P rating histories of roughly 6600 US corporate bonds from December 1980 until December 2002. The results are given in figure 2.2.

If one wants to perform this kind of analysis on sub-samples the problem may arise that there is not enough data available. Nickell et al. (1998) suggest to use an ordered probit model to test for significance of different exogenous factors (such as the business cycle) on transition matrices for quite specific

	AAA	AA	A	BBB	BB	B	CCC	CC	C	D
AAA	93.98	5.40	0.50	0.04	0.08	0.00	0.00	0.00	0.00	0.00
AA	0.58	91.78	6.78	0.61	0.10	0.12	0.01	0.00	0.00	0.01
A	0.07	2.05	91.81	5.38	0.48	0.17	0.01	0.02	0.00	0.02
BBB	0.04	0.22	4.53	89.63	4.40	0.80	0.13	0.01	0.02	0.21
BB	0.03	0.07	0.47	6.06	83.40	8.15	0.88	0.03	0.00	0.91
B	0.00	0.09	0.25	0.36	4.89	84.77	4.89	0.54	0.04	4.17
CCC	0.07	0.00	0.14	0.43	0.93	5.14	78.87	1.36	0.14	12.92
CC	0.00	0.00	0.00	0.00	0.77	0.77	3.08	83.85	0.77	10.77
C	0.00	0.00	0.00	0.00	0.00	0.00	12.50	0.00	50.00	37.50
D	0.00	0.00	0.00	0.00	0.00	0.00	0.00	0.00	0.00	100.00

	AAA	AA	A	BBB	BB	B	CCC	CC	C	D
AAA	92.73	6.73	0.36	0.09	0.09	0.00	0.00	0.00	0.00	0.00
AA	0.60	93.39	5.14	0.66	0.03	0.13	0.03	0.00	0.00	0.03
A	0.07	2.15	93.21	4.13	0.28	0.15	0.00	0.00	0.00	0.02
BBB	0.04	0.31	5.47	89.53	3.56	0.85	0.07	0.00	0.00	0.16
BB	0.07	0.07	0.36	7.13	84.75	6.53	0.66	0.00	0.00	0.43
B	0.00	0.03	0.41	0.41	5.42	86.55	3.45	0.38	0.06	3.30
CCC	0.00	0.00	0.18	0.55	0.91	6.03	83.18	0.37	0.18	8.59
CC	0.00	0.00	0.00	0.00	0.00	0.00	0.00	90.63	3.13	6.25
C	0.00	0.00	0.00	0.00	0.00	0.00	33.33	0.00	33.33	33.33
D	0.00	0.00	0.00	0.00	0.00	0.00	0.00	0.00	0.00	100.00

	AAA	AA	A	BBB	BB	B	CCC	CC	C	D
AAA	95.02	3.98	0.87	0.00	0.12	0.00	0.00	0.00	0.00	0.00
AA	0.57	90.42	7.82	0.88	0.13	0.18	0.00	0.00	0.00	0.00
A	0.07	1.63	91.34	5.92	0.72	0.28	0.02	0.00	0.00	0.02
BBB	0.00	0.18	3.90	89.99	4.66	0.94	0.18	0.03	0.00	0.12
BB	0.00	0.08	0.58	5.49	85.27	6.73	0.87	0.08	0.00	0.91
B	0.00	0.07	0.07	0.40	5.16	84.83	5.26	0.44	0.00	3.78
CCC	0.00	0.00	0.18	0.36	0.89	5.35	78.43	1.43	0.18	13.19
CC	0.00	0.00	0.00	0.00	1.39	1.39	5.56	77.78	0.00	13.89
C	0.00	0.00	0.00	0.00	0.00	0.00	0.00	0.00	75.00	25.00
D	0.00	0.00	0.00	0.00	0.00	0.00	0.00	0.00	0.00	100.00

	AAA	AA	A	BBB	BB	B	CCC	CC	C	D
AAA	95.04	4.76	0.20	0.00	0.00	0.00	0.00	0.00	0.00	0.00
AA	0.54	90.24	8.99	0.00	0.23	0.00	0.00	0.00	0.00	0.00
A	0.07	2.52	89.48	7.31	0.51	0.04	0.00	0.07	0.00	0.00
BBB	0.10	0.10	3.51	89.26	5.79	0.45	0.20	0.00	0.10	0.50
BB	0.00	0.07	0.51	4.71	77.16	14.21	1.38	0.00	0.00	1.96
B	0.00	0.26	0.26	0.20	3.15	80.62	7.42	1.12	0.07	6.90
CCC	0.34	0.00	0.00	0.34	1.02	3.07	71.67	3.07	0.00	20.48
CC	0.00	0.00	0.00	0.00	0.00	0.00	0.00	92.31	0.00	7.69
C	0.00	0.00	0.00	0.00	0.00	0.00	0.00	0.00	0.00	100.00
D	0.00	0.00	0.00	0.00	0.00	0.00	0.00	0.00	0.00	100.00

Fig. 2.2. Transition matrices estimated from US corporate bond rating histories from December 1980 until December 2002. From above: Unconditional transition matric, conditional transition matrix: business cycle peak, conditional transition matrix: business cycle nnormal, conditional transition matrix: business cycle trough.

Table 2.11. Conditional transition matrix based on the time frame [1970,1997] and Moody's unsecured long-term corporate and sovereign bond ratings, business cycle normal (entries in %).

Initial Rating	Rating at year end								
	Aaa	*Aa*	*A*	*Baa*	*Ba*	*B*	*Caa*	*Ca/C*	*D*
Aaa	92.2	7.4	0.3	–	0.1	–	–	–	–
Aa	1.5	87.5	10.1	0.7	0.2	–	–	–	–
A	0.0	1.8	91.7	5.4	0.8	0.2	0.0	–	–
Baa	0.1	0.2	5.2	88.1	4.9	1.2	0.0	–	0.2
Ba	0.1	0.0	0.3	5.4	85.7	6.7	0.2	0.0	1.5
B	0.1	0.1	0.4	0.8	6.6	83.6	1.6	0.3	6.6
Caa	–	–	–	–	2.8	9.3	59.8	8.4	19.6
Ca/C	–	–	–	–	–	8.3	8.3	70.8	12.5
D	–	–	–	–	–	–	–	–	100.0

Source: (Nickell et al. 1998).

Table 2.12. Conditional transition matrix based on the time frame [1970,1997] and Moody's unsecured long-term corporate and sovereign bond ratings, business cycle peak (entries in %).

Initial Rating	Rating at year end								
	Aaa	*Aa*	*A*	*Baa*	*Ba*	*B*	*Caa*	*Ca/C*	*D*
Aaa	89.7	8.5	1.8	–	–	–	–	–	–
Aa	0.8	93.2	5.6	0.3	0.1	0.1	–	–	–
A	0.0	2.3	93.4	3.9	0.3	0.1	–	–	–
Baa	–	0.2	4.4	92.2	2.8	0.3	0.1	–	0.1
Ba	–	0.0	0.6	4.8	88.5	5.0	0.3	–	0.7
B	–	–	0.1	0.3	7.2	85.8	2.0	0.1	4.5
Caa	–	–	–	1.8	2.7	5.4	76.6	0.9	12.6
Ca/C	–	–	–	–	2.0	4.1	24.5	46.9	22.4
D	–	–	–	–	–	–	–	–	100.0

Source: (Nickell et al. 1998).

obligor categories (for a definition and explanation of the ordered probit model see equations 2.1 - 2.3). E.g., they consider US banking and US industrial bonds as sub-samples - both for different states of the business cycle (see tables 2.13 - 2.16). Such a model implicitly assumes independence across obligors within a year, given the state of the business cycle. Correlation is induced only by changes of the state of the business cycle.

We can make several interesting observations:

Table 2.13. Conditional transition matrix based on the time frame [1970,1997] and Moody's unsecured long-term US banking bond ratings, business cycle trough (entries in %).

Initial Rating	Rating at year end								
	Aaa	*Aa*	*A*	*Baa*	*Ba*	*B*	*Caa*	*Ca/C*	*D*
Aaa	83.1	15.0	1.9	–	–	–	–	–	–
Aa	0.4	84.6	14.1	0.7	0.2	–	–	–	–
A	0.0	2.0	92.0	5.3	0.5	0.1	–	–	–
Baa	–	0.7	11.1	86.3	1.7	0.2	0.0	–	0.0
Ba	–	–	0.8	7.9	86.6	4.0	0.1	–	0.6
B	–	–	0.4	1.1	10.2	82.9	1.3	0.3	3.8
Caa	–	–	–	–	1.7	6.2	66.5	4.3	21.3
Ca/C	–	–	–	–	–	0.7	4.2	48.8	46.3
D	–	–	–	–	–	–	–	–	100.0

Source: (Nickell et al. 1998).

Table 2.14. Conditional transition matrix based on the time frame [1970,1997] and Moody's unsecured long-term US banking bond ratings, business cycle peak (entries in %).

Initial Rating	Rating at year end								
	Aaa	*Aa*	*A*	*Baa*	*Ba*	*B*	*Caa*	*Ca/C*	*D*
Aaa	87.7	11.1	1.1	–	–	–	–	–	–
Aa	0.9	90.0	8.7	0.3	0.1	–	–	–	–
A	0.0	2.0	91.9	5.4	0.5	0.1	–	–	–
Baa	–	0.7	10.8	86.5	1.8	0.2	0.0	–	0.0
Ba	–	–	1.1	9.7	85.6	3.1	0.1	–	0.4
B	–	–	0.6	1.5	12.1	81.7	1.0	0.2	2.9
Caa	–	–	–	–	3.7	10.6	69.8	3.1	12.7
Ca/C	–	–	–	–	–	2.0	8.5	58.6	30.9
D	–	–	–	–	–	–	–	–	100.0

Source: (Nickell et al. 1998).

- In a trough, highly-rated banks are much more subject to downgrades than industrials.
- There is less of a difference for lower credit quality bank and industrial obligors.

We did the same analysis on sub-samples of our own database, i.e. on S&P rating histories of roughly 6600 US corporate bonds from December 1980 until December 2002. We considered two sub-samples: bonds from the financial

Table 2.15. Conditional transition matrix based on the time frame [1970,1997] and Moody's unsecured long-term US industrial bond ratings, business cycle trough (entries in %).

Initial Rating	Rating at year end								
	Aaa	*Aa*	*A*	*Baa*	*Ba*	*B*	*Caa*	*Ca/C*	*D*
Aaa	89.0	10.0	0.9	–	–	–	–	–	–
Aa	0.6	87.8	10.9	0.5	0.1	–	–	–	–
A	0.1	2.3	92.4	4.7	0.4	0.1	–	–	–
Baa	–	0.2	4.6	89.5	4.8	0.7	0.1	–	0.1
Ba	–	–	0.2	3.5	85.7	8.5	0.3	–	1.8
B	–	–	0.2	0.5	5.7	83.5	2.1	0.5	7.5
Caa	–	–	–	–	2.2	7.5	68.1	3.9	18.3
Ca/C	–	–	–	–	–	3.9	13.1	61.8	21.2
D	–	–	–	–	–	–	–	–	100.0

Source: (Nickell et al. 1998).

Table 2.16. Conditional transition matrix based on the time frame [1970,1997] and Moody's unsecured long-term US industrial bond ratings, business cycle peak (entries in %).

Initial Rating	Rating at year end								
	Aaa	*Aa*	*A*	*Baa*	*Ba*	*B*	*Caa*	*Ca/C*	*D*
Aaa	92.4	7.1	0.5	–	–	–	–	–	–
Aa	1.4	91.9	6.5	0.2	0.1	–	–	–	–
A	0.1	2.3	92.3	4.8	0.5	0.1	–	–	–
Baa	–	0.2	4.5	89.5	4.9	0.7	0.1	–	0.1
Ba	–	–	0.3	4.4	86.7	7.0	0.2	–	1.3
B	–	–	0.2	0.6	7.0	83.9	1.8	0.4	6.0
Caa	–	–	–	–	4.8	12.3	69.7	2.8	10.5
Ca/C	–	–	–	–	–	8.8	20.5	59.4	11.4
D	–	–	–	–	–	–	–	–	100.0

Source: (Nickell et al. 1998)

industry and insurance companies on the one hand, and bonds from all other industries on the other hand. The results are given in figures 2.3 and 2.4.

To allow for the influence of the business cycle we extend our Markov chain approach to time non-homogeneity.

The Time Non-Homogeneous Case. In analyzing transition rate data Altman & Kao (1992a), Altman & Kao (1992c), Carty & Fons (1993), Altman (1998), Nickell, Perraudin & Varotto (2000), Bangia et al. (26), and Lando & Skodeberg (2002) have shown the presence of time non-homogeneity such as

	AAA	AA	A	BBB	BB	B	CCC	CC	C	D
AAA	95.24	4.41	0.21	0.07	0.07	0.00	0.00	0.00	0.00	0.00
AA	0.50	92.82	6.22	0.37	0.03	0.00	0.03	0.00	0.00	0.03
A	0.07	2.83	93.53	3.15	0.29	0.10	0.00	0.00	0.00	0.02
BBB	0.06	0.91	5.95	87.66	3.79	0.79	0.51	0.00	0.00	0.34
BB	0.00	0.30	1.64	10.76	76.83	7.62	1.49	0.00	0.00	1.35
B	0.00	0.28	0.84	1.68	7.56	77.59	7.56	0.28	0.00	4.20
CCC	0.53	0.00	0.00	0.00	2.65	2.12	85.71	1.59	0.00	7.41
CC	0.00	0.00	0.00	0.00	0.00	0.00	0.00	87.50	0.00	12.50
C	-	-	-	-	-	-	-	-	-	-
D	0.00	0.00	0.00	0.00	0.00	0.00	0.00	0.00	0.00	100.00

	AAA	AA	A	BBB	BB	B	CCC	CC	C	D
AAA	95.13	4.41	0.15	0.15	0.15	0.00	0.00	0.00	0.00	0.00
AA	0.56	94.37	4.52	0.35	0.07	0.00	0.07	0.00	0.00	0.07
A	0.05	3.22	94.33	2.20	0.05	0.10	0.00	0.00	0.00	0.05
BBB	0.12	1.09	7.41	87.73	2.19	0.73	0.36	0.00	0.00	0.36
BB	0.00	0.35	1.75	14.39	75.79	5.96	1.05	0.00	0.00	0.70
B	0.00	0.00	1.29	2.58	9.03	80.00	5.16	0.65	0.00	1.29
CCC	0.00	0.00	0.00	0.00	1.20	3.61	87.95	0.00	0.00	7.23
CC	0.00	0.00	0.00	0.00	0.00	0.00	0.00	100.00	0.00	0.00
C	-	-	-	-	-	-	-	-	-	-
D	0.00	0.00	0.00	0.00	0.00	0.00	0.00	0.00	0.00	100.00

	AAA	AA	A	BBB	BB	B	CCC	CC	C	D
AAA	95.74	3.87	0.39	0.00	0.00	0.00	0.00	0.00	0.00	0.00
AA	0.28	92.49	6.67	0.56	0.00	0.00	0.00	0.00	0.00	0.00
A	0.07	2.44	94.56	2.72	0.14	0.07	0.00	0.00	0.00	0.00
BBB	0.00	0.96	4.82	88.10	4.50	0.80	0.48	0.00	0.00	0.32
BB	0.00	0.40	1.61	10.89	79.44	5.24	0.81	0.00	0.00	1.61
B	0.00	0.00	0.72	1.45	8.70	77.54	6.52	0.00	0.00	5.07
CCC	0.00	0.00	0.00	0.00	2.94	1.47	88.24	1.47	0.00	5.88
CC	0.00	0.00	0.00	0.00	0.00	0.00	0.00	83.33	0.00	16.67
C	-	-	-	-	-	-	-	-	-	-
D	0.00	0.00	0.00	0.00	0.00	0.00	0.00	0.00	0.00	100.00

	AAA	AA	A	BBB	BB	B	CCC	CC	C	D
AAA	94.49	5.51	0.00	0.00	0.00	0.00	0.00	0.00	0.00	0.00
AA	0.81	89.00	10.18	0.00	0.00	0.00	0.00	0.00	0.00	0.00
A	0.15	2.52	89.04	6.81	1.33	0.15	0.00	0.00	0.00	0.00
BBB	0.00	0.31	4.36	86.60	6.54	0.93	0.93	0.00	0.00	0.31
BB	0.00	0.00	1.47	2.94	74.26	15.44	3.68	0.00	0.00	2.21
B	0.00	1.56	0.00	0.00	1.56	71.88	15.63	0.00	0.00	9.38
CCC	2.63	0.00	0.00	0.00	5.26	0.00	76.32	5.26	0.00	10.53
CC	0.00	0.00	0.00	0.00	0.00	0.00	0.00	75.00	0.00	25.00
C	-	-	-	-	-	-	-	-	-	-
D	0.00	0.00	0.00	0.00	0.00	0.00	0.00	0.00	0.00	100.00

Fig. 2.3. Transition matrices estimated from US bond rating histories (financial industry and insurance companies) from December 1980 until December 2002. From above: Unconditional transition matric, conditional transition matrix: business cycle peak, conditional transition matrix: business cycle nnormal, conditional transition matrix: business cycle trough.

	AAA	AA	A	BBB	BB	B	CCC	CC	C	D
AAA	92.14	6.84	0.92	0.00	0.10	0.00	0.00	0.00	0.00	0.00
AA	0.64	90.95	7.23	0.80	0.16	0.21	0.00	0.00	0.00	0.00
A	0.07	1.70	91.03	6.40	0.56	0.20	0.01	0.02	0.00	0.01
BBB	0.04	0.07	4.22	90.07	4.53	0.80	0.05	0.01	0.02	0.19
BB	0.03	0.05	0.34	5.54	84.11	8.21	0.81	0.03	0.00	0.86
B	0.00	0.08	0.22	0.30	4.76	85.11	4.76	0.55	0.04	4.17
CCC	0.00	0.00	0.17	0.50	0.66	5.61	77.81	1.32	0.17	13.78
CC	0.00	0.00	0.00	0.00	0.88	0.88	3.51	83.33	0.88	10.53
C	0.00	0.00	0.00	0.00	0.00	0.00	12.50	0.00	50.00	37.50
D	0.00	0.00	0.00	0.00	0.00	0.00	0.00	0.00	0.00	100.00

	AAA	AA	A	BBB	BB	B	CCC	CC	C	D
AAA	89.16	10.16	0.68	0.00	0.00	0.00	0.00	0.00	0.00	0.00
AA	0.63	92.58	5.65	0.91	0.00	0.23	0.00	0.00	0.00	0.00
A	0.07	1.63	92.67	5.06	0.39	0.17	0.00	0.00	0.00	0.00
BBB	0.03	0.14	5.03	89.94	3.88	0.88	0.00	0.00	0.00	0.11
BB	0.07	0.04	0.22	6.37	85.69	6.59	0.62	0.00	0.00	0.40
B	0.00	0.03	0.36	0.30	5.25	86.86	3.37	0.36	0.06	3.40
CCC	0.00	0.00	0.22	0.65	0.86	6.47	82.33	0.43	0.22	8.84
CC	0.00	0.00	0.00	0.00	0.00	0.00	0.00	88.46	3.85	7.69
C	0.00	0.00	0.00	0.00	0.00	0.00	33.33	0.00	33.33	33.33
D	0.00	0.00	0.00	0.00	0.00	0.00	0.00	0.00	0.00	100.00

	AAA	AA	A	BBB	BB	B	CCC	CC	C	D
AAA	93.73	4.18	1.74	0.00	0.35	0.00	0.00	0.00	0.00	0.00
AA	0.84	88.52	8.86	1.18	0.25	0.34	0.00	0.00	0.00	0.00
A	0.07	1.22	89.72	7.52	1.01	0.38	0.03	0.00	0.00	0.03
BBB	0.00	0.00	3.69	90.42	4.70	0.98	0.11	0.04	0.00	0.07
BB	0.00	0.05	0.46	4.87	85.93	6.90	0.87	0.09	0.00	0.83
B	0.00	0.07	0.04	0.35	4.99	85.18	5.20	0.46	0.00	3.72
CCC	0.00	0.00	0.20	0.41	0.61	5.88	77.08	1.42	0.20	14.20
CC	0.00	0.00	0.00	0.00	1.52	1.52	6.06	77.27	0.00	13.64
C	0.00	0.00	0.00	0.00	0.00	0.00	0.00	0.00	75.00	25.00
D	0.00	0.00	0.00	0.00	0.00	0.00	0.00	0.00	0.00	100.00

	AAA	AA	A	BBB	BB	B	CCC	CC	C	D
AAA	95.60	4.00	0.40	0.00	0.00	0.00	0.00	0.00	0.00	0.00
AA	0.37	90.99	8.27	0.00	0.37	0.00	0.00	0.00	0.00	0.00
A	0.05	2.52	89.62	7.47	0.24	0.00	0.00	0.10	0.00	0.00
BBB	0.12	0.06	3.35	89.76	5.65	0.35	0.06	0.00	0.12	0.53
BB	0.00	0.08	0.40	4.91	77.47	14.08	1.13	0.00	0.00	1.93
B	0.00	0.21	0.27	0.21	3.22	81.00	7.06	1.17	0.07	6.79
CCC	0.00	0.00	0.00	0.39	0.39	3.53	70.98	2.75	0.00	21.96
CC	0.00	0.00	0.00	0.00	0.00	0.00	0.00	95.45	0.00	4.55
C	0.00	0.00	0.00	0.00	0.00	0.00	0.00	0.00	0.00	100.00
D	0.00	0.00	0.00	0.00	0.00	0.00	0.00	0.00	0.00	100.00

Fig. 2.4. Transition matrices estimated from US corporate bond rating histories (all industries but financial industry and insurance companies) from December 1980 until December 2002. From above: Unconditional transition matric, conditional transition matrix: business cycle peak, conditional transition matrix: business cycle nnormal, conditional transition matrix: business cycle trough.

ratings drift and sensitivity to the business cycle. If we assume that the transition probabilities are changing over time, hence relaxing the assumption of time homogeneity the multi-period matrix must be indexed by the start and end date of the period over which we consider the transitions.

Definition 2.3.2.
The transition matrix describing the probability distribution of ratings at time t given the rating at time t_0 $(t_0 < t)$ is denoted by

$$Tr\,(t_0, t) = \left(tr_{R,\check{R}}\,(t_0, t)\right)_{R,\check{R}},$$

where $tr_{R,\check{R}}\,(t_0, t)$ is the probability of transition from rating R at time t_0 to rating $\check{R}$ at time t.

If there are discrete points in time

$$t_0 < t_1 < .... < t_n < t_{n+1} < ...,$$

and the transition matrices for the time periods $[t_{i-1}, t_i]$ are given by $Tr\,(t_{i-1}, t_i)\,,\ i = 1, ..., n$, then by the Markov property the multi-period matrix $Tr\,(t_0, t_n)$ is given by

$$Tr\,(t_0, t_n) = \prod_{i=1}^{n} Tr\,(t_{i-1}, t_i)\,.$$

Given this structure, we can calculate the probability of default occurring after time t_n for a counterparty with initial rating $R \in \{1, 2, ..., K-1\}$ as

$$\begin{aligned}\sum_{\check{R} \neq K} tr_{R,\check{R}}\,(t_0, t_n) &= 1 - tr_{R,K}\,(t_0, t_n) \\ &= 1 - \left(\prod_{i=1}^{n} Tr\,(t_{i-1}, t_i)\right)_{R,K}.\end{aligned}$$

Example 2.3.2.
Let $t_i - t_{i-1} = 1$ year for all $i = 1, 2, ...$ We assume that the one-year transition probabilities are only dependent on some underlying driving state variable such as the business cycle. In addition, we assume that this state variable forms a Markov process X_{t_i}, $i = 0, 1, 2, ..., n, ...$. Then we can write for all i

$$Tr\,(t_{i-1}, t_i) = Tr\left(X_{t_{i-1}}\right)$$

and get for all n

$$Tr\,(t_0, t_n) = \prod_{i=1}^{n} Tr\left(X_{t_{i-1}}\right),$$

i.e. the multiyear transition probabilities depend on the path taken by the Markov process X.

Detecting Non-Markovian Behavior. Even if we consider changing economic conditions by using a time non-homogeneous model the Markov chain assumption is still problematic. As, e.g., Altman & Kao (1992a), Altman & Kao (1992c), Carthy & Fons (1994), Jonsson & Fridson (1996), Helwege & Kleiman (1996), Kavvathas (2000), and Lando & Skodeberg (2002) point out, there is dependence of transition probabilities on the duration in a rating class and the age of the specific defaultable financial instrument, sometimes called aging effect. E.g., Altman & Kao (1992a) track over time ratings for individual bond issues. They find that default risk is increasing in the first three or four years of an issue's life although the effect disappears thereafter. In addition, as Carthy & Fons (1994), Behar & Nagpal (1999), Kavvathas (2000), and Lando & Skodeberg (2002) show, there is momentum in ratings transition data, i.e. for obligors of certain ratings, the prior rating is a relevant determinant of the likelihood of a down- or upgrade. In addition, obligors of the same rating may still have different credit qualities.

Transition Probabilities and Aging Effects. Carthy & Fons (1994) use a rating database containing data from May 1923 through June 1993 and a corporate bond default database covering years from 1970 through 1992 to analyze the average length of time that an issuer holds a specific rating. They find that a Weibull distribution[10] most closely models the life span characteristics of bond ratings. Using maximum likelihood techniques they estimate the parameters of the distribution for each of Moody's long- and short-term rating categories. From the estimated distributions it is easy to calculate the mean lifetime of each rating category. Tables (2.17) and (2.18) show the results of the parameter estimations. In general, the shape and the scale parameters and the length of time that an issuer holds a specific rating decrease as the credit quality decreases.

Rating Momentum. Rating momentum means that prior rating changes may have predictive power for the direction of future rating changes. For example, downgrade rating momentum suggests that an obligor downgraded to B is more likely to be further downgraded than upgraded within the next year. Therefore, the Markov property fails, as the current rating does not fully determine the transition probabilities.

For Moody's investment grade letter rating categories Carthy & Fons (1994) test the following hypotheses:

- Hypothesis 1 (H1) : There is no upgrade momentum (i.e. the probability of an upgrade, given the previous rating change was an upgrade, is less

[10] A random variable X has the Weibull distribution if it has a probability density function of the form

$$w(x, a, b) = \left(\frac{b}{a}\right)\left(\frac{x}{a}\right)^{b-1} e^{-\left(\frac{x}{a}\right)^{b}},$$

where $a > 0$ and $b \geq 0$. a is called shape parameter and b is called scale parameter.

Table 2.17. Estimated scale and shape parameters for the Weibull distribution for long-term rating categories.

Rating	Shape	Scale	Mean
Aaa	1.10	10.29	9.93
Aa	1.16	8.90	8.45
A	1.02	9.99	9.91
Baa	0.99	7.28	7.31
Ba	1.06	5.04	4.93
B	0.91	3.90	4.08
Caa	0.57	2.27	3.66

Source: Carthy & Fons (1994) and own calculations.

Table 2.18. Estimated scale and shape parameters for the Weibull distribution for short-term rating categories.

Rating	Shape	Scale	Mean
Aaa	1.38	6.85	6.26
Aa1	1.05	7.76	7.61
Aa2	1.25	3.66	3.41
Aa3	1.12	5.30	5.08
A1	1.16	4.02	3.82
A2	1.09	3.83	3.71
A3	1.09	3.04	2.94
Baa1	1.13	2.90	2.77
Baa2	1.05	2.66	2.61
Baa3	1.15	2.29	2.18
Ba1	1.10	2.62	2.53
Ba2	1.10	2.47	2.38
Ba3	1.20	3.08	2.90
B1	1.03	2.64	2.61
B2	1.06	2.14	2.09
B3	0.82	1.76	1.96
Caa	0.70	1.16	1.47

Source: Carthy & Fons (1994) and own calculations.

than or equal to the probability of a downgrade, given the previous rating change was an upgrade).

- Hypothesis 2 (H2) : There is no downgrade momentum (i.e. the probability of a downgrade, given the previous rating change was a downgrade, is less than or equal to the probability of a upgrade, given the previous change was a downgrade).

Therefore, they only consider obligors with at least one rating change and group them according to whether the rating change is an upgrade or downgrade and according to the rating category.

For the obligors of each group i, j $(i \in \{Aa, A, Baa, Ba, B\}, \ j \in \{u, d\})$, $u = up$, $d = down$, we define the following variables:

$$\delta_{i,j} = \begin{cases} 1 \ , & \text{in case of an upgrade within 1 year,} \\ -1 \ , & \text{in case of no rating change within 1 year,} \\ 0 \ , & \text{in case of a downgrade within 1 year,} \end{cases}$$

and

$$\begin{aligned} p_{i,j,u} &= \text{ the probability that } \delta_{i,j} = 1, \\ p_{i,j,d} &= \text{ the probability that } \delta_{i,j} = -1, \\ p_{i,j,0} &= \text{ the probability that } \delta_{i,j} = 0. \end{aligned}$$

Let $n_{i,j}$ be the number of obligors in group i, j and $\delta_{i,j,k}$ the realization of random variable $\delta_{i,j}$ according to the behavior of the $k-$th obligor in group i, j. As the sample sizes are very large, it is assumed that the sample means

$$Z_{i,j} = \frac{1}{n_{i,j}} \sum_{k=1}^{n_{i,j}} \delta_{i,j,k}, \ i \in \{Aa, A, Baa, Ba, B\}, \ j \in \{u, d\},$$

are asymptotically normal with mean $p_{i,j,u} - p_{i,j,d}$ and variance $p_{i,j,u} + p_{i,j,d} - (p_{i,j,u} - p_{i,j,d})^2$. Then for each rating class i, H1 and H2 can be expressed as $p_{i,u,u} - p_{i,u,d} \leq 0$ and $p_{i,d,d} - p_{i,d,u} \leq 0$, respectively. If the t-statistics for H1 and H2,

$$\frac{\sqrt{n_{i,u}} \left(p_{i,u,u} - p_{i,u,d}\right)}{\sqrt{p_{i,u,u} + p_{i,u,d} - \left(p_{i,u,u} - p_{i,u,d}\right)^2}} \text{ and } \frac{\sqrt{n_{i,d}} \left(p_{i,d,d} - p_{i,d,u}\right)}{\sqrt{p_{i,d,u} + p_{i,d,d} - \left(p_{i,d,u} - p_{i,d,d}\right)^2}},$$

are large positive values, H1 and H2 have to be rejected, respectively. The results of the tests are given in table 2.19.

For a 5% confidence interval we can summarize the results of Carthy & Fons (1994) as follows:

- Besides for rating category B there is no upward momentum.
- The probability of a downgrade following a downgrade exceeds that of an upgrade following a downgrade. The hypothesis that there is a downgrade momentum can't be rejected.

Transition Probabilities and Industry/Domicile Effects. Depending on the specific purpose, worldwide transition data is not the right data to use. Country or region specific data can vary extremely from the average worldwide data. Tables 2.20, 2.21, and 2.22 show region specific transition matrices for the U.S. & Canada, Europe and Malaysia, respectively. Especially the Malaysian bond market is quite different from international default experience.

Table 2.19. Results of the tests for upgrade and downgrade momentum of Moody's investment grade ratings.

Rating	Momentum	$p_{\cdot,\cdot,u}$	$p_{\cdot,\cdot,d}$	t-Statistic	Reject
Aa	Up	1.32%	3.62%	−1.81	No
Aa	Down	1.52%	5.56%	2.16	Yes
A	Up	1.08%	2.37%	−1.50	No
A	Down	1.19%	6.96%	4.62	Yes
Baa	Up	2.87%	3.16%	−0.22	No
Baa	Down	1.85%	11.47%	6.64	Yes
Ba	Up	7.66%	7.66%	0.00	No
Ba	Down	4.20%	15.38%	6.25	Yes
B	Up	7.69%	0.00%	2.06	Yes
B	Down	5.71%	23.82%	8.74	Yes

Source: Carthy & Fons (1994)

Table 2.20. S&P's historical U.S. & Canadian corporate average one-year transition and default rates (in %, based on the time frame [1985,2002]).

	Rating at year end								
I.R.	*AAA*	*AA*	*A*	*BBB*	*BB*	*B*	*CCC*	*D*	*NR*
AAA	91.59	4.13	0.32	0.09	0.09	0.00	0.00	0.00	3.77
AA	0.48	87.31	7.22	0.68	0.05	0.14	0.03	0.02	4.08
A	0.05	1.66	87.74	5.19	0.52	0.21	0.06	0.07	4.51
BBB	0.02	0.24	4.04	84.79	4.26	0.83	0.23	0.36	5.22
BB	0.04	0.07	0.38	5.90	76.86	6.60	0.86	1.25	8.04
B	0.00	0.09	0.23	0.33	4.83	74.24	4.36	6.18	9.74
CCC	0.12	0.00	0.35	0.71	1.77	8.83	49.35	27.09	11.78

Source: Standard & Poor's (*Special Report: Ratings Performance 2002* 2003).

All tables clearly support a correlation between ratings and default occurrences. High ratings correspond to lower default experience. Obligors further down the credit curve have a higher tendency to be downgraded compared to issuers further up the curve. Credit migration matrices are said to be diagonally dominant, meaning that most of the probability mass resides along the diagonal. Most of the time there is no migration. Bangia et al. (26) show that diagonal elements are usually estimated with a high precision whereas the estimation precision decreases as one moves away from the diagonal.

Nickell et al. (1998) quantify the dependence of rating transition probabilities on the industry and domicile of the obligor. Therefore, they use a non-parametric approach which simply consists of estimating transition probabilities based on relative frequencies for separate data sets corresponding to obligors of different types. In a second approach they employ a parametric ordered probit model (see, e.g., equations 2.1 - 2.3). As data they use the uni-

Table 2.21. S&P's historical European corporate average one-year transition and default rates (in %, based on the time frame [1980,2001]).

I.R.	Rating at year end								
	AAA	*AA*	*A*	*BBB*	*BB*	*B*	*CCC*	*D*	*NR*
AAA	91.20	6.21	0.26	0.00	0.00	0.00	0.00	0.00	2.33
AA	0.25	88.57	7.99	0.37	0.00	0.00	0.00	0.00	2.83
A	0.00	2.21	88.47	4.47	0.29	0.05	0.00	0.00	4.51
BBB	0.00	0.26	5.13	85.88	0.77	0.13	0.13	0.13	7.57
BB	0.00	0.00	0.30	3.34	78.42	5.78	1.82	0.91	9.42
B	0.00	0.00	0.74	0.37	3.68	69.12	6.25	7.35	12.50
CCC	0.00	0.00	0.00	0.00	0.00	15.22	63.04	15.22	6.52

Source: Standard & Poor's (*Special Report: Ratings Performance 2001* 2002).

Table 2.22. RAM's historical Malaysian corporate average one-year transition and default rates (in %, based on the time frame [1992,2002]).

I.R.	Rating at year end								
	AAA	*AA*	*A*	*BBB*	*BB*	*B*	*CCC*	*D*	*NR*
AAA	93.30	3.30	0.00	0.00	0.00	0.00	0.00	0.00	3.30
AA	0.60	86.90	5.10	2.90	0.00	0.00	0.60	0.00	4.00
A	0.00	2.10	81.20	11.20	1.20	0.60	0.00	0.60	3.20
BBB	0.00	0.30	3.10	72.40	9.80	1.60	1.60	0.80	10.60
BB	0.00	0.00	0.60	6.50	58.00	10.10	5.30	3.60	16.00
B	0.00	0.00	0.00	0.00	8.30	73.30	5.00	1.70	11.70
CCC	0.00	0.00	0.00	0.00	0.00	0.00	40.00	8.00	52.00

Source: RAM (Berhad 2003).

verse of notional unsecured Moody's long-term corporate and sovereign bond ratings from December 1970 to December 1997 - a sample of 6534 obligor rating histories. The results for the banking and industrial bonds are given in tables 2.23 and 2.24. The main findings are:

- The volatility of rating transitions is higher for banks than for industrials: the probabilities of remaining in the same rating are consistently lower for banks independent of the rating category.
- Large movements in ratings are just as likely or more likely for industrials than for banks: the distribution of changes in credit quality is relatively fat-tailed for industrials.
- The banking transition matrix is significantly different from the unconditional transition matrix 2.9, especially for highly-rated banks.
- The industrial transition matrix is very similar to the unconditional transition matrix 2.9.

Table 2.23. Conditional transition matrix based on the time frame [1970,1997] and notional unsecured Moody's long-term banking bond ratings (entries in %).

	Rating at year end								
I.R.	*Aaa*	*Aa*	*A*	*Baa*	*Ba*	*B*	*Caa*	*Ca/C*	*D*
Aaa	84.7	15.0	0.3	–	–	–	–	–	–
Aa	0.4	87.8	11.5	0.3	–	–	–	–	–
A	–	2.7	90.0	6.4	0.7	0.2	–	–	–
Baa	–	0.9	16.4	75.1	5.8	1.8	–	–	–
Ba	–	–	4.3	10.3	76.2	5.9	0.5	–	2.7
B	–	–	–	2.7	13.4	78.6	0.9	–	4.5
Caa	–	–	–	–	50.0	–	–	–	50.0
Ca/C	–	–	–	–	–	–	–	–	–
D	–	–	–	–	–	–	–	–	100.0

Source: (Nickell et al. 1998)

Table 2.24. Conditional transition matrix based on the time frame [1970,1997] and notional unsecured Moody's long-term industrial bond ratings (entries in %).

	Rating at year end								
I.R.	*Aaa*	*Aa*	*A*	*Baa*	*Ba*	*B*	*Caa*	*Ca/C*	*D*
Aaa	91.6	7.8	0.7	–	–	–	–	–	–
Aa	1.1	89.3	9.1	0.3	0.2	0.0	–	–	–
A	0.1	1.9	92.4	4.8	0.6	0.2	–	–	0.0
Baa	0.0	0.1	3.9	89.9	4.9	0.8	0.1	–	0.2
Ba	0.0	0.1	0.4	3.4	87.0	7.4	0.2	0.0	1.5
B	0.0	0.1	0.2	0.5	6.2	84.0	1.9	0.4	6.8
Caa	–	–	–	0.8	2.1	7.5	68.2	3.8	17.6
Ca/C	–	–	–	–	1.4	6.8	20.5	56.2	15.1
D	–	–	–	–	–	–	–	–	100.0

Source: (Nickell et al. 1998)

The results for US, UK and Japanese bonds are given in tables 2.25, 2.26, and 2.27. The main findings are:

- The US and UK transition matrices closely resemble the unconditional transition matrix 2.9.
- The Japanese transition matrix differs significantly from the whole-sample matrix 2.9.
- Lowly-rated Japanese obligors show little volatility compared to US obligors.
- Highly-rated Japanese obligors possess more volatile ratings than US obligors.

Table 2.25. Conditional transition matrix based on the time frame [1970,1997] and notional unsecured Moody's long-term US bond ratings (entries in %).

	Rating at year end								
I.R.	*Aaa*	*Aa*	*A*	*Baa*	*Ba*	*B*	*Caa*	*Ca/C*	*D*
Aaa	91.9	6.9	1.1	–	0.1	–	–	–	–
Aa	1.2	89.3	8.8	0.5	0.2	0.0	–	–	–
A	0.1	2.3	92.0	4.9	0.6	0.2	0.0	–	0.0
Baa	0.0	0.2	5.5	88.9	4.5	0.6	0.1	–	0.1
Ba	0.0	0.1	0.5	5.4	85.5	6.9	0.3	0.0	1.4
B	0.0	0.1	0.2	0.7	6.5	82.9	1.9	0.5	7.2
Caa	–	–	–	1.0	2.5	7.6	67.3	3.5	18.1
Ca/C	–	–	–	–	1.0	5.7	14.3	58.1	21.0
D	–	–	–	–	–	–	–	–	100.0

Source: (Nickell et al. 1998)

Table 2.26. Conditional transition matrix based on the time frame [1970,1997] and notional unsecured Moody's long-term UK bond ratings (entries in %).

	Rating at year end								
I.R.	*Aaa*	*Aa*	*A*	*Baa*	*Ba*	*B*	*Caa*	*Ca/C*	*D*
Aaa	90.4	8.9	0.7	–	–	–	–	–	–
Aa	0.3	88.2	11.0	0.5	–	–	–	–	–
A	–	3.4	94.1	2.5	–	–	–	–	–
Baa	–	–	11.9	86.4	1.7	–	–	–	–
Ba	–	–	–	16.0	76.0	8.0	–	–	–
B	–	–	–	11.1	5.6	83.3	–	–	–
Caa	–	–	–	–	–	–	–	–	–
Ca/C	–	–	–	–	–	–	–	–	–
D	–	–	–	–	–	–	–	–	100.0

Source: (Nickell et al. 1998)

Summary of Fundamental Problems with the Historical Approach. The historical method has a lot of shortcomings:

- The default and transition rates can only be applied to firms with a known current rating. Where a firm is not rated, its financial data must be used to calculate key accounting ratios (usually those used by rating agencies). Then, these ratios can be compared with the corresponding median figures for rated firms. Based on such comparisons equivalent ratings and corresponding default probabilities can be assigned to the firms.
- Using historical default rate estimates as approximations for future default rates neglects default rate volatility, especially at low rating levels, and the correlation to the business cycle. Actual transition and default probabilities are very dynamic and can vary depending on general economic

Table 2.27. Conditional transition matrix based on the time frame [1970,1997] and notional unsecured Moody's long-term Japanese bond ratings (entries in %).

	Rating at year end								
I.R.	*Aaa*	*Aa*	*A*	*Baa*	*Ba*	*B*	*Caa*	*Ca/C*	*D*
Aaa	86.9	12.1	1.0	–	–	–	–	–	–
Aa	0.3	88.9	10.5	0.3	–	–	–	–	–
A	–	0.8	95.2	4.0	–	–	–	–	–
Baa	–	–	1.2	96.9	1.6	–	0.3	–	–
Ba	–	–	–	3.5	94.4	2.1	–	–	–
B	–	–	–	–	9.5	90.5	–	–	–
Caa	–	–	–	–	–	–	–	–	–
Ca/C	–	–	–	–	–	–	–	–	–
D	–	–	–	–	–	–	–	–	100.0

Source: (Nickell et al. 1998)

conditions. Hence, the historical observations may not reflect the current credit environment.

- A number of studies performed by rating agencies and academics have documented not only the volatility of default rates but the dependence of ratings transition matrices on the industry of the obligor, the time since the issuance of the specific financial instrument and the stage of the business cycle. Prominent among these are Lucas & Lonski (1992), Altman & Kao (1992b), Altman & Kao (1992a), Carty & Fons (1993), Altman (1997), and Carty (1997). Nickell et al. (1998) show that there are significant differences in the transition matrices of banks and industrials, US and non-US obligors, business cycle peaks and troughs. Cross-country differences are especially important for highly-rated obligors, but appear to be less important for non investment-grade issuers. Business cycle effects are especially important for low ratings.
- Despite the large number of observations, some measurements have low statistical significance.
- The rating of a debtor can be a too rough estimate of its actual quality. Within a rating class there are great variations in default rates. Especially in the high-yield debt market, stability provided by a broad categorization scheme is an illusion.
- Rating Agencies used to be too slow to change their ratings. Hence, the historical average probability of staying in the same rating class overestimates the true figure.
- The differences between mean and median default rates within a rating class shows that the historical average default probability overestimates the true default probability.

- If rating agencies overestimate probabilities of staying in the same rating class and default probabilities, they must underestimate at least some of the transition probabilities to the other credit states.

Conclusion 2.3.1
The ratings from rating agencies appear to be a too broad categorization method. Hence, one shouldn't only rely on historical default and transition rates but use alternative methods as well.

2.3.2 Excursus: Some Fundamental Mathematics

In the following, for modeling purposes we fix a terminal time horizon $T^* \in \mathbb{R}$ and let $\left(\Omega, \mathcal{F}, (\mathcal{F}_t)_{0 \leq t \leq T^*}, \mathbf{P}\right)$ be a filtered probability space[11] where $\mathbf{P}$ is some subjective probability measure. We assume that $\mathcal{F}_0$ is trivial, i.e. $\mathcal{F}_0 = \{\emptyset, \Omega\}$, that $\mathcal{F}_{T^*} = \mathcal{F}$ and that the filtration $\mathbf{F} = (\mathcal{F}_t)_{0 \leq t \leq T^*}$ satisfies the usual conditions[12]. The filtration $\mathbf{F}$ represents the arrival of information over time. Recovery rates, transition and default probabilities are modelled in this fixed probability space. We claim that the event of default is measurable and that the information whether an obligor defaults or doesn't default is included in the set of measurable events.

Because there are some specific classes of stochastic processes[13] that play an essential part in the theory of credit risk modeling we give some technical definitions first:

Definition 2.3.3.

1. *A stochastic process M on a filtered probability space $\left(\Omega, \mathcal{F}, (\mathcal{F}_t)_{t \geq 0}, \mathbf{P}\right)$ is called a martingale with respect to a filtration $(\mathcal{F}_t)_{t \geq 0}$ and with respect to $\mathbf{P}$ if*
   - *M_t is $\mathcal{F}_t$−measurable for all $t \geq 0$,*
   - *$E^{\mathbf{P}}[|M_t|] < \infty$ for all $t \geq 0$,*
   - *$E^{\mathbf{P}}[M_s | \mathcal{F}_t] = M_t$ for all $s \geq t \geq 0$.*

[11] Jacod & Shiryaev (1987), p. 2: A stochastic basis or a filtered probability space is a probability space $(\Omega, \mathcal{F}, \mathbf{P})$ equipped with a filtration $\mathbf{F} = (\mathcal{F}_t)_{t \geq 0}$; here filtration means an increasing and right-continuous family of sub-σ-fields of $\mathcal{F}$.

[12] Jacod & Shiryaev (1987), p.2: The stochastic basis $\left(\Omega, \mathcal{F}, (\mathcal{F}_t)_{t \geq 0}, \mathbf{P}\right)$ is called complete, or equivalently is said to satisfy the usual conditions if the σ-field $\mathcal{F}$ is $\mathbf{P}$−complete and if every $\mathcal{F}_t$ contains all $\mathbf{P}$−null sets of $\mathcal{F}$. It is always possible to complete a given stochastic basis.

[13] Jacod & Shiryaev (1987), p. 3: A stochastic process (or, an E-valued process) is a family $X = \{X_t\}_{t \geq 0}$ of mappings from Ω into some set E endowed with a σ−algebra $\mathcal{E}$ defined on a probability space $(\Omega, \mathcal{F}, \mathbf{P})$. Unless otherwise stated, E will be $\mathbb{R}^d$ for some $d \in \mathbb{N}$. We shall say indifferently: the stochastic process X, or $\{X_t\}_{t \geq 0}$, or $\{X(t)\}_{t \geq 0}$.

2. *$\mathcal{M}$ is the class of all uniformly integrable martingales, i.e. of all martingales M such that the family of random variables $\{M_t\}_{t\geq 0}$ is uniformly integrable.*
3. *A local martingale is a process that belongs to the localized class $\mathcal{M}_{loc}$ constructed from $\mathcal{M}$. A process M belongs to the localized class $\mathcal{M}_{loc}$ if and only if there exists an increasing sequence $(T_n)_{n\in\mathbf{N}}$ of stopping times*[14] *such that $\lim_{n\to\infty} T_n = \infty$ a.s. and that each stopped process*[15] *$M_{T_n\wedge t}$ belongs to $\mathcal{M}$.*
4. *$\mathcal{L}$ denotes the set of all local martingales M such that $M_0 = 0$.*
5. *We denote by $\mathcal{V}$ the set of all real-valued processes A that are right continuous with existing left limits, $\mathcal{F}_t$-adapted*[16]*, with $A_0 = 0$, and each path $t \rightsquigarrow A_t(\omega)$ is of finite variation*[17]*.*
6. *A semimartingale is a process X of the form $X = X_0 + M + A$ where X_0 is finite-valued and $\mathcal{F}_0$-measurable, where $M \in \mathcal{L}$, and where $A \in \mathcal{V}$.*

Definition 2.3.4.
A Brownian motion is a real-valued, continuous stochastic process $\{X_t\}_{t\geq 0}$, with independent and stationary increments. In other words:

- *continuity: $\mathbf{P}$ a.s. the map $s \longmapsto X_s(\omega)$ is continuous.*
- *independent increments: for all $s \leq t$, $X_t - X_s$ is independent of $\mathcal{F}_s = \sigma(X_u, u \leq s)$, i.e. the σ-algebra generated by $\{X_u, u \leq s\}$.*
- *stationary increments: if $s \leq t$, $X_t - X_s$ and $X_{t-s} - X_0$ have the same probability law.*

Remark 2.3.2.

1. If $\{X_t\}_{t\geq 0}$ is a Brownian motion, then $X_t - X_0$ is a normal random variable with mean rt and variance $\sigma^2 t$, where r and σ are constant real numbers. For a proof consult the book by Gihman & Skorohod (1980).
2. A Brownian motion is standard if $X_0 = 0$ $\mathbf{P}$ a.s., $E[X_t] = 0$, $E[X_t^2] = t$.

---

[14] Lamberton & Lapeyre (1996), p. 30: τ is a stopping time with respect to the filtration $\mathbf{F} = (\mathcal{F}_t)_{t\geq 0}$ if τ is a random variable in $\mathbf{R}^+ \cup \{+\infty\}$, such that for any $t \geq 0 : \{\tau \leq t\} \in \mathcal{F}_t$.

[15] Jacod & Shiryaev (1987), p. 3: If X is a process and if T is a mapping: $\Omega \to \mathbf{R}^+ \cup \{+\infty\}$, we define the process stopped at time T by $X_{T\wedge t}$.

[16] Jacod & Shiryaev (1987), p. 3: An E-valued stochastic process X endowed with a σ-algebra $\mathcal{E}$ is adapted to the filtration $\mathbf{F}$ (or equivalently is $\mathcal{F}_t$-adapted), if for all t, X_t is $\mathcal{F}_t$-$\mathcal{E}$-measurable.

[17] Jacod & Shiryaev (1987), p. 28: The stochastic process X is of finite variation if almost all of the paths of X are of finite variation on each finite interval $[0,t]$, $t \geq 0$. The variation process $VAR(X)$ (sometimes denoted by $|X|$) of X is defined by

$$VAR(X)_t(\omega) = |X|_t = \lim_{n\to\infty} \sum_{1\leq k\leq n} \left| X_{tk/n}(\omega) - X_{t(k-1)/n}(\omega) \right|.$$

If X is of finite variation, then $VAR(X)_t(\omega) = |X|_t < \infty$.

3. A standard Brownian motion is a martingale.

Finally, we give some technical definitions of some stochastic processes describing default and transition probabilities.

Definition 2.3.5.
For each $0 \leq T \leq T^*$ *we let* $\{p^s(t,T)\}_{0\leq t\leq T}$ $\left(\{p^d(t,T)\}_{0\leq t\leq T}\right)$ *denote an* $\mathcal{F}_t$*-adapted stochastic process where* $p^s(t,T)$ $\left(p^d(t,T)\right)$ *is the conditional probability at time t, given all available information at that time, of survival to (default by) time* T*. Obviously, for all t, T:* $p^d(t,T) = 1 - p^s(t,T)$*. In addition, we denote the probability of surviving to time* T*, given survival to time t, by* $p^s(T|t)$*.*

Based on these definitions we can introduce the concept of forward default probabilities:

Definition 2.3.6.
The forward default probability is defined as the probability of default by time T*, given survival to time t, and denoted by* $p^d(T|t)$ *(obviously, for all t, T:* $p^d(T|t) = 1 - p^s(T|t)$*).*

We want to emphasize that $p^d(T|t)$ and $p^s(T|t)$ are only conditioned on survival up to time t, whereas $p^d(t,T)$ and $p^s(t,T)$ are conditioned on all information available at time t.

Definition 2.3.7.
Let there be a finite state space of ratings $\{1,2,...,K\}$*. These states represent the various credit classes, with state* 1 *being the highest and state* K *being bankruptcy. For each* $0 \leq T \leq T^*$ *and* $R, \check{R} \in \{1,2,...,K\}$ *we let* $\{p_{R,\check{R}}(t,T)\}_{0\leq t\leq T}$ *denote an* $\mathcal{F}_t$*-adapted stochastic process, where* $p_{R,\check{R}}(t,T)$ *is the conditional probability that a counterparty who is in rating class* R *at time t, migrates to rating class* $\check{R}$ *until time* T*.*

2.3.3 The Asset Based Method

Merton's Default Model. The asset based or firm value approach[18] for modeling default probabilities goes back to the initial proposal of Merton (1974), where he assumes that a firm is only financed by equity and one type of zero coupon debt with promised terminal constant payoff $L > 0$ at maturity time T. The firm is able to redeem its debt if the total market value of its assets V_T at time T is worth more than the amount L owed to the creditor. Otherwise the obligor defaults on its obligation and the bondholders receive V_T. In other words the payoff to the bondholders equals the minimum of L and V_T. Because this approach explicitly models the firm's capital structure, it is sometimes called structural approach. Merton assumes a Black-Scholes type frictionless market[19]. By the given assumptions, the $\mathbf{F}-$adapted stochastic

[18] Equivalently firm value, Merton-based or structural approach
[19] Some of the most important assumptions are:

process $\{p^d(t,T)\}_{0\leq t\leq T}$ in the filtered probability space $(\Omega,\mathcal{F},\mathbf{F},\mathbf{P})$, which describes the conditional default probability until time T, given all information available at time t, is implicitly defined by

$$\begin{aligned} p^d(t,T) &= \mathbf{P}(V_T < L|\mathcal{F}_t) \\ &= \mathbf{P}(\ln V_T < \ln L|\mathcal{F}_t) \\ &= \mathbf{P}(l_T < 0|\mathcal{F}_t), \end{aligned}$$

where the log-ratio process l_t is defined by $l_t = \ln\left(\frac{V_t}{L}\right),\ 0 \leq t \leq T$. If the dynamics of the firm value under the probability measure $\mathbf{P}$ are specified as the $\mathbf{F}$−adapted stochastic process $\{V_t\}_{0\leq t\leq T^*}$ with

$$\frac{dV_t}{V_t} = \mu_V dt + \sigma_V dW_V(t),\ 0 \leq t \leq T, \tag{2.5}$$

where μ_V is the unknown instantaneous expected rate of return on the firm per unit of time, σ_V is the constant volatility parameter, and W_V is a standardized Brownian motion (i.e. the firm value is treated as a log-normal diffusion), easy calculations yield

$$p^d(t,T) = \Phi(-d_t),\ 0 \leq t \leq T,$$

where Φ denotes the operator of the standard normal distribution and

$$d_t = \frac{l_t + \left(\mu_V - \frac{1}{2}\sigma_V^2\right)(T-t)}{\sigma_V\sqrt{T-t}}$$

is the so called distance to default. Hence, the default probability $p^d(t,T)$ can be obtained from the probability distribution of the firm values at maturity time T (see figure 2.5).
Because the instantaneous default probability of a healthy firm is zero under a continuous firm value process which in general does not seem to be very realistic, Zhou (1997) models the evolution of the firm value as a jump-diffusion process. Formally,

$$\frac{dV_t}{V_t} = (\mu_V - \lambda_V\nu_V)\,dt + \sigma_V dW_V(t) + (\Pi_V - 1)\,dU_V(t),\ 0 \leq t \leq T, \tag{2.6}$$

where λ_V and ν_V are constants, $\Pi_V > 0$ is the i.i.d. log-normal distributed jump amplitude, such that $\ln(\Pi_V) \sim N(\pi_\Pi, \sigma_\Pi)$ for some constant $\sigma_\Pi > 0$

- the market is complete;
- there is no arbitrage in the market;
- trading takes place continuously in time;
- traded assets are infinitely divisible;
- unrestricted borrowing and lending is possible at the same constant interest rate;
- no restrictions on short-sellings;
- no transaction costs and taxes;
- bankruptcy is costless.

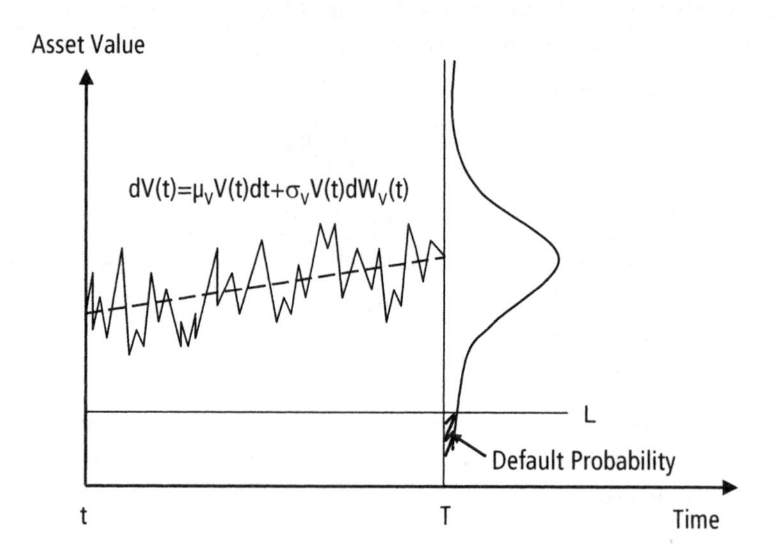

Fig. 2.5. Merton's model of default.

and $\pi_{\Pi} = \nu_V + 1$, U_V is a Poisson process[20] with intensity parameter λ_V, and $dW_V(t)$, $dU_V(t)$ and Π_V are mutually independent for all $0 \leq t \leq T$. N denotes the normal distribution. Because the process V is a semimartingale (for a proof see, e.g., Protter (1992), pages 47-48) we apply the semimartingale version of Itô's lemma[21] to equation (2.6) and get

[20] Ross (1996), p.59: A stochastic process $\{X_t\}_{t\geq 0}$ is said to be a counting process if X_t represents the total number of events that have occurred up to time t. Hence, a counting process must satisfy: $X_t \geq 0$, X_t is integer valued, and if $s < t$, $X_t - X_s$ equals the number of events that have occurred in the interval $(s, t]$.

A counting process $\{X_t\}_{t\geq 0}$ is said to be a Poisson process with intensity λ, $\lambda > 0$, if $X_0 = 0$, the process has independent increments, and the number of events in any interval of length t is Poisson distributed with mean λt.

[21] Protter (1992), p.71: Let X be a semimartingale and let f be a twice continously differentiable real function. Then $f(X)$ is again a semimartingale , and the following formula holds:

$$f(X_t) - f(X_0) = \int_{0+}^{t} f'(X_{s-})\, dX_s + \frac{1}{2}\int_{0+}^{t} f''(X_{s-})\, d[X,X]_s^c + \sum_{0<s\leq t} \{f(X_s) - f(X_{s-}) - f'(X_{s-})\, \Delta X_s\},$$

where $[X,X]^c$ denotes the path by path continuous part of $[X,X]$ and $\Delta X_s = X_s - X_{s-}$. The quadratic variation of X is defined as

$$d\ln(V_t) = \left(\mu_V - \frac{1}{2}\sigma_V^2 - \lambda_V \nu_V\right) dt + \sigma_V dW_V(t) + \ln(\Pi_V)\, dU_V(t).$$

Hence, the conditional default probability for $0 \leq t \leq T$ is given by[22]

$$p^d(t,T) = \mathbf{P}\left(l_T \leq 0 \middle| \mathcal{F}_t\right).$$

Let $U_V(t,T)$ be the total number of jumps from time t to time T. Then we have

$$\begin{aligned} &\ln(V_T)|\left(V_t, U_V(t,T) = i\right) \\ &\sim N\left(\ln(V_t) + \left(\mu_V - \frac{1}{2}\sigma_V^2 - \lambda_V \nu_V\right)(T-t) + i\pi_\Pi, \sigma_V^2(T-t) + i\cdot\sigma_\Pi^2\right), \end{aligned}$$

and

$$\begin{aligned} p^d(t,T) = \sum_{i=0}^{\infty} &\frac{\exp(-\lambda_V(T-t))(\lambda_V(T-t))^i}{i!} \\ &\cdot\Phi\left(-\frac{l_t + \left(\mu_V - \frac{1}{2}\sigma_V^2 - \lambda_V\nu_V\right)(T-t) + i\mu_\Pi}{\sqrt{\sigma_V^2(T-t) + i\cdot\sigma_\Pi^2}}\right). \end{aligned}$$

Asset from Equity Values in Merton's Default Model. In general, asset values are not observable. Therefore, we have to show how asset values can be derived from observable equity values. Equity is given by the firm's market capitalization, i.e. by the numbers of shares times the value of one share. Again, we assume a Black-Scholes type frictionless market and assume that the asset value process is defined by equation (2.5). The firm is only financed by equity and one type of zero coupon debt with promised terminal constant payoff $L > 0$ at maturity time T. The firm is able to redeem its debt if the total market value of its assets V_T at time T is worth more than the amount L owed to the creditor. Otherwise the obligor defaults on its obligation and the bondholders receive V_T, which means that the bondholders are the owners of the company. In the first case the value of the shareholders' equity at time T is $V_T - L$ in the second case 0. In other words the value of the stockholder's equity at time T equals $\max(0, V_T - L)$. But this is the payoff of a European call option on the asset values of the firm. Therefore, the market value of the equity equals the value of call option with underlying V, strike L and maturity T. Then, by the well-known Black-Scholes formula (see, e.g., Lamberton & Lapeyre (1996)), the value of the equity at time t $(0 \leq t \leq T)$ is given by

$$[X]_t = [X,X]_t = X_t^2 - 2\int_0^t X_{u-}dX_u,\ \forall\, t \geq 0.$$

[22] For a detailed proof see the paper by Zhou (1997), page 26.

$$E_t = V_t \Phi \left(\frac{l_t + \left(r + \frac{1}{2}\sigma_V^2\right)(T-t)}{\sigma_V \sqrt{T-t}} \right) \tag{2.7}$$

$$-L \cdot e^{-r(T-t)} \Phi \left(\frac{l_t + \left(r - \frac{1}{2}\sigma_V^2\right)(T-t)}{\sigma_V \sqrt{T-t}} \right),$$

where r is the instantaneous interest rate and assumed to be a non-negative constant. Applying Itô's lemma to equation (2.7) yields

$$\frac{dE_t}{E_t} = rdt + \frac{\Phi\left(\frac{l_t + \left(r + \frac{1}{2}\sigma_V^2\right)(T-t)}{\sigma_V\sqrt{T-t}}\right) V_t \sigma_V}{E_t} dW_V(t).$$

Hence,

$$\sigma_E = \frac{\Phi\left(\frac{l_t + \left(r + \frac{1}{2}\sigma_V^2\right)(T-t)}{\sigma_V\sqrt{T-t}}\right) V_t \sigma_V}{E_t}. \tag{2.8}$$

Using the observable equity values and parameters, we can solve the system of equations (2.7) and (2.8) for V_t and σ_V numerically.

First-Passage Default Model. First-passage time models generalize the traditional asset based model, such that a default can occur not only at maturity of the debt contract but at any point of time, and assume that bankruptcy occurs if the firm value V_t hits a specified possibly stochastic boundary or default point D_t such as the current value of the firm's liabilities. However, empirical studies show that in general firms do not default when their firm value falls below the book value of their total liabilities. Rather the default point lies somewhere between short-term and total liabilities. Therefore, it can be important to classify the debt outstanding by, for example, short-term, medium-term and long-term debt.

First-passage time models were introduced by Black & Cox (1976) who modified Merton's default model to facilitate the modeling of safety covenants in the indenture provisions. A safety covenant allows the bondholders to force bankruptcy if certain conditions are met. The aim of such provisions is to protect bondholders from further devaluations of the firm value.

Now let T^d represent the time when a default occurs the first time.

Definition 2.3.8.
For $\mathbf{F}-$adapted firm value and boundary processes V_t and D_t we define the $\mathbf{F}-$adapted log-ratio process $\{l_t\}_{0 \le t \le T^*}$ by

$$l_t = \ln\left(\frac{V_t}{D_t}\right), \quad 0 \le t \le T^*.$$

The first passage time or hitting time for V_t to hit D_t is

$$T^d = \inf\left\{0 \le t \le T^* | l_t \le 0\right\}.$$

Let us now consider a quite general version of the firm value and the barrier processes. Assume that their dynamics are given by

$$\frac{dV_t}{V_t} = \mu_V dt + \sigma_V dW_V(t), \ V_0 > 0, \tag{2.9}$$

$$\frac{dD_t}{D_t} = \mu_D dt + \sigma_D dW_D(t), \ D_0 > 0, \tag{2.10}$$

where W_V and W_D are n-dimensional standard Brownian motions with respect to the underlying filtration $\mathbf{F}$, $\mu_V, \mu_D \in \mathbb{R}$ are constants, $\sigma_V, \sigma_D \in \mathbb{R}^n$ are constant vectors, $V_0 > D_0 > 0$, and the SDEs (2.9) and (2.10) posses unique, strong solutions. Then the dynamics of l_t are given by

$$\begin{aligned} dl_t &= \left(\mu_V - \mu_D + \frac{1}{2}\|\sigma_D\|^2 - \frac{1}{2}\|\sigma_V\|^2\right)dt + \sqrt{\left(\|\sigma_V\|^2 + \|\sigma_D\|^2\right)}dW_l(t) \\ &= \mu_l dt + \sigma_l dW_l(t), \end{aligned}$$

where W_l is a standard one-dimensional Brownian motion with respect to $\mathbf{F}$, and it is easy to show that for all $0 \le t < T \le T^*$

$$\begin{aligned} p^d(t,T) &= \mathbf{P}\left(T^d \le T | \mathcal{F}_t\right) \\ &= \Phi\left(\frac{-l_t - \mu_l(T-t)}{\sigma_l\sqrt{T-t}}\right) + e^{-2\frac{\mu_l}{\sigma_l^2}l_t}\Phi\left(\frac{-l_t + \mu_l(T-t)}{\sigma_l\sqrt{T-t}}\right). \end{aligned}$$

The following two examples follow directly from this result.

Example 2.3.3.
The easiest generalization of Merton's default model to a first-passage default model is the additional assumption that default occurs at the first time at which the log-normally distributed asset value (given by equation (2.5)) hits an arbitrary but constant default boundary D (which need not be the face value of debt). Then, using

$$\frac{dV_t}{V_t} = \mu_V dt + \sigma_V dW_V(t), \ V(0) > 0, \ l_t = \ln\left(\frac{V_t}{D}\right)$$

and the result of Harrison (1990) on the probability law of the first passage time, for $0 \le t \le T$,

$$\begin{aligned} p^d(t,T) &= \mathbf{P}\left(T^d \le T | \mathcal{F}_t\right) \\ &= \Phi\left(\frac{-l_t - \left(\mu_V - \frac{1}{2}\|\sigma_V\|^2\right)(T-t)}{\|\sigma_V\|\sqrt{T-t}}\right) \\ &\quad + \left(\frac{V_t}{D}\right)^{\left(1-\frac{2\mu_V}{\|\sigma_V\|^2}\right)}\Phi\left(\frac{-l_t + \left(\mu_V - \frac{1}{2}\|\sigma_V\|^2\right)(T-t)}{\|\sigma_V\|\sqrt{T-t}}\right). \end{aligned}$$

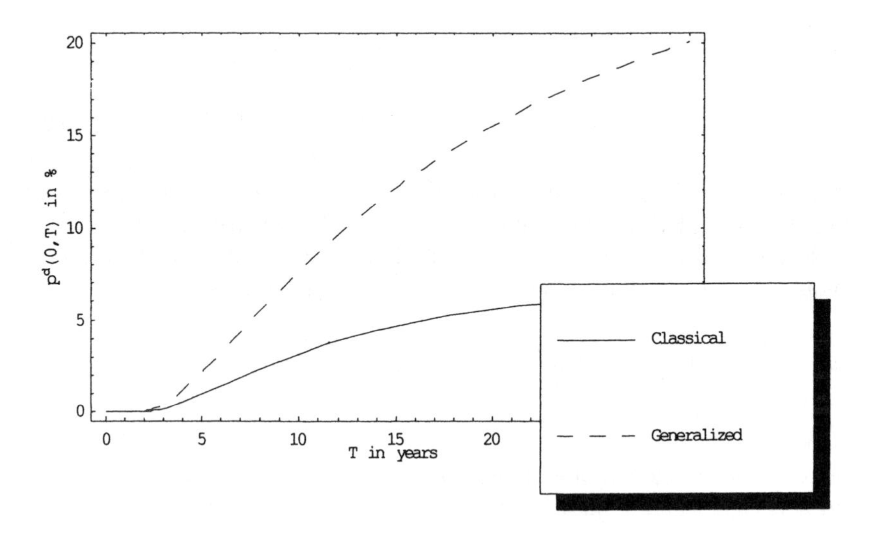

Fig. 2.6. Term structure of default probabilities for classical and generalized Merton model with parameter values $n = 1$, $t = 0$, $l(0) = \ln 2$, $\mu_V = 3\%$, $\sigma_V = 15\%$.

This result is used, e.g., in Leland & Toft (1996). In the following we show how the default probabilities in the classical Merton model differ from the ones implied by this generalized model. Therefore, we assume the following parameter specifications: $n = 1$, $t = 0$, $l_0 = \ln 2$, $\mu_V = 3\%$, $\sigma_V = 15\%$. Obviously, the default probabilities in the generalized Merton model are always higher than in the classical model. With growing T the difference between the two default probabilities in the two models even increases.

Example 2.3.4.
For fixed $a, b > 0$, let the barrier process D be defined as

$$D_t = a \cdot e^{-b(T-t)},\ t \geq 0,$$

which means that

$$\frac{dD_t}{D_t} = b \cdot dt,\ D_0 = a \cdot e^{-bT}.$$

Then, using

$$\frac{dV_t}{V_t} = \mu_V dt + \sigma_V dW_V(t),\ V_0 > 0,\ l_t = \ln\left(\frac{V_t}{D_t}\right),$$

for $0 \leq t \leq T$,

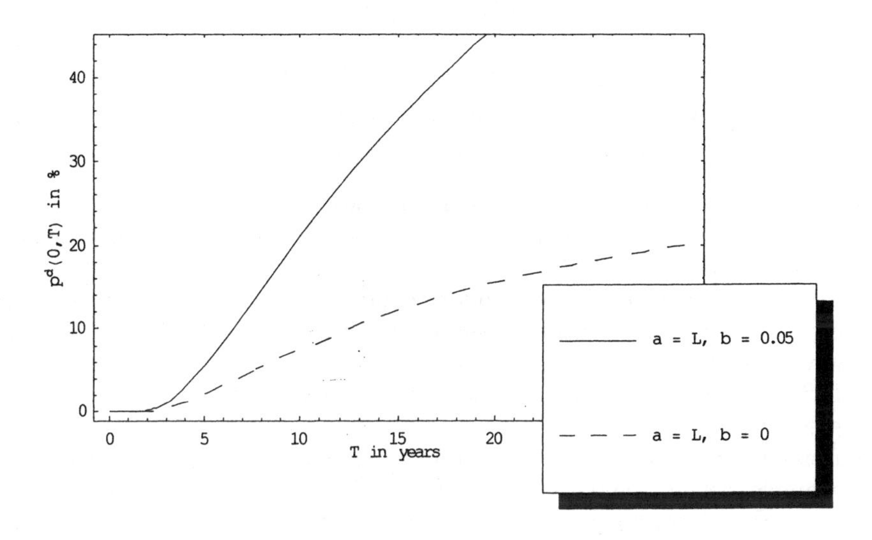

Fig. 2.7. Term structure of default probabilities for examples with parameter values $n = 1$, $t = 0$, $l(0) = \ln 2$, $\mu_V = 3\%$, $\sigma_V = 15\%$, $a = L$, $b = 0$ und $b = 0.05$.

$$\begin{aligned} p^d(t,T) &= \mathbf{P}\left(T^d \le T | \mathcal{F}_t\right) \\ &= \Phi\left(\frac{-l_t - \left(\mu_V - b - \frac{1}{2}\|\sigma_V\|^2\right)(T-t)}{\|\sigma_V\|\sqrt{T-t}}\right) \\ &\quad + \left(\frac{V_t}{D_t}\right)^{\left(1-\frac{2\mu_V+b}{\|\sigma_V\|^2}\right)} \Phi\left(\frac{-l_t + \left(\mu_V - b - \frac{1}{2}\|\sigma_V\|^2\right)(T-t)}{\|\sigma_V\|\sqrt{T-t}}\right). \end{aligned}$$

This result is used, e.g., in Black & Cox (1976). Note that the previous example is a special case of this example if we set $b = 0$ and $a = D$. In the following we show how the default probabilities in the previous example differ from the ones implied by this model when $b > 0$. Therefore, we assume the following parameter specifications: $n = 1$, $t = 0$, $l_0 = \ln 2$, $\mu_V = 3\%$, $\sigma_V = 15\%$, $a = D = L$ (face value of the bond), $b = 0.05$.
Because we assume that l_0 equals $\ln 2$ in both examples $V_0 - D_0 = L \cdot e^{-bT}$. With growing values for b the difference $V_0 - D_0$ decreases and therefore the default probabilities increase.

Except for some very special processes (such as the ones we have discussed right now), explicit solutions for first passage times are not known[23]. In this

[23] For a survey of the first passage time problem see Abrahams (1986).

case the conditional default probability

$$p^d(t,T) = \mathbf{P}\left(T^d \leq T|\mathcal{F}_t\right), \ 0 \leq t \leq T,$$

can be determined, e.g., by Monte Carlo simulation.

Summary of Fundamental Problems of the Asset Based Approach. Asset based and first-passage time models have some shortcomings:

- Usually firm values are not easily observable.
- Analyzing corporate-level data issuer by issuer is impractical to empirically study a structural model.
- Complicated balance sheets cannot be captured easily by structural models, e.g., there may be too many classes of claims on the assets of the firm that have to be priced.
- Different accounting standards in different countries complicate the whole process.
- Default can be determined by factors other than assets and liabilities (for example, default could occur for reasons of illiquidity, or, as it is sometimes the case with sovereign debt, for general economic and political considerations).
- Often it is difficult to define a meaningful firm value process, e.g., in case of sovereign debt.
- To generate realistic levels of default probabilities (especially short term default probabilities) one has to extend the basic Merton approach in many ways, hence, work with a complicated model. For example, if we model the firm value process only with a continuous diffusion the probability of a firm to default in the next instant is zero. Hence, firms would never default unexpectedly.

2.3.4 The Intensity Based Method

The Default Intensity. Intensity based methods take an approach completely different from asset based methods. They define the event of default as a stopping time T^d with respect to the filtration $\mathbf{F}$ with an exogenously given probability distribution. One may associate with the stopping time T^d the first jump time of a point or counting process[24] $\{A_t\}_{0 \leq t \leq T^*}$ defined by

[24] Brémaud (1981), pp.18-19: A realization of a point process over $[0, \infty)$ can be described by a sequence $(T_n)_{n \in \mathbb{N}}$ in $[0, \infty]$ such that $T_0 = 0$, $T_n < \infty \Longrightarrow T_n < T_{n+1}$. This realization is, by definition, called nonexplosive iff $T_\infty = \lim_{n \to \infty} T_n = \infty$. To each realization T_n corresponds a counting function A_t defined by

$$A_t = \begin{cases} n & , \text{ if } t \in [T_n, T_{n+1}), \ n \geq 0, \\ +\infty & , \text{ if } t \geq T_\infty. \end{cases}$$

If the above T_n are random variables defined on some probability space $(\Omega, \mathcal{F}, \mathbf{P})$, the sequence $(T_n)_{n \in \mathbb{N}}$ is called a point process. The associated count-

$A_t = \sum_{n \geq 1} 1_{\{T_n \leq t\}}$. The sequence T_n is a sequence of jump times with T_n defined as the time of occurrence of the nth jump. If we define H_t as the process A_t stopped at time T^d, then H_t can be interpreted as the default indicator function which equals 0 before default and 1 if default has already occurred, i.e. $H_t = 1_{\{T^d \leq t\}}$. Now we are able to introduce the concept of stochastic intensity like, e.g., in Brémaud (1981), page 27:

Definition 2.3.9.
Let $\{A_t\}_{t \geq 0}$ be a point process adapted to some filtration $\mathbf{F}$, and let λ_t be a nonnegative $\mathcal{F}_t$–progressive[25] *process such that for all $t \geq 0$*

$$\int_0^t \lambda_s ds < \infty \ \mathbf{P} - a.s.$$

If for all nonnegative $\mathcal{F}_t$–predictable[26] *processes C_t, the equality*

$$E^{\mathbf{P}} \left[\int_0^\infty C_s dA_s \right] = E^{\mathbf{P}} \left[\int_0^\infty C_s \lambda_s ds \right]$$

is verified, then we say: A_t admits the $(\mathbf{P}, \mathcal{F}_t)$ –intensity λ_t.

Remark 2.3.3.
A stochastic integral is defined with respect to a specific class of integrators (for possible choices such as right-continuous, square integrable martingales[27] see, e.g., Karatzas (1988), page 131). This integrator class defines which type of stochastic processes (e.g., $\mathcal{F}_t$–progressive processes or only the smaller class of $\mathcal{F}_t$–predictable processes) can be used as integrands of the stochastic integral (for details see, e.g., Karatzas (1988), page 131, and Protter (1992), pages 48-49). E.g., if the stochastic integral is defined for semimartingale integrators one can use $\mathcal{F}_t$–predictable integrands.

ing process A_t is also called a point process, by abuse of notation. Moreover, if the condition $E^{\mathbf{P}}[A_t] < \infty$, $t \geq 0$, holds, the point process A_t is said to be integrable.

[25] Brémaud (1981), p. 288: Let $\mathbf{F} = (\mathcal{F}_t)_{t \geq 0}$ be a filtration defined on the probability space $(\Omega, \mathcal{F}, \mathbf{P})$. An E–valued process X endowed with a σ–algebra $\mathcal{E}$ is said to be $\mathcal{F}_t$–progressive (or $\mathcal{F}_t$–progressively measurable) iff for all $t \geq 0$ the mapping $(t, \omega) \rightarrow X_t(\omega)$ from $[0,t] \times \Omega \rightarrow E$ is $(\mathcal{B}([0,t]) \otimes \mathcal{F}_t) - \mathcal{E}-$ measurable.

[26] Brémaud (1981), p. 8: Let $\mathbf{F} = (\mathcal{F}_t)_{t \geq 0}$ be a filtration defined on the probability space $(\Omega, \mathcal{F}, \mathbf{P})$, and define $\mathcal{P}(\mathcal{F}_t)$ to be the σ–field over $(0, \infty) \times \Omega$ generated by the rectangles of the form

$$(s,t] \times A; \ 0 \leq s \leq t, \ A \in \mathcal{F}_s.$$

$\mathcal{P}(\mathcal{F}_t)$ is called the $\mathcal{F}_t$–predictable σ–field over $(0,\infty) \times \Omega$. A real-valued process X such that X_0 is $\mathcal{F}_0$–measurable and the mapping $(t, \omega) \rightarrow X_t(\omega)$ defined from $(0, \infty) \times \Omega$ into $\mathbf{R}$ is $\mathcal{P}(\mathcal{F}_t)$ –measurable is said to be $\mathcal{F}_t$–predictable.

[27] Karatzas & Shreve (1988), p.30: A martingale is called square integrable if $E^{\mathbf{P}}(X_t^2) < \infty$ for all $t \geq 0$.

To be able to give an intuitive interpretation of the notion of intensity (in a setting where we want to model the arrival of default) we need to mention some properties of stochastic intensities:

Theorem 2.3.1 (Integration Theorem).
If A_t admits the $(\mathbf{P},\mathcal{F}_t)-$intensity λ_t, then A_t is nonexplosive and

1. *$M_t = A_t - \int_0^t \lambda_s ds$ is an $\mathcal{F}_t-$local martingale. We call A_t local compensated jump martingale.*
2. *If X is an $\mathcal{F}_t-$predictable process such that $E^{\mathbf{P}}\left[\int_0^t |X_s| \lambda_s ds\right] < \infty$, $t \geq 0$, then $\int_0^t X_s dM_s$ is an $\mathcal{F}_t-$martingale.*
3. *If X is an $\mathcal{F}_t-$predictable process such that $\int_0^t |X_s| \lambda_s ds < \infty$, $\mathbf{P}-a.s.$, $t \geq 0$, then $\int_0^t X_s dM_s$ is an $\mathcal{F}_t-$local martingale.*

Proof.
Brémaud (1981), page 27.

Theorem 2.3.2 (Martingale Characterization of Intensity).
Let A_t be a nonexplosive point process adapted to $\mathcal{F}_t$, and suppose that for some nonnegative $\mathcal{F}_t-$progressive process λ_t and for all $n \geq 1$, $A_{t\wedge T_n} - \int_0^{t\wedge T_n} \lambda_s ds$ is a $(\mathbf{P},\mathcal{F}_t)-$martingale. Then λ_t is the $(\mathbf{P},\mathcal{F}_t)-$intensity of A_t.

Proof.
Brémaud (1981), page 28.

If the intensity is constrained to be predictable, it is essentially unique:

Theorem 2.3.3 (Uniqueness of Predictable Intensities).
Let A_t be a point process adapted to the filtration $\mathbf{F}$, and let λ_t and $\widetilde{\lambda_t}$ be two $(\mathbf{P},\mathcal{F}_t)-$intensities of A_t which are $\mathcal{F}_t-$predictable. Then $\lambda_t(\omega) = \widetilde{\lambda_t}(\omega)$ $\mathbf{P}(d\omega)\, dA_t(\omega) - a.e.$

Proof.
Brémaud (1981), page 31.

Also, one can always find such a predictable version of the intensity:

Theorem 2.3.4 (Existence of Predictable Versions of the Intensity).
Let A_t be a point process with an $(\mathbf{P},\mathcal{F}_t)-$intensity λ_t. Then one can find an $(\mathbf{P},\mathcal{F}_t)-$intensity $\widetilde{\lambda_t}$ that is $\mathcal{F}_t-$predictable.

Proof.
Brémaud (1981), page 31.

To show how default probabilities can be modeled using intensities we make the following assumption.

Assumption 2.3.1
The default indicator function H_t admits a $(\mathbf{P},\mathcal{F}_t)$ −intensity λ_t.

Thus,

$$\begin{aligned} M_{t\wedge T^d}(t) &= H_t - \int_0^{T^d \wedge t} \lambda_s ds \\ &= H_t - \int_0^t \lambda_s 1_{\{s\leq T^d\}} ds \end{aligned}$$

is an $\mathcal{F}_t$−local martingale and λ_t is a nonnegative $\mathcal{F}_t$−progressive process with $\int_0^t \lambda_s ds < \infty$ $\mathbf{P} - a.s.$ for all $t \geq 0$. Hence, the conditional expectation $E^{\mathbf{P}}[H_{t+\varepsilon} - H_t|\mathcal{F}_t]$, $\varepsilon > 0$, is given by

$$\begin{aligned} &E^{\mathbf{P}}[H_{t+\varepsilon} - H_t|\mathcal{F}_t] \qquad (2.11) \\ &= E^{\mathbf{P}}[M_{t\wedge T^d}(t+\varepsilon) - M_{t\wedge T^d}(t)|\mathcal{F}_t] + E^{\mathbf{P}}\left[\int_t^{t+\varepsilon} \lambda_s 1_{\{s\leq T^d\}} ds|\mathcal{F}_t\right] \\ &= M_{t\wedge T^d}(t) - M_{t\wedge T^d}(t) + E^{\mathbf{P}}\left[\int_t^{t+\varepsilon} \lambda_s 1_{\{s\leq T^d\}} ds|\mathcal{F}_t\right] \\ &= E^{\mathbf{P}}\left[\int_t^{t+\varepsilon} \lambda_s 1_{\{s\leq T^d\}} ds|\mathcal{F}_t\right] \end{aligned}$$

and by application of the Lebesgue averaging theorem and the Lebesgue dominated convergence theorem, successively, in case of a right-continuous intensity λ_t, we get

$$\begin{aligned} &\lim_{\varepsilon\to 0+} \frac{E^{\mathbf{P}}[H_{t+\varepsilon} - H_t|\mathcal{F}_t]}{\varepsilon} \\ &= \lim_{\varepsilon\to 0+} \frac{E^{\mathbf{P}}\left[\int_t^{t+\varepsilon} \lambda_s 1_{\{s\leq T^d\}} ds|\mathcal{F}_t\right]}{\varepsilon} \\ &= \lambda_t 1_{\{t\leq T^d\}} \ \mathbf{P} - a.s. \end{aligned}$$

On the other hand,

$$\begin{aligned} &E^{\mathbf{P}}[H_{t+\varepsilon} - H_t|\mathcal{F}_t] \\ &= 0\cdot \mathbf{P}(T^d \leq t|\mathcal{F}_t) + 1\cdot \mathbf{P}(t < T^d \leq t+\varepsilon|\mathcal{F}_t) + 0\cdot \mathbf{P}(T^d > t+\varepsilon|\mathcal{F}_t) \\ &= p^d(t, t+\varepsilon) \end{aligned}$$

and

$$\lim_{\varepsilon\to 0+} \frac{E^{\mathbf{P}}[H_{t+\varepsilon} - H_t|\mathcal{F}_t]}{\varepsilon} = \lim_{\varepsilon\to 0+} \frac{p^d(t, t+\varepsilon)}{\varepsilon}.$$

Hence,

$$\lambda_t 1_{\{t \leq T^d\}} = \lim_{\varepsilon \to 0+} \frac{p^d(t, t+\varepsilon)}{\varepsilon},$$

i.e. given all information available at time t, the intensity λ_t can be interpreted as the arrival rate of default at time t.

Conclusion 2.3.2
The probability of default over the next infinitesimal time interval of length $\varepsilon > 0$ is approximately $\lambda_t \cdot \varepsilon$. This interpretation links the notion of intensity to the problem of estimating default probabilities. Sometimes λ_t is called the intensity of T^d.

In addition, using equation (2.11), we have found the following result:

Proposition 2.3.1.
$p^d(t, T)$, $t < T$, is given by

$$p^d(t, T) = E^{\mathbf{P}}\left[\int_t^T \lambda_s 1_{\{s \leq T^d\}} ds | \mathcal{F}_t\right].$$

This proposition gives us a general relationship between the conditional probability of default $p^d(t, T)$ and the intensity process λ. Imposing some strict assumptions on the intensity process λ, Madan & Unal (1994) show that the probability of default can be related to a discount factor in which discounting is done at the intensity λ. Hence, a striking similarity to default-free interest rate modeling is found. Duffie (1998c) shows the same result under more general assumptions:

Proposition 2.3.2.
Suppose T^d is a stopping time with a bounded intensity process[28] λ. Fixing some time $T > 0$, let

$$Y_t = E^{\mathbf{P}}\left[\exp\left(-\int_t^T \lambda_s ds\right) | \mathcal{F}_t\right], \quad t \leq T.$$

If the jump

$$\Delta Y_{T^d} = Y_{T^d} - Y_{T^d-}$$

is zero almost surely, then

$$p^d(t, T) = 1 - Y_t, \; t < T^d.$$

Proof.
See Duffie (1998c), pages 4-5.

[28] The assumption that λ is bounded can be replaced by some integrability conditions. For details see Duffie (1998c), p.5.

Remark 2.3.4.
Given any predictable nonnegative process λ such that $\int_0^t \lambda_s ds < \infty$ $\mathbf{P}-a.s.$ for all $t \geq 0$, and defining the stopping time T^d by

$$T^d = \inf\left\{t \geq 0 : \int_0^t \lambda_s ds = \Lambda\right\},$$

where Λ is an exponentially distributed random variable with mean 1, independent of Y defined in proposition 2.3.2, then λ is the intensity of T^d and $\Delta Y_\tau = 0$ a.s.[29] Such a model is called a Cox process or doubly stochastic Poisson process model.

Conclusion 2.3.3
As soon as we have specified the intensity process λ, the conditional default and survival probabilities are given by proposition 2.3.2. The following examples summarize analytically tractable choices of λ.

Examples of Intensity Models.

Constant intensity models. $\lambda_t = \lambda$ if $0 \leq t < T^d$ and 0 otherwise, where $\lambda > 0$ is some constant. Under this assumption $p^s(t,T) = e^{-\lambda(T-t)}$ and $p^d(t,T) = 1 - e^{-\lambda(T-t)}$. The time to default is exponentially distributed, the expected time to default is $\frac{1}{\lambda}$. We associate with the arrival of default the first event of a Poisson arrival process, at some constant mean arrival rate λ. Figure 2.8 shows the term structure of default probabilities in the constant intensity model for different values of λ.

Deterministically time-varying intensity models. λ_t is a deterministically time-varying function. Under this assumption $p^s(t,T) = e^{-\int_t^T \lambda_s ds}$ and $p^d(t,T) = 1 - e^{-\int_t^T \lambda_s ds}$.

Affine jump diffusion intensity models:. λ_t is an affine function of the factors X, i.e. $\lambda_t = l_0(t) + l_1(t) \cdot X_t$ for bounded continuous functions l_0 and l_1 on $[0,\infty)$, where X is a strong $n-$dimensional Markov process[30] valued in $\mathbb{R}^n$ uniquely solving the stochastic differential equation

[29] For details see, e.g., Lando (1994), Lando (1996), Lando (1998), Duffie (1998c).

[30] The intuitive meaning of the Markov property is that the future behaviour of the process X after t depends only on the value X_t and is not influenced by the history of the process before t. Mathematically speaking, an $\mathcal{F}_t-$adapted process X satisfies the Markov property if, for any bounded Borel function f and for any s and t such that $s \leq t$, we have $E^{\mathbf{P}}\left[f(X_t)|\mathcal{F}_s\right] = E^{\mathbf{P}}\left[f(X_t)|X_s\right]$. Roughly, the strong Markov property states that a relation of the previous form continues to hold if the time s is replaced by a stopping time τ. Mathematically speaking, an $\mathcal{F}_t-$adapted process X satisfies the strong Markov property if, for any bounded Borel function f and for any stopping time τ with respect to the filtration$\{\mathcal{F}_t\}_{t\geq 0}$, $\tau < \infty$ a.s., we have $E^{\mathbf{P}}\left[f(X_t)|\mathcal{F}_\tau\right] = E^{\mathbf{P}}\left[f(X_t)|X_\tau\right]$. For further details see, e.g., Lamberton & Lapeyre (1996), pp. 54 ff., Oksendal (1998), pp. 107 ff., and Brémaud (1981), pp. 290 ff.

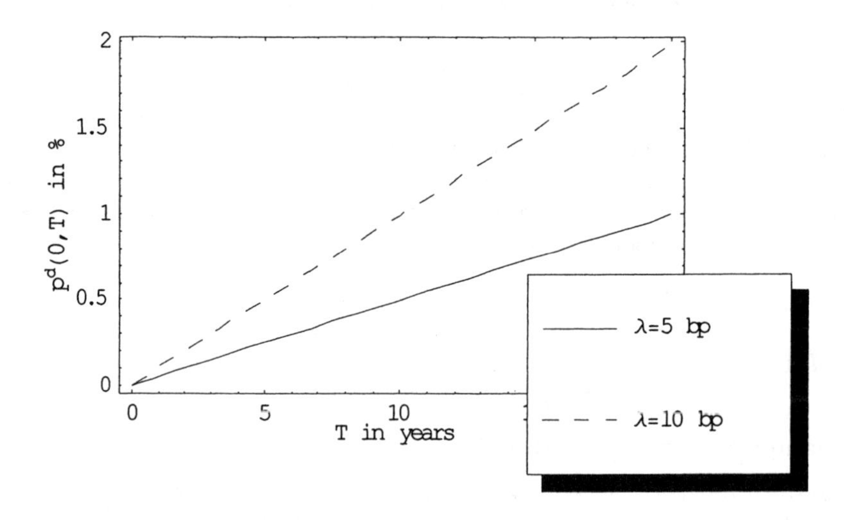

Fig. 2.8. Constant intensity model: Term structure of default probabilities for different constant intensity values.

$$dX_t = \mu\left(X_t, t\right) dt + \sigma\left(X_t, t\right) dW_t + dZ_t.$$

W is an $\mathbf{F}$−adapted n−dimensional standard Brownian motion in $\mathbb{R}^n$, $\mu : \Theta \to \mathbb{R}^n$, $\sigma : \Theta \to \mathbb{R}^{n \times n}$, $\Theta \subset \mathbb{R}^n \times [0, \infty)$, Z is a pure jump process whose jump-counting process A_t admits a stochastic intensity $\gamma : \Theta \to [0, \infty)$ depending on X with a jump-size distribution ν only depending on t and for each t, $\{x : (x, t) \in \Theta\}$ containing an open subset of $\mathbb{R}^n$. Then, under certain restrictions[31] on the parameters of the process X and a specific choice of Θ the conditional survival probabilities are of the form

$$p^s\left(t, T\right) = e^{\alpha(t,T) + \beta^T(t,T) \cdot X_t},$$

where $\alpha\left(t, T\right) \in \mathbb{R}$ and $\beta\left(t, T\right) \in \mathbb{R}^n$ are coefficients, depending on t and T that are given explicitly. Each model with survival probabilities of this form is called affine. All parameters of λ may be inferred from market data. Therefore, one can use methods similar to those described in section 6.6. Examples for empirical studies and parameter estimation of intensity based models include those of Duffee (1996a), and Düllmann & Windfuhr (2000).

As affine models are very tractable and flexible, they are pretty popular. They were first applied to (non-defaultable) interest rate modeling. Famous

[31] See, e.g., Duffie & Kan (1996).

examples are the term structure models of Vasicek (1977), Cox et al. (1985), Longstaff & Schwartz (1992), and Hull & White (1993). Brown & Schaefer (1994) were the first ones to consider affine models as an own category. Their theory was further developed by Duffie & Kan (1994), Duffie & Kan (1996), and Duffie, Pan & Singleton (1998). Dai & Singleton (1998) provided a classification of affine models.

It turns out that imposing the simple relationship

$$p^s(t,T) = e^{\alpha(t,T)+\beta^T(t,T)\cdot X_t}$$

between survival probabilities and state variables imposes very much tractability on the model. As for $t < T^d$

$$\alpha(t,t) = 0,\ \beta_i(t,t) = 0,\ i = 1,...,n,$$

we find

$$\begin{aligned}
\lambda_t &= -\lim_{\varepsilon\to 0}\frac{\alpha(t,t+\varepsilon)}{\varepsilon} - \lim_{\varepsilon\to 0}\frac{\beta^T(t,t+\varepsilon)}{\varepsilon}X_t \\
&= -\lim_{\varepsilon\to 0}\frac{\alpha(t,t+\varepsilon)-\alpha(t,t)}{\varepsilon} - \lim_{\varepsilon\to 0}\frac{\beta^T(t,t+\varepsilon)-\beta^T(t,t)}{\varepsilon}X_t \\
&= -\left.\frac{\partial\alpha(t,T)}{\partial T}\right|_{T=t} - \left.\frac{\partial\beta^T(t,T)}{\partial T}\right|_{T=t} X_t. \qquad (2.12)
\end{aligned}$$

Once the processes for the state variables X_t have been specified, equation (2.12) is sufficient to establish survival probabilities in the model. $\alpha(t,T)$ and $\beta(t,T)$ are determined by the specific choice for X. But it is not possible to choose X arbitrarily. In case of affine diffusion intensity models Duffie & Kan (1996) show that the process for X must be of the form

$$dX_t = (aX_t + b)\,dt + \sigma(X_t,t)\,dW_t,$$

where a is a constant $n\times n$ matrix, b is a constant n-dimensional vector, W is a $n-$dimensional standard Brownian motion and $\sigma(X_t,t)$ is a $n\times n$ diagonal matrix of the following type:

$$\sigma(X_t,t) = \sigma\cdot\Sigma(X_t,t),$$

where σ is a constant $n\times n$ matrix and

$$\Sigma(X_t) = \begin{pmatrix} \sqrt{c_1+d_1X_t} & & 0 \\ & \ddots & \\ 0 & & \sqrt{c_n+d_nX_t} \end{pmatrix},$$

where $c_i\in\mathbb{R}$, $d_i\in\mathbb{R}^n$, $i = 1,...,n$, are constants (with mild regularity conditions to ensure that $c_i + d_iX_t$, $i = 1,...,n$, remain positive). Let

$$c = \begin{pmatrix} c_1 & & 0 \\ & \ddots & \\ 0 & & c_n \end{pmatrix} \text{ and } d = \begin{pmatrix} d_1 & & 0 \\ & \ddots & \\ 0 & & d_n \end{pmatrix}.$$

To calculate survival probabilities we have to solve for $\alpha(t,T)$ and $\beta(t,T)$. As

$$p^s(t,T) = E^{\mathbf{P}}\left[\exp\left(-\int_t^T \lambda_s ds\right)|\mathcal{F}_t\right],$$

by applying the Feynman-Kac formula (see, e.g., Duffie (1992)) $p^s(t,T)$ satisfies the differential equation

$$\begin{aligned} 0 = & \frac{\partial p^s(t,T)}{\partial t} + \sum_{i=1}^n (a_i X_t + b_i) \frac{\partial p^s(t,T)}{\partial X_i} \\ & + \frac{1}{2}\sum_{i,j=1}^n \left(\sigma \Sigma^2(X_t)\sigma^T\right)_{ij} \frac{\partial^2 p^s(t,T)}{\partial X_i \partial X_j} - \lambda_t p^s(t,T), \end{aligned}$$

where a_i is the i-th row of matrix a. Using

$$\begin{aligned} \frac{\partial p^s(t,T)}{\partial t} &= p^s(t,T)\left(\frac{\partial \alpha(t,T)}{\partial t} + \frac{\partial \beta^T(t,T)}{\partial t} X_t\right), \\ \frac{\partial p^s(t,T)}{\partial X_i} &= p^s(t,T)\,\beta_i(t,T), \\ \frac{\partial^2 p^s(t,T)}{\partial X_i \partial X_j} &= p^s(t,T)\,\beta_i(t,T)\,\beta_j(t,T), \end{aligned}$$

and equation (2.12) yields

$$\begin{aligned} 0 = & \left(\left.\frac{\partial \alpha(t,T)}{\partial T}\right|_{T=t} + \frac{\partial \alpha(t,T)}{\partial t} + b^T\beta(t,T) + \frac{1}{2}\beta^T(t,T)\sigma c \sigma^T \beta(t,T)\right) \\ & + \left(\left.\frac{\partial \beta^T(t,T)}{\partial T}\right|_{T=t} + \frac{\partial \beta^T(t,T)}{\partial t} + a^T\beta(t,T) \right. \\ & \left. + \frac{1}{2}\beta^T(t,T)\sigma d \sigma^T \beta(t,T)\right) X_t \end{aligned}$$

and therefore

$$\begin{aligned} \frac{\partial \alpha(t,T)}{\partial t} &= - \left.\frac{\partial \alpha(t,T)}{\partial T}\right|_{T=t} - b^T\beta(t,T) - \frac{1}{2}\beta^T(t,T)\sigma c \sigma^T \beta(t,T), \\ \frac{\partial \beta(t,T)}{\partial t} &= - \left.\frac{\partial \beta(t,T)}{\partial T}\right|_{T=t} - a^T\beta(t,T) - \frac{1}{2}\beta^T(t,T)\sigma d \sigma^T \beta(t,T). \end{aligned}$$

Together with the boundary conditions $\alpha(t,t) = 0$, $\beta_i(t,t) = 0$, $i = 1, ..., n$, these Riccati equations may sometimes have explicit formulae. Even if numerical solutions are necessary there are often easy methods available.

In the following we want to distinguish between the following types of affine models:

- (Extended) Gaussian affine diffusion intensity models:

$$\lambda_t = l_0 + l_1 \cdot X_t,$$

where l_0 is a constant and l_1 a n-dimensional constant, and

$$dX_t = (aX_t + b)\, dt + \sigma dW_t,$$

where a is a constant $n \times n$ matrix, σ is a constant diagonal $n \times n$ matrix, and b is a constant n-dimensional vector. Such a model is called extended if some of its parameters are deterministic functions of time. Gaussian affine models permit intensities to become zero with non-zero probability which is forbidden in a sensible model. But for models fitted to ordinary market data the risk of actually getting negative intensities is very low. Gaussian affine diffusion intensity models are very popular in iterest rate modeling. Examples of one-factor Gaussian interest rate models are the models of Vasicek (1977), Hull & White (1990), and Hull & White (1993). Steeley (1991) works with a two-factor model. Chen & Yang (1996) provide a three-factor model. Beaglehole & Tenney (1991), Babbs (1993), Nunes (1998), and Babbs & Nowman (1999) consider the general n-dimensional case.

- (Extended) Gaussian affine jump diffusion intensity models:

$$\lambda_t = l_0 + l_1 \cdot X_t,$$

where l_0 is a constant and l_1 a n-dimensional constant, and

$$dX_t = (aX_t + b)\, dt + \sigma dW_t + dZ_t,$$

where a is a constant $n \times n$ matrix, σ is a constant diagonal $n \times n$ matrix, b is a constant n-dimensional vector, and Z is a pure jump process whose jump-counting process A_t admits a stochastic intensity $\gamma : \Theta \to [0, \infty)$ depending on X with a jump-size distribution ν only depending on t. Such a model is called extended if some of its parameters are deterministic functions of time.

- (Extended) CIR affine diffusion intensity models:

$$\lambda_t = l_0 + l_1 \cdot X_t,$$

where l_0 is a constant and l_1 a n-dimensional constant, and

$$dX_t = (aX_t + b)\, dt + \sigma \Sigma^{CIR}(X_t)\, dW_t,$$

where a is a constant $n \times n$ matrix, σ is a constant diagonal $n \times n$ matrix, b is a constant n-dimensional vector, and

$$\Sigma^{CIR}(X_t) = \begin{pmatrix} \sqrt{X_{1,t}} & & 0 \\ & \ddots & \\ 0 & & \sqrt{X_{n,t}} \end{pmatrix}.$$

Such a model is called extended if some of its parameters are deterministic functions of time. Examples of discussions of one-factor CIR interest rate models are the works of Cox et al. (1985), Hull & White (1990), Jamshidian (1995), Pelsser (1996). Among others, Richard (1978), Longstaff & Schwartz (1992), and Chen & Scott (1992) work with CIR two-factor models. Beaglehole & Tenney (1991), Jamshidian (1996), and Scott (1995) discuss the general theory.

- (Extended) CIR affine jump diffusion intensity models:

$$\lambda_t = l_0 + l_1 \cdot X_t,$$

where l_0 is a constant and l_1 a n-dimensional constant, and

$$dX_t = dX_t = (aX_t + b)\,dt + \sigma\Sigma^{CIR}(X_t)\,dW_t + dZ_t,$$

where a is a constant $n \times n$ matrix, σ is a constant diagonal $n \times n$ matrix, b is a constant n-dimensional vector,

$$\Sigma^{CIR}(X_t) = \begin{pmatrix} \sqrt{X_{1,t}} & & 0 \\ & \ddots & \\ 0 & & \sqrt{X_{n,t}} \end{pmatrix}.$$

and Z is a pure jump process whose jump-counting process A_t admits a stochastic intensity $\gamma : \Theta \to [0, \infty)$ depending on X with a jump-size distribution ν only depending on t. Such a model is called extended if some of its parameters are deterministic functions of time.

- Affine diffusion or jump diffusion intensity models that mix Gaussian and CIR type state variables:
The most famous family of affine diffusion models that mix Gaussian and CIR type state variables is the so called three-factor affine family. The idea behind this family is to allow for a stochastic mean and/or a stochastic volatility making sure that volatilities stay positive:

$$\lambda_t = X_{1,t},$$

with

$$dX_t = a\,(b_t - X_t)\,dt + \sigma\Sigma(X_t)\,dW_t,$$

where

$$a = \begin{pmatrix} a_1 & 0 & 0 \\ 0 & a_2 & 0 \\ 0 & 0 & a_3 \end{pmatrix}, \quad b_t = \begin{pmatrix} X_{2t} \\ b_2 \\ b_3 \end{pmatrix},$$

$$\sigma = \begin{pmatrix} 1 & 0 & 0 \\ 0 & \sigma_2 & 0 \\ 0 & 0 & \sigma_3 \end{pmatrix}, \Sigma(X_t) = \begin{pmatrix} \sqrt{X_{3,t}} & 0 & 0 \\ 0 & X_{2,t}^{\gamma} & 0 \\ 0 & 0 & \sqrt{X_{3,t}} \end{pmatrix},$$

and $\gamma \in \left\{0, \frac{1}{2}\right\}$. a_i, $i = 1, 2, 3$, and b_i, σ_i, $i = 2, 3$, are constants. E.g., Balduzzi, Das, Foresi & Sundaram (1996), Chen (1996) and Rhee (1999) work with this three-factor affine family and apply it to interest rate modeling.

When choosing a specific model, one should pose the following questions:

- Does the model imply positive intensities, i.e. $\lambda_t > 0$ for all t ?
- What distribution does the model imply for the intensity ? Is this distribution empirically validated ?
- Are default and survival probabilities explicitly computable from the dynamics ?
- Is the model mean reverting ?
- Are the default and survival probabilities implied by the model always between 0 and 1 ?
- How suited is the model for Monte Carlo simulation ?
- Does the chosen dynamics allow for historical estimation techniques to be used for parameter estimation ?

In the following we give some examples of affine survival/default probability term structure models:

Example 2.3.5.
The Vasicek model (see, e.g., Vasicek (1977)) applied to intensity modeling: The intensity λ follows an Ornstein-Uhlenbeck process, i.e.

$$d\lambda_t = a\left(b - \lambda_t\right) dt + \sigma dW_t, \tag{2.13}$$

where W is a one-dimensional standard Brownian motion and the initial value λ_0, and the parameter values $a > 0$, b, and $\sigma > 0$ are constants. The process

$$\lambda_t = e^{-at}\lambda_0 + \left(1 - e^{-at}\right) b + \sigma e^{-at} \int_0^t e^{au} dW_u \tag{2.14}$$

is the unique [32] λ that solves the stochastic equation (2.13) with initial value λ_0 (for a proof see, e.g., Nielsen (1999), *p.* 107). λ_t conditional on $\mathcal{F}_0$ is

[32] Let $W = (W_1, ..., W_m)'$ be a $m-$dimensional standard Brownian motion, $m \in \mathbb{N}$. A stochastic process X is called an Itô process if for all $t \geq 0$ we have

$$X_t = X_0 + \int_0^t \mu_s ds + \int_0^t \sigma_s dW_s,$$

where X_0 is $\mathcal{F}_{0-}-$ measurable, $\mu = \mu_t$ and $\sigma = \sigma_t$ are progressively measurable stochastic processes with

$$\int_0^t \left|\mu_s\right| ds < \infty$$

and

$$\int_0^t \sigma_i^2 \left(s\right) ds < \infty \; \mathbf{P}-\mathrm{a.s.}$$

for all $t \geq 0$, $i = 1, ..., m$.

Korn & Korn (1999), p.36: If all paths of a stochastic process are right continuous the process is progressively measurable.

normally distributed with mean and variance given respectively by

$$E^{\mathbf{P}}\left[\lambda_t \middle| \mathcal{F}_0\right] = e^{-at}\lambda_0 + \left(1 - e^{-at}\right) b, \tag{2.15}$$

and

$$Var^{\mathbf{P}}\left[\lambda_t \middle| \mathcal{F}_0\right] = \frac{\sigma^2}{2a}\left(1 - e^{-2at}\right). \tag{2.16}$$

This implies that $\mathbf{P}(\lambda_t < 0) > 0$, which is not very satisfactory from a theoretical and practical point of view and is a major drawback of using the Vasicek model for specifying intensity processes. As a consequence of equations (2.15) and (2.16) the intensity λ is mean reverting, since the expected intensity tends, for t going to infinity, to the value b. Using proposition 2.3.2 and equation (2.14)

$$\begin{aligned}
&p^d(t,T) \\
&= 1 - E^{\mathbf{P}}\left[e^{-\int_t^T \lambda_s ds} \middle| \mathcal{F}_t\right] \\
&= 1 - E^{\mathbf{P}}\left[e^{-(T-t)b - \frac{1}{a}\left(1-e^{-a(T-t)}\right)(\lambda_t - b) - \frac{\sigma}{a}\int_t^T \left(1-e^{-a(T-u)}\right)dW_u} \middle| \mathcal{F}_t\right] \\
&= 1 - e^{-(T-t)b - \frac{1}{a}\left(1-e^{-a(T-t)}\right)(\lambda_t - b) + \frac{\sigma^2}{4a^3}\left(4e^{-a(T-t)} - e^{-2a(T-t)} + 2a(T-t) - 3\right)}.
\end{aligned}$$

Through this formula, the model gives a complete description of the term structure of default probabilities at any given point in time t. The term structure of survival is given by

$$p^s(t,T) = e^{\alpha(t,T) + \beta(t,T)\lambda_t},$$

where

$$\begin{aligned}
\beta(t,T) &= -\frac{1}{a}\left(1 - e^{-a(T-t)}\right), \\
\alpha(t,T) &= \left(b - \frac{\sigma^2}{2a^2}\right)\left(\beta(t,T) - T + t\right) - \frac{\sigma^2}{4a}\beta^2(t,T).
\end{aligned}$$

Figure 2.10 shows two term structures of default probabilities conditional on the parameter set $a = 0.25$, $b = 5$ basis points, $\sigma = 3$ basis points, and two different initial intensities, $\lambda_0 = 5$ basis points and $\lambda_0 = 10$ basis points. The upper curve corresponds to an initial intensity of 10 basis points, the lower curve corresponds to an initial value of 5 basis points. In addition, figure 2.10 illustrates the fact that in this model the default probability tends to zero as the time to maturity approaches zero. Since λ_t is normally distributed, survival probabilities are lognormally distributed:

$$\begin{aligned}
d\ln p^s(t,T) &= d\alpha(t,T) + \beta(t,T)\, d\lambda_t + \lambda_t d\beta(t,T) \\
&= \left(\lambda_t - \frac{\sigma^2}{2a^2}\left(1 - e^{-a(T-t)}\right)^2\right) dt \\
&\quad - \frac{\sigma}{a}\left(1 - e^{-a(T-t)}\right) dW_t.
\end{aligned}$$

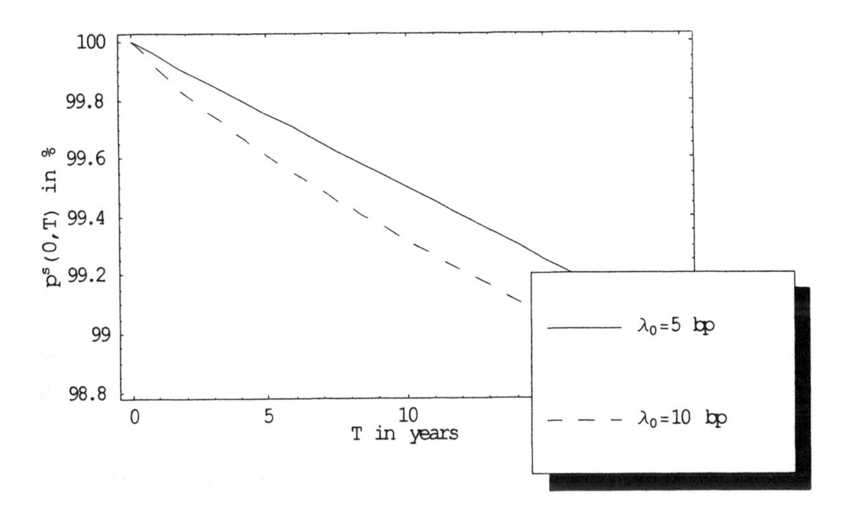

Fig. 2.9. λ follows an Ornstein-Uhlenbeck process: Term structure of survival probabilities for different initial intensity values.

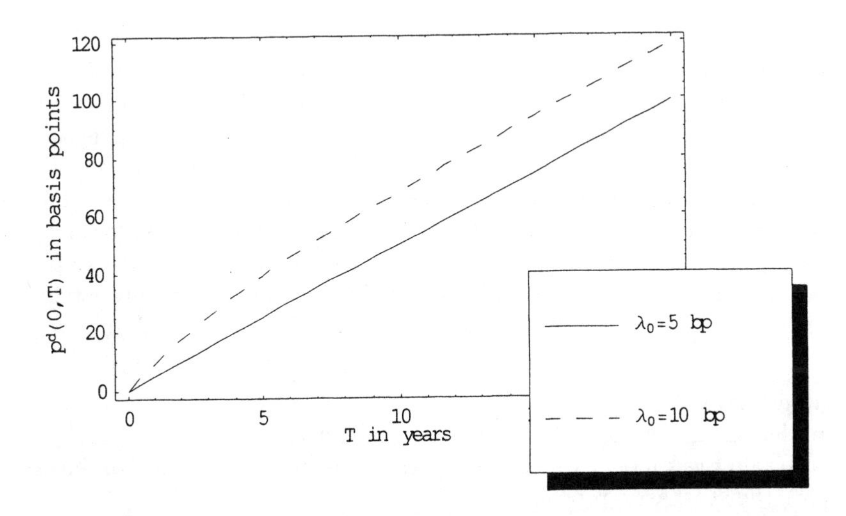

Fig. 2.10. λ follows an Ornstein-Uhlenbeck process: Term structure of default probabilities for different initial intensity values.

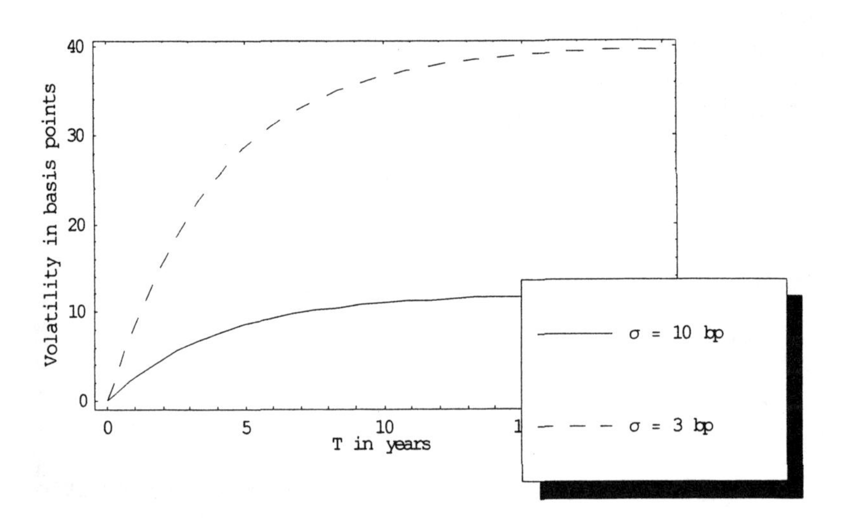

Fig. 2.11. Volatility of the survival probability $p^s(0,T)$ for different values of σ.

A positive shock of W makes the default probability and the intensity go up and the survival probability go down. The volatility of the survival probability is

$$\frac{\sigma}{a}\left(1-e^{-a(T-t)}\right).$$

It is a deterministic function of t. It tends to zero as t tends to T and to $\frac{\sigma}{a}$ as T tends to infinity. It is dependent on the speed of mean reversion and the time horizon but independent of the current intensity value λ_0 and the mean reversion level b of the intensity process. Figure 2.11 shows the term structures of the survival probability volatilities for different values of σ. The upper curve corresponds to $\sigma = 3$ basis points and the lower curve corresponds to $\sigma = 10$ basis points. The parameter value of a is like in the previous calculations: $a = 0.25$. The volatility of the survival probability is an increasing function of $T-t$ and converges to 0 as $T-t \to 0$ and to $\frac{\sigma}{a}$ as $T-t \to \infty$.

Example 2.3.6.
The Merton model applied to intensity modeling:
It is the simplest model which doesn't show mean reversion. The intensity λ follows the stochastic process

$$d\lambda_t = adt + \sigma dW_t, \tag{2.17}$$

where W is a one-dimensional standard Brownian motion and the initial value λ_0, and the parameter values $a > 0$,and $\sigma > 0$ are constants. The process

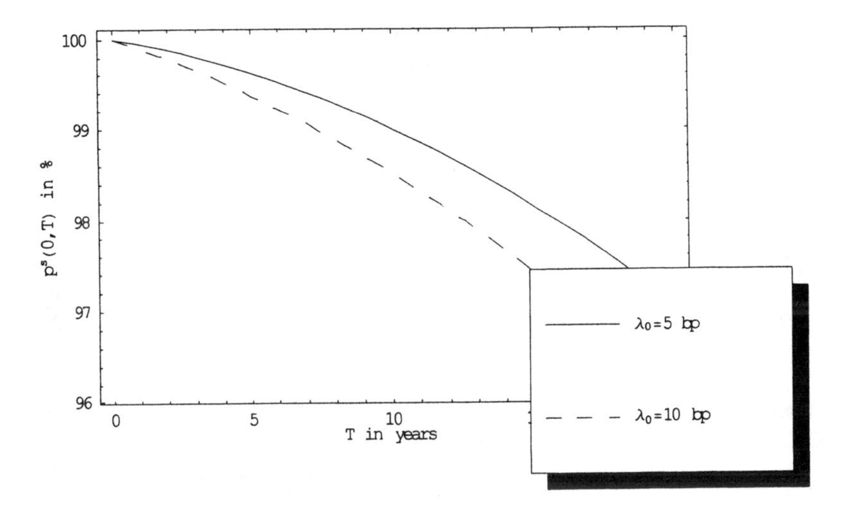

Fig. 2.12. λ follows a Merton process: Term structure of survival probabilities for different initial intensity values.

$$\lambda_t = \lambda_0 + at + \sigma W_t \tag{2.18}$$

is the unique Itô process λ that solves the stochastic equation (2.17) with initial value λ_0. Using proposition 2.3.2 and equation (2.18)

$$\begin{aligned} p^d(t,T) &= 1 - E^{\mathbf{P}}\left[e^{-\int_t^T \lambda_s ds} \middle| \mathcal{F}_t \right] \\ &= 1 - e^{-\lambda_t(T-t) - \frac{1}{2}a(T-t)^2 + \frac{\sigma^2}{6}(T-t)^3}. \end{aligned}$$

Through this formula, the model gives a complete description of the term structure of default probabilities at any given point in time t. The term structure of survival is given by

$$p^s(t,T) = e^{\alpha(t,T) + \beta(t,T)\lambda_t},$$

where

$$\begin{aligned} \beta(t,T) &= -(T-t), \\ \alpha(t,T) &= -\frac{1}{2}a(T-t)^2 + \frac{\sigma^2}{6}(T-t)^3. \end{aligned}$$

Figure 2.12 shows two term structures of default probabilities conditional on the parameter set $a = 1$ basis point, $\sigma = 3$ basis points, and two different

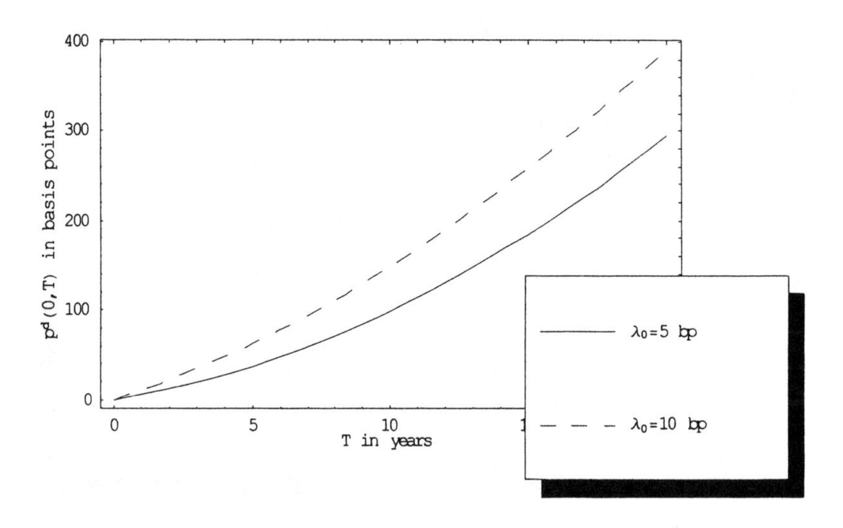

Fig. 2.13. λ follows a Merton process: Term structure of default probabilities for different initial intensity values.

initial intensities, $\lambda_0 = 5$ basis points and $\lambda_0 = 10$ basis points. The upper curve corresponds to an initial intensity of 10 basis points, the lower curve corresponds to an initial value of 5 basis points. In addition, figure 2.13 illustrates the fact that in this model the default probability tends to zero as the time to maturity approaches zero. Although figures 2.12 and 2.13 seem plausible the model has the annoying feature that

$$p^s(t,T) \to \infty \text{ as } T \to \infty.$$

This implies that the model can't be used to model survival probabilities for long time horizons. For some configurations of the parameters, in particular for low values of σ, this may not be a serious problem. However, Figure 2.14 shows an example where we have chosen a high value for σ of 40 basis points. The other parameter values are like in the previous example: $a = 1$ basis point and $\lambda_0 = 5$ basis points. For longer time horizons the model produces survival probabilities which does not make sense. Especially, for $T > 26$ years the values for $p^s(t,T)$ become greater than 100%.
Since λ_t is normally distributed, survival probabilities are lognormally distributed:

$$\begin{aligned} d\ln p^s(t,T) &= d\alpha(t,T) + \beta(t,T)\,d\lambda_t + \lambda_t d\beta(t,T) \\ &= \left(\lambda_t - \frac{\sigma^2}{2}(T-t)^2\right)dt - \sigma(T-t)\,dW_t. \end{aligned}$$

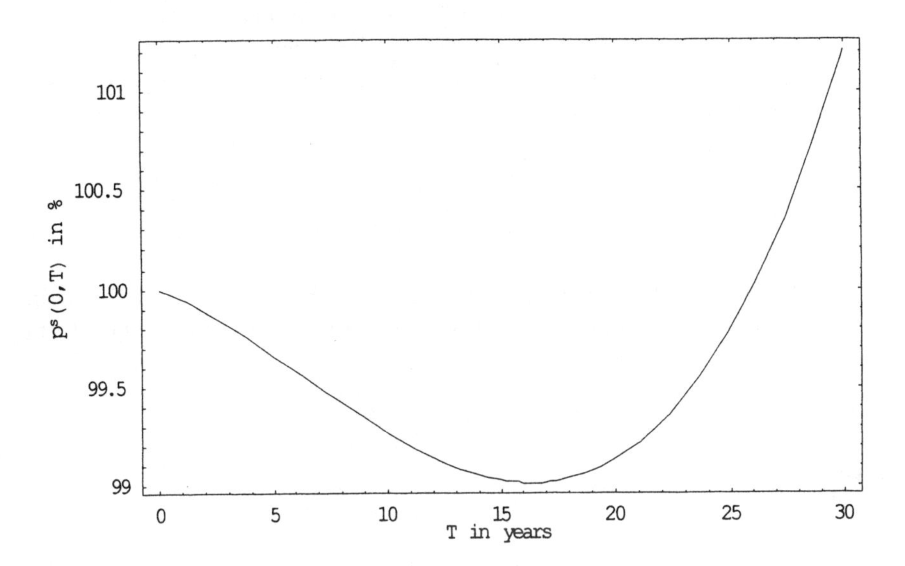

Fig. 2.14. λ follows a Merton process: Term structure of survival probabilities for a high value of σ. The model produces survival probabilities that does not make sense.

A positive shock of W makes the default probability and the intensity go up and the survival probability go down. The volatility of the survival probability is

$$\sigma\,(T-t)\,.$$

It is a deterministic function of t and tends to zero as t tends to T.

Example 2.3.7.
The continuous-time Ho-Lee model (see, e.g., Ho & Lee (1986)) applied to intensity modeling:
In the continuous-time Ho-Lee model the intensity λ follows the stochastic process

$$d\lambda_t = a_t dt + \sigma dW_t \tag{2.19}$$

with initial value $\lambda_0 > 0$, where W is a one-dimensional standard Brownian motion, a is a deterministic function of time such that for all $t \leq T^*$

$$\int_0^t |a_s|\, ds < \infty,$$

and $\sigma > 0$ is a constant. Then the process

$$\lambda_t = \lambda_0 + \int_0^t a_s ds + \sigma W_t \tag{2.20}$$

is the unique Itô process λ that solves the stochastic equation (2.19) with initial value λ_0. The Ho-Lee model is like the Merton model, except that we allow the parameter a to be a deterministic function of time which allows us to fit the initial term structure of default probabilities. Using proposition 2.3.2 and equation (2.20) we find

$$\begin{aligned} p^d(t,T) &= 1 - E^{\mathbf{P}}\left[e^{-\int_t^T \lambda_s ds} \middle| \mathcal{F}_t \right] \\ &= 1 - e^{-\lambda_t (T-t) - \int_t^T \left(\int_t^s a_u du \right) ds + \frac{\sigma^2}{6} (T-t)^3} . \end{aligned}$$

Through this formula, the model gives a complete description of the term structure of default probabilities at any given point in time t. The term structure of survival is given by

$$p^s(t,T) = e^{\alpha(t,T) + \beta(t,T)\lambda_t},$$

where

$$\begin{aligned} \beta(t,T) &= -(T-t), \\ \alpha(t,T) &= -\int_t^T \left(\int_t^s a_u du \right) ds + \frac{\sigma^2}{6} (T-t)^3 . \end{aligned}$$

Example 2.3.8.
The Cox, Ingersoll and Ross model (see, e.g., Cox et al. (1985)) applied to intensity modeling:
The general equilibrium approach developed by Cox et al. (1985) led to the introduction of a square-root term in the diffusion coefficient of the dynamics proposed by Vasicek (1977). The resulting model has been a benchmark for many years as it combines two important features: it is very well analytically tractable and the stochastic process is always positive. The intensity λ follows the stochastic process

$$d\lambda_t = a(b - \lambda_t)\, dt + \sigma \sqrt{\lambda_t} dW_t, \tag{2.21}$$

with initial value $\lambda_0 > 0$, and where W is a one-dimensional standard Brownian motion and $a, b, \sigma > 0$ are constants. The condition

$$2ab > \sigma^2$$

has to be imposed to ensure that the origin is inaccessible to the process (2.21) so that λ remains positive. The process λ features a noncentral χ^2-squared distribution, where the density function f_{λ_t} of λ_t is given by

$$f_{\lambda_t}(x) = c_t f_{\chi^2(v,\mu_t)}(c_t x),$$

where

$$c_t = \frac{4a}{\sigma^2 \left(1 - e^{-at}\right)},$$
$$\upsilon = \frac{4ab}{\sigma^2},$$
$$\mu_t = c_t \lambda_0 e^{-at},$$

and

$$f_{\chi^2(\upsilon,\mu)}(y) = \sum_{i=0}^{\infty} \frac{e^{-\frac{\mu}{2}} \left(\frac{\mu}{2}\right)^i}{i!} \frac{\left(\frac{1}{2}\right)^{i+\frac{\upsilon}{2}}}{\Gamma\left(i + \frac{\upsilon}{2}\right)} y^{i-1+\frac{\upsilon}{2}} e^{-\frac{y}{2}},$$

where Γ is Euler's gamma function[33]. Hence, the mean and the variance of λ_t conditional on $\mathcal{F}_0$ are given by

$$E^{\mathbf{P}}\left[\lambda_t \middle| \mathcal{F}_0\right] = e^{-at}\lambda_0 + \left(1 - e^{-at}\right) b,$$

and

$$Var^{\mathbf{P}}\left[\lambda_t \middle| \mathcal{F}_0\right] = \frac{\sigma^2}{a}\left(e^{-at} - e^{-2at}\right)\lambda_0 + \frac{\sigma^2}{2a}\left(1 - e^{-at}\right)^2 b.$$

The term structure of default and survival probabilities are given by

$$p^d(t,T) = 1 - p^s(t,T),$$
$$p^s(t,T) = e^{\alpha(t,T) + \beta(t,T)\lambda_t},$$

where

$$\beta(t,T) = -\frac{2\left(e^{(T-t)\sqrt{a^2+2\sigma^2}} - 1\right)}{2\sqrt{a^2+2\sigma^2} + \left(a + \sqrt{a^2+2\sigma^2}\right)\left(e^{(T-t)\sqrt{a^2+2\sigma^2}} - 1\right)},$$

$$\alpha(t,T) = \frac{2ab}{\sigma^2} \ln\left(\frac{2\sqrt{a^2+2\sigma^2}\, e^{\frac{\left(a+\sqrt{a^2+2\sigma^2}\right)(T-t)}{2}}}{2\sqrt{a^2+2\sigma^2} + \left(a + \sqrt{a^2+2\sigma^2}\right)\left(e^{(T-t)\sqrt{a^2+2\sigma^2}} - 1\right)}\right).$$

Transition Intensities. So far, we have defined an obligor specific default intensity λ_t with the following property

$$\lambda_t 1_{\{t \leq T^d\}} = \lim_{\varepsilon \to 0+} \frac{p^d(t, t+\varepsilon)}{\varepsilon}, \text{ for all } t \geq 0. \tag{2.22}$$

Sometimes it might be more convenient to consider and model the migration and default behavior of rating classes rather than the behavior of single counterparties. Therefore, starting from condition (2.22) we generalize the default

[33] Euler's gamma function is defined by the integral

$$\Gamma(z) = \int_0^\infty t^{z-1} e^{-t} dt, \ \mathrm{Re}(z) > 0.$$

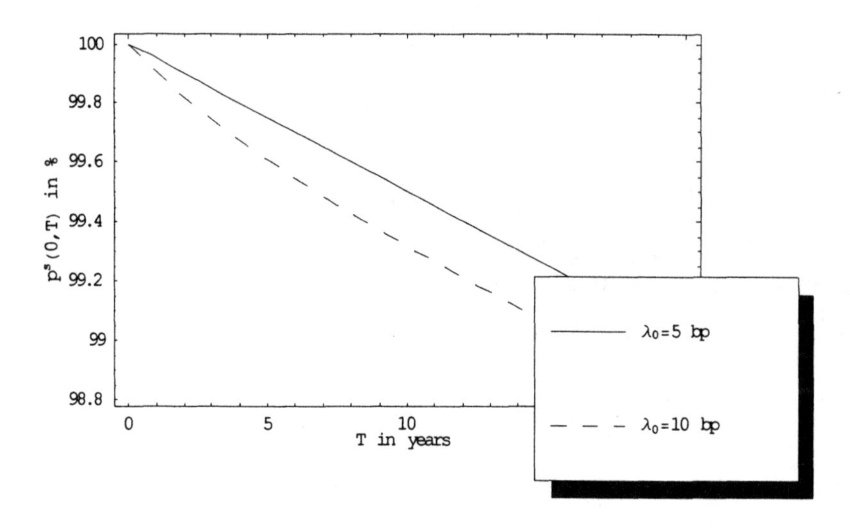

Fig. 2.15. λ follows a CIR process: Term structure of survival probabilities for different initial intensity values.

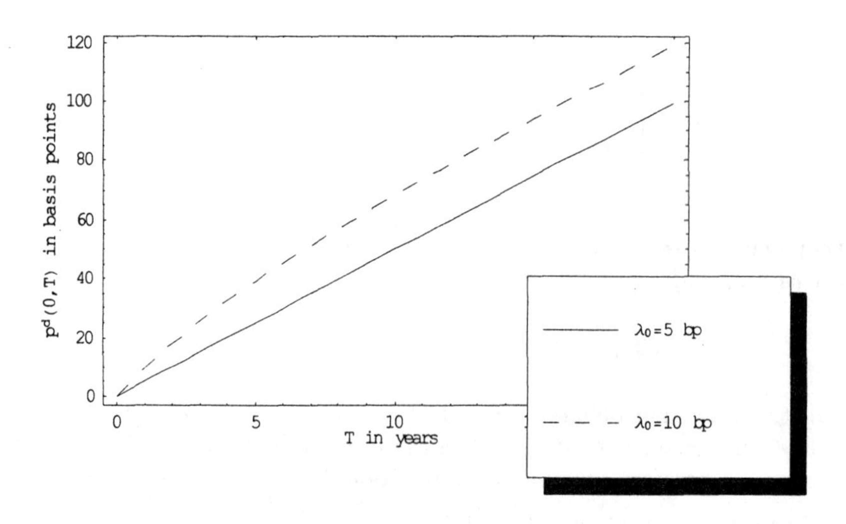

Fig. 2.16. λ follows a CIR process: Term structure of default probabilities for different initial intensity values.

intensity approach to a general transition intensity approach. We assume that there exist transition intensities for each type of transition, i.e. for all $t \geq 0$ and each pair of ratings $R \neq \check{R}$ $\left(R \in \{1,2,...,K-1\},\ \check{R} \in \{1,2,...,K\}\right)$ the following limit exists:

$$\lambda_{R,\check{R}}(t) = \lim_{\varepsilon \to 0+} \frac{p_{R,\check{R}}(t, t+\varepsilon)}{\varepsilon},$$

and as K is an absorbing state, for all $t \geq 0$:

$$\lambda_{K,\check{R}}(t) = 0.$$

If these limits exist for all transitions, we can define for all $R \in \{1,2,...,K-1\}$

$$\lambda_{R,R}(t) = \sum_{\check{R} \neq R} \lambda_{R,\check{R}}(t).$$

Note, that $\lambda_{R,\check{R}}(t) \geq 0$ for all $R, \check{R} \in \{1,2,...,K\}$.

Conclusion 2.3.4 *The probability of leaving state R over the next infinitesimal time interval of length* $\varepsilon > 0$ *is approximately* $\lambda_{R,R}(t) \cdot \varepsilon$. *This interpretation links the notion of transition intensity to the problem of estimating transition probabilities.*

Rating transitions as continuous time, time homogeneous Markov chains. If we assume that the rating process is a continuous time, time-homogeneous Markov chain with transition matrix $Tr(\tau)$ with respect to the time horizon τ, then, $\lambda_{R,\check{R}}(t) \equiv \lambda_{R,\check{R}}$ for all $t \geq 0$, and the $K \times K$ dimensional matrix Λ defined by

$$\Lambda = \begin{pmatrix} -\lambda_{1,1} & \lambda_{1,2} & \cdots & \cdots & \lambda_{1,K} \\ \lambda_{2,1} & -\lambda_{2,2} & \cdots & \cdots & \lambda_{2,K} \\ \vdots & \vdots & \ddots & \vdots & \vdots \\ \lambda_{K,1} & \lambda_{K,2} & \cdots & \cdots & -\lambda_{K,K} \end{pmatrix}$$

is the generator matrix of the Markov chain, i.e.

$$Tr(\tau) = \exp(\Lambda\tau) \text{ for all } \tau \geq 0,$$

where

$$\exp(\Lambda\tau) \equiv \sum_{k=0}^{\infty} \frac{(\Lambda\tau)^k}{k!}.$$

Especially, the one-year transition matrix is given by $Tr(1) = \exp(\Lambda)$. The waiting time for leaving state R has an exponential distribution with mean $\frac{1}{\lambda_R}$. Once the Markov chain leaves state R it migrates to state $\check{R} \neq R$ with probability $\frac{\lambda_{R,\check{R}}}{\lambda_R}$. The maximum-likelihood estimator of $\lambda_{R,\check{R}}$ based on observations between times t_1 and t_2 $(t_1 < t_2)$ is

$$\widehat{\lambda}_{R,\check{R}} = \frac{M_{R,\check{R}}(t_1,t_2)}{\int_{t_1}^{t_2} N_R(s)\,ds}, \tag{2.23}$$

where $M_{R,\check{R}}(t_1,t_2)$ counts the total number of transactions from state R to $\check{R}$ in the time interval $[t_1,t_2]$ and $N_R(s)$ is the number of counterparties in credit class R at time s. E.g., the maximum-likelihood estimator for the one-year transition matrix $Tr(1)$ is given by

$$\widehat{Tr}(1) = \exp\left(\widehat{\Lambda}\right).$$

Example 2.3.9.
Suppose there is a rating system with three non-defafult rating categories A, B, C, and the default category D. Over one year we observe a universe of 30 firms. At the beginning of the observation period each rating category contains 10 firms.

- After one months one B rated firm migrates to C and one B rated firm migrates to A and stay there for the rest of the year.
- After two months two A rated firms migrate to rating class B and stay there for the rest of the year.
- After six months one C rated firm migrates to B and stays there for the rest of the year. In addition, one C rated firm defaults.
- After eight months a C rated firm defaults.

Using estimator (2.23),

$$\begin{aligned}
\widehat{\lambda}_{A,B} &= \frac{2}{8\cdot\frac{12}{12}+1\cdot\frac{11}{12}+2\cdot\frac{2}{12}} = 0.21622,\\
\widehat{\lambda}_{B,A} &= \frac{1}{8\cdot\frac{12}{12}+2\cdot\frac{1}{12}+2\cdot\frac{10}{12}+1\cdot\frac{6}{12}} = 0.09677,\\
\widehat{\lambda}_{B,C} &= \frac{1}{8\cdot\frac{12}{12}+2\cdot\frac{1}{12}+2\cdot\frac{10}{12}+1\cdot\frac{6}{12}} = 0.09677,\\
\widehat{\lambda}_{C,B} &= \frac{1}{7\cdot\frac{12}{12}+1\cdot\frac{11}{12}+2\cdot\frac{6}{12}+1\cdot\frac{8}{12}} = 0.10435,\\
\widehat{\lambda}_{C,D} &= \frac{2}{7\cdot\frac{12}{12}+1\cdot\frac{11}{12}+2\cdot\frac{6}{12}+1\cdot\frac{8}{12}} = 0.20870,\\
\widehat{\lambda}_{A,C} &= \widehat{\lambda}_{A,D} = \widehat{\lambda}_{B,D} = \widehat{\lambda}_{C,A} = 0.
\end{aligned}$$

Hence, the maximum-likelihood estimates for the generator matrix and the one year transition matrix are given by

$$\widehat{\Lambda} = \begin{pmatrix} -0.21622 & 0.21622 & 0 & 0\\ 0.09677 & -0.19354 & 0.09677 & 0\\ 0 & 0.10435 & -0.31305 & 0.20870\\ 0 & 0 & 0 & 0 \end{pmatrix}$$

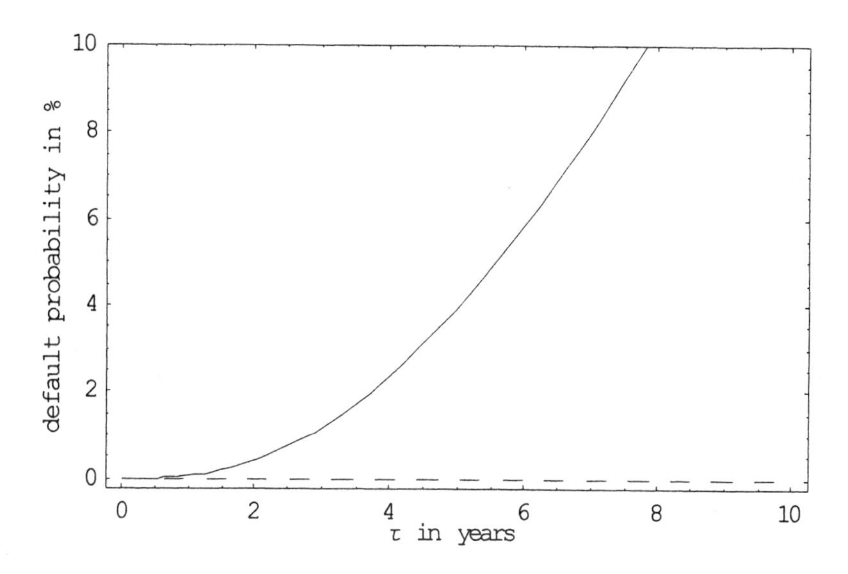

Fig. 2.17. Term structure of default probability directly from first rating class (bottom graph) and term structure of true default probability of first rating class (top graph).

and

$$\widehat{Tr}\,(1) = \begin{pmatrix} 0.81407 & 0.17707 & 0.00825 & 0.00061 \\ 0.07925 & 0.83663 & 0.07556 & 0.00857 \\ 0.00398 & 0.08148 & 0.73507 & 0.17948 \\ 0 & 0 & 0 & 1 \end{pmatrix}, \tag{2.24}$$

respectively. Examining the first row of $\widehat{\Lambda}$, the probability of staying in rating category 1 over a small time period $\varepsilon > 0$ is approximately $1 - 0.21622 \cdot \varepsilon$. The transition rate from the first rating category to default is 0. One could then easily draw the inappropriate conclusion that the probability of defaulting between time 0 and time τ is equal to $1 - \exp(0 \cdot \tau) = 0$. But this does not take into account the possibility of rating migrations and subsequent default. In figure 2.17 we show the difference of the rate of default directly from the first rating category and the true default probability as a function of time. Especially for higher-rated firms it is important to estimate risks of downgrading and risks of default in lower rating categories. The two top rating classes have the same risk of direct default (the rate is 0 in both cases). Nevertheless, as figure 2.18 shows, the probability of default for the second rating category as represented by the middle graph is higher than that of the first rating category, because it is closer to the third rating category. The top graph represents the default probabilities of the third rating category. Let

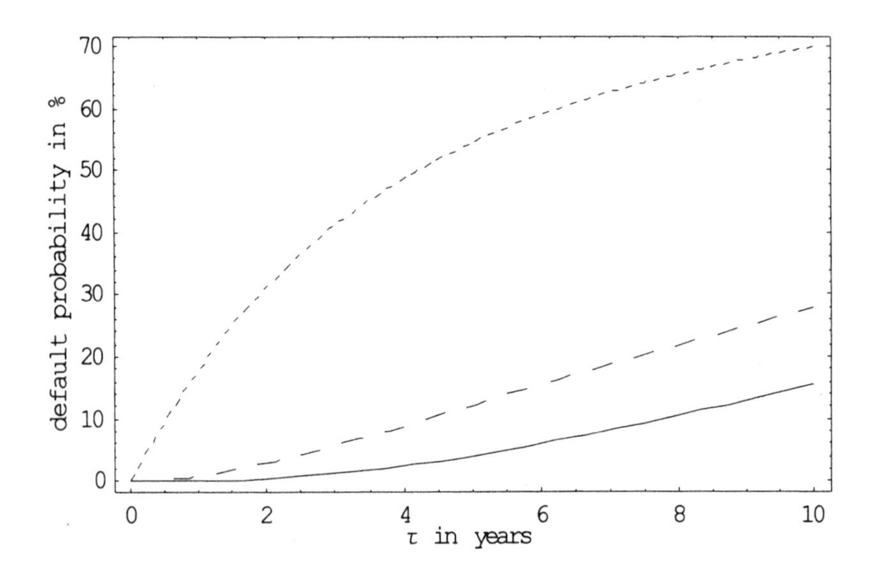

Fig. 2.18. Term structures of default probabilities of rating categories 1 (bottom graph), 2 (middle graph) and 3 (top graph).

us finally mention that if we had used the method as described in definition 2.3.1 we would have got

$$\widehat{Tr}\,(1) = \begin{pmatrix} 0.8 & 0.2 & 0 & 0 \\ 0.1 & 0.8 & 0.1 & 0 \\ 0 & 0.1 & 0.7 & 0.2 \\ 0 & 0 & 0 & 1 \end{pmatrix}. \tag{2.25}$$

Because there is no counterparty directly defaulting from rating classes A and B, matrix (2.25) does not capture default risk in these rating categories. However, the entries of matrix (2.24) are strictly positive for all rows but the last one. Especially, the default probabilities for all rating classes are positive, although there is no default observed for rating categories A and B.

Using $\exp(\Lambda) \approx I + \Lambda$, one can use the approximation

$$\widehat{\Lambda} \approx \widehat{Tr}\,(1) - I$$

to estimate the generator matrix from a given one-year transition matrix. Unfortunately, sometimes this approximation doesn't provide a fit of sufficient quality to the given data. Then the following algorithm can be applied to the transition matrix. First, we assume that every obligor has made either one or no transition throughout one year and that $\lambda_{R,R} \neq 0$ for $R \in \{1, 2, ..., K-1\}$. Then it can be shown that

$$\exp(\Lambda) \approx \begin{pmatrix} e^{\lambda_{1,1}} & \frac{\lambda_{1,2}\left(e^{\lambda_{1,1}}-1\right)}{\lambda_{1,1}} & \cdots\cdots & \frac{\lambda_{1,K}\left(e^{\lambda_{1,1}}-1\right)}{\lambda_{1,1}} \\ \frac{\lambda_{2,1}\left(e^{\lambda_{2,2}}-1\right)}{\lambda_{2,2}} & e^{\lambda_{22}} & \cdots\cdots & \frac{\lambda_{2,K}\left(e^{\lambda_{2,2}}-1\right)}{\lambda_{2,2}} \\ \vdots & \vdots \ddots & \vdots & \vdots \\ \frac{\lambda_{K-1,1}\left(e^{\lambda_{K-1,K-1}}-1\right)}{\lambda_{K-1,K-1}} & \frac{\lambda_{K-1,2}\left(e^{\lambda_{K-1,K-1}}-1\right)}{\lambda_{K-1,K-1}} & \cdots \ddots & \frac{\lambda_{K-1,K}\left(e^{\lambda_{K-1,K-1}}-1\right)}{\lambda_{K-1,K-1}} \\ 0 & 0 \cdots & 0 & 1 \end{pmatrix}$$

follows immediately. To get $\widehat{\Lambda}$ we set $\widehat{Tr}(1)$ equal to the right side of the equation above, i.e.,

$$\widehat{tr}_{R,R}(1) = e^{\widehat{\lambda}_{R,R}}, \ R = 1, ..., K-1,$$

$$\widehat{tr}_{R,\check{R}}(1) = \frac{\widehat{\lambda}_{R,\check{R}}\left(e^{\widehat{\lambda}_{R,R}} - 1\right)}{\widehat{\lambda}_{R,R}}, \ R \neq \check{R}, \ R, \check{R} \in \{1, ..., K-1\}.$$

We solve the system for $\widehat{\lambda}_{R,R}$ and $\widehat{\lambda}_{R,\check{R}}$ and get

$$\widehat{\lambda}_{R,R} = \ln\left(\widehat{tr}_{R,R}(1)\right), \ R = 1, ..., K-1, \tag{2.26}$$

$$\widehat{\lambda}_{R,\check{R}} = \frac{\widehat{tr}_{R,\check{R}}(1)\ln\left(\widehat{tr}_{R,R}(1)\right)}{\left(\widehat{tr}_{R,R}(1) - 1\right)}, \ R \neq \check{R}, \ R, \check{R} \in \{1, ..., K-1\}. \tag{2.27}$$

Example 2.3.10.
We determine the generator matrix of the transition matrix (2.3). Therefore, we calculate all estimates of the entries of the generator matrix $\widehat{\lambda}_{R,R}$, $R = 1, ..., K-1$, and $\widehat{\lambda}_{R,\check{R}}$, $R \neq \check{R}$ and $R, \check{R} \in \{1, ..., K-1\}$

$$\widehat{\Lambda} = \begin{pmatrix} -0.0719 & 0.0652 & 0.0047 & 0.0015 & 0.0006 & 0.0000 & 0.0000 & 0.0000 \\ 0.0062 & -0.0944 & 0.0795 & 0.0064 & 0.0006 & 0.0012 & 0.0002 & 0.0001 \\ 0.0005 & 0.0221 & -0.0896 & 0.0589 & 0.0049 & 0.0020 & 0.0004 & 0.0005 \\ 0.0003 & 0.0024 & 0.0470 & -0.1168 & 0.0498 & 0.0101 & 0.0030 & 0.0041 \\ 0.0004 & 0.0010 & 0.0048 & 0.0666 & -0.1896 & 0.0866 & 0.0134 & 0.0168 \\ 0.0000 & 0.0009 & 0.0031 & 0.0045 & 0.0586 & -0.1977 & 0.0540 & 0.0766 \\ 0.0013 & 0.0000 & 0.0041 & 0.0083 & 0.0207 & 0.1316 & -0.5830 & 0.4168 \\ 0.0000 & 0.0000 & 0.0000 & 0.0000 & 0.0000 & 0.0000 & 0.0000 & 0.0000 \end{pmatrix}$$

according to equations (2.26) and (2.27).

Rating Transitions as Continuous Time, Time-Non-Homogeneous Markov Chains. Now we consider the more general case of a time-non-homogeneous, continuous time Markov process and present the so called product-limit estimator (see, e.g., Aalen & Johansen (1978)) for $Tr(s,t)$. Therefore, we define the cumulative intensity function for a transition from credit state R to $\check{R}$ as

$$\Xi_{R,\check{R}}(t) = \int_0^t \lambda_{R,\check{R}}(s)\,ds, \;\; R \neq \check{R},$$
$$\Xi_{R,R}(t) = -\sum_{\check{R}\neq R} \Xi_{R,\check{R}}(t).$$

Then, the transition matrix for a non-homogeneous, continuous time Markov chain is given by

$$\begin{aligned} Tr(s,t) &= \prod_{[s,t]} (I + d\Xi) \\ &= \lim_{\max|t_i - t_{i-1}|\to 0} \prod_i \left(I + \Xi_{t_i} - \Xi_{t_{i-1}}\right), \end{aligned} \tag{2.28}$$

where I is the identity matrix, $\Xi_t = \left(\Xi_{R,\check{R}}(t)\right)_{R,\check{R}}$, and $0 \leq s = t_0 < t_1 < \ldots. < t_n = t$ is a partition of the time interval $(s,t]$. The product-limit estimator directly uses the relationship (2.28) by estimating the increments of the individual intensity functions. These increments are computed from observed transitions divided by the number of exposed firms. All the cumulative intensities together produce the estimator for the transition probabilities. Given a sample of m transitions over the period from s to t, one can consistently estimate $Tr(s,t)$ as

$$\widehat{Tr}(s,t) = \prod_{i=1}^{m} \left(I + \Delta\widehat{\Xi}_{T_i}\right),$$

where T_i is a transition time in the interval $(s,t]$ and

$$\Delta\widehat{\Xi}_{T_i} = \begin{pmatrix} -\frac{\Delta M_1(T_i)}{N_1(T_i)} & \frac{\Delta M_{1,2}(T_i)}{N_1(T_i)} & \frac{\Delta M_{1,3}(T_i)}{N_1(T_i)} & \cdots & \frac{\Delta M_{1,K}(T_i)}{N_1(T_i)} \\ \frac{\Delta M_{2,1}(T_i)}{N_2(T_i)} & -\frac{\Delta M_2(T_i)}{N_2(T_i)} & \frac{\Delta M_{2,3}(T_i)}{N_2(T_i)} & \cdots & \frac{\Delta M_{2,K}(T_i)}{N_2(T_i)} \\ \vdots & \vdots & \ddots & \cdots & \vdots \\ \frac{\Delta M_{K-1,1}(T_i)}{N_{K-1}(T_i)} & \frac{\Delta M_{K-1,1}(T_i)}{N_{K-1}(T_i)} & \cdots & -\frac{\Delta M_{K-1}(T_i)}{N_{K-1}(T_i)} & \frac{\Delta M_{K-1,K}(T_i)}{N_{K-1}(T_i)} \\ 0 & 0 & 0 & \cdots & 0 \end{pmatrix}.$$

Hereby, the notation is used as follows:

- $\Delta M_{j,k}(T_i)$, $1 \leq j \leq K-1$, $1 \leq k \leq K$, $j \neq k$, is the number of transitions from rating class j to rating class k at date T_i.
- $\Delta M_j(T_i)$, $1 \leq j \leq K-1$ is the number of transitions away from rating class j at date T_i. Obviously, for all $j \in \{1, ..., K-1\}$,

$$\Delta M_j(T_i) = \sum_{k\neq j} \Delta M_{j,k}(T_i).$$

- $N_j(T_i)$, $j \in \{1, ..., K-1\}$ is the number of obligors in rating class j right before date T_i.

Therefore, the matrix elements have a clear interpretation:

- The absolute value of the diagonal element in row j is the fraction of obligors of rating j just prior to date T_i that leaves that rating category at time T_i.
- The off-diagonal element in row j and column k is the fraction of obligors of rating j just prior to date T_i that migrates to rating category k at time T_i.

Note, that each row of $\Delta\widehat{\Xi}_{T_i}$ sums to zero and that the bottom row is zero because we assume that default is an absorbing state.

Example 2.3.11.
We use the same example as in the time homogeneous case. As there are four dates of rating migrations, $m = 4$. $T_1 = \frac{1}{12}$, $T_2 = \frac{2}{12}$, $T_3 = \frac{6}{12}$, $T_4 = \frac{8}{12}$.

$$\Delta\widehat{\Xi}_{T_1} = \begin{pmatrix} 0 & 0 & 0 & 0 \\ \frac{1}{10} & -\frac{2}{10} & \frac{1}{10} & 0 \\ 0 & 0 & 0 & 0 \\ 0 & 0 & 0 & 0 \end{pmatrix}, \quad \Delta\widehat{\Xi}_{T_2} = \begin{pmatrix} -\frac{2}{11} & \frac{2}{11} & 0 & 0 \\ 0 & 0 & 0 & 0 \\ 0 & 0 & 0 & 0 \\ 0 & 0 & 0 & 0 \end{pmatrix},$$

$$\Delta\widehat{\Xi}_{T_3} = \begin{pmatrix} 0 & 0 & 0 & 0 \\ 0 & 0 & 0 & 0 \\ 0 & \frac{1}{11} & -\frac{2}{11} & \frac{1}{11} \\ 0 & 0 & 0 & 0 \end{pmatrix}, \quad \Delta\widehat{\Xi}_{T_4} = \begin{pmatrix} 0 & 0 & 0 & 0 \\ 0 & 0 & 0 & 0 \\ 0 & 0 & -\frac{1}{9} & \frac{1}{9} \\ 0 & 0 & 0 & 0 \end{pmatrix}.$$

Hence,

$$\begin{aligned} \widehat{Tr}(0,1) &= \prod_{i=1}^{4} \left(I + \Delta\widehat{\Xi}_{T_i}\right) \\ &= \begin{pmatrix} 0.81818 & 0.18182 & 0 & 0 \\ 0.08182 & 0.82727 & 0.07273 & 0.01818 \\ 0 & 0.09091 & 0.72727 & 0.18182 \\ 0 & 0 & 0 & 1 \end{pmatrix}. \end{aligned}$$

Advantages of Continuous Time Estimators. As e.g. Christensen & Lando (2002) and Lando & Skodeberg (2002) mention, using the continuous-time estimator has a lot of advantages to using the discrete-time estimator:

- The continuous-time estimator yields non-zero estimates for the transition probabilities, i.e. it yields meaningful probabilities even for rare events.
- Based on the estimator for the generator matrix transition matrices for arbitrary horizons can be obtained.
- In the discrete setting, the estimator can't use all available information, e.g., the exact date within a period that a counterparty changed its rating.

- The continuous-time estimator improves confidence set estimations.
- The continuous-time framework permits a rigorous formulation and testing of non-Markov type behavior.
- The dependence on external covariates and changes in regimes can be formulated, tested and quantified.
- Censoring can be handled easily within the continuous-time framework.

Some Fundamental Problems. Although intensity based methods are very flexible, nevertheless they have several shortcomings:

- The intensity process is purely exogenously specified.
- There is no link between the causes of default (e.g., low asset values) and the default event.
- "Default is never expected".

2.3.5 Adjusted Default Probabilities

The previous sections describe the determination of default probabilities by using stochastic models. Young & Bhagat (2000) from Goldman Sachs argue that

> "in certain circumstances, the purely stochastic approach overlooks valuable credit information".

Hence, the quantitative approach should always be linked to and be based on a strong fundamental credit analysis. This can be done by making mathematical models use and take advantage of the experience of credit analysts. They usually have a very good objective and intuitive feeling on how counterparties are going to perform on their contracts. One possibility to close this gap between quantitative modeling and fundamental analysis is the concept of default probability adjustment. Basically, for each counterparty the credit analyst adjusts the values of the default probabilities over time which were derived implicitly from public traded debt by using a quantitative method or from rating agencies. So the unique characteristics of the counterparties can be taken into consideration. These adjustments should be done such that they stay within a target range of the original default probabilities and take into account events that can have an impact on the default probabilities' term structure such as changes in regulatory obligations, changes in the stability of the industry, contract expirations or debt maturities.

2.4 Modeling Recovery Rates

2.4.1 Definition of Recovery Rates

In the real world payoffs of defaulted securities are usually greater than zero. The recovery rate (given default), denoted by w, is defined as the extent to which the value of an obligation can be recovered once the obligor has defaulted, i.e. the recovery rate is a measure for the expected fractional recovery in case of default and as such takes any value in the interval $[0, 1]$. If the defaultable security under consideration lives during the time interval $[t, T]$, $0 \leq t < T \leq T^*$, we allow for the occurrence of default at any time in the interval $[t, T]$. Recovery rates are of natural interest to investors who wish to estimate potential credit losses. The loss rate (given default), is defined as 1 minus recovery rate. Let us stress that recovery rates are as important as the probabilities of default in estimating potential credit losses. A proportional error in either the probability of default or recovery rate affects potential credit losses identically. There are different concepts of expressing and modeling recovery rates:

- **Fractional recovery of par:** it is assumed that there is compensation in terms of cash (invested in the non-defaultable money market account[34]), and the recovery rate is expressed as a fraction of par. The model has been applied, e.g., by Duffie (1998b).
- **Fractional recovery of market value:** it is assumed that there is compensation in terms of equivalent defaultable bonds, which have not defaulted yet, i.e. the recovery rate is expressed as a fraction of the market value of the defaulted bond just prior to default. By equivalent we mean bonds with the same maturity, quality and face value. This model was mainly developed by Duffie & Singleton (1997) and applied, e.g., by Schönbucher (2000).
- **Fractional recovery of a default-free but otherwise equivalent bond (equivalent recovery model):** it is assumed that there is compensation in terms of (the value of) non-defaultable bonds, i.e. at the time of default T^d a defaultable bond has the payoff of w equivalent non-defaultable bonds. By equivalent we mean bonds with the same maturity and face value. Several authors have proposed this model, e.g., Jarrow & Turnbull (1995), and Madan & Unal (1998).

There are some basic observations about recovery conventions:

- Different recovery models can be transformed into each other. Basically, the value of the defaultable security at default is expressed using different numéraires.

[34] See, e.g., definition 6.2.1.

- The equivalent recovery model is relatively more complex than the other two models and requires the knowledge of the non-defaultable interest rate term structure.
- Fractional recovery of par and the equivalent recovery model don't assume recovery of the coupons.

A firm in distress usually experiences four different stages (see, e.g., Schuermann (2003)):

- The last cash paid date is of course only known ex post but obviously is the start date of the severe problems of the firm.
- For bonds default occurs typically approximately six months later than the last cash paid date. This is quite obvious as default is usually defined as the first date a coupon is not paid and coupons are very often paid semiannually.
- Bankruptcy is declared any time between default and a year later. A firm can default and still not declare bankruptcy depending on the negotiations with the creditors.
- Emergence from bankruptcy proceedings.

Cash flows from distressed instruments may occur any time throughout this process. That makes recovery rates really hard to estimate - all the cash flows subsequent to a default must be discounted back to the time of default. Obviously, this cash flow information is usually not available. Most of the firms in financial distress are not liquidated but restructured, which makes it even harder to find good recovery rate estimates. The time spent in bankruptcy can dramatically reduce the recovery rate. On average this time is approximately two years (see, e.g., Helwege (1999), Eberhart, Altman & Aggarwal (1998),Gupton, Gates & Carty (Moody's Special Comment), and Garbade (2001)). Eberhart & Sweeney (1992) and Wagner (1996) find in bond-only studies a longer time of 2.5 years. Obviously, recovery values can vary substantially. They are dependent on the type (loan or bond - recovery rates for loans are usually higher than for bonds - see, e.g., Van de Castle, Keisman & Yang (2000)) and seniority of the risky asset (see figure 2.19), the rating of the issuer before default (e.g., recovery rates are of second order consideration for defaultable bonds of good quality and even irrelevant for treasuries but they are crucial for the high-yield debt market), the collateral guaranteed by the asset, the business cycle and the specific industry. Typically, they are estimated by appealing to historical averages although the extreme range of the historical data makes one doubt about its use.

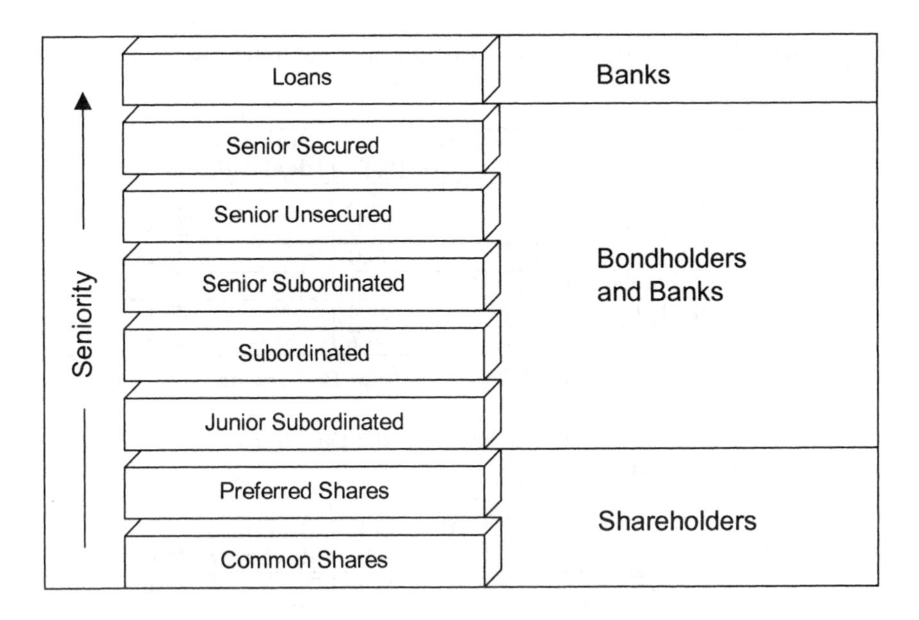

Fig. 2.19. Typical capital structure of a firm arranged according to seniority.

2.4.2 The Impact of Seniority

Figure 2.19 (from Schuermann (2003)) shows the typical capital structure of a firm. According to the absolute priority rule the value of a bankrupt firm has to be distributed to shareholders, bondholders and banks such that senior creditors are paid before any money is paid to more junior ones, and they are paid off before the shareholders. However, this rule is more a theoretical than a practical one. E.g., Eberhart & Weiss (1998) find that in 65% – 80% of all bankruptcies the absolute priority rule is violated and Eberhart & Sweeney (1992) show that the bond markets price these violations are already in. Altman & Eberhart (1994) examine recovery rates on defaulted debt by seniority at the default date. Gupton et al. (Moody's Special Comment) find that the average syndicated loan recovery rate for senior secured debt is on average 70% whereas for senior unsecured debt the average is 52%. Thornburn (2000) show in case of Swedish small companies that senior claims recover on average 69% whereas junior claims have recovery rates of only 2%. Altman & Kishore (1996) and Altman & Kishore (1997) show weighted average recovery rates on defaulted debt by seniority, industry and original rating from 1978 until 1996. They find an average recovery rate on a sample of over 750 defaulting bond issues of 40.11%. Carty & Lieberman (1998) find that the average bond recovery rate for the period 1970-1995 is 41.25% with a high standard deviation of 26.55%. Carty & Lieberman (1996) assessed the recovery rate on a small sample of senior unsecured defaulted bank loans and found it to be on average about 71%. Asarnow & Edwards

Table 2.28. Average recovery rates based on the time frame [1988, 3rd quarter 2002] as reported by S&P.

Type	Recovery rate (%)	Standard deviation (%)
Bank debt	81.6	27.7
Senior secured notes	67.0	32.8
Senior unsecured notes	46.0	36.1
Senior subordinated notes	32.4	33.3
Subordinated notes	31.2	35.1
Junior subordinated notes	18.7	29.9

Source: Standard & Poor's (*Special Report: Ratings Performance 2001* 2002)

Table 2.29. Average recovery rates based on the time frame [1998, 3rd quarter 2002] as reported by S&P.

Type	Recovery rate (%)	Standard deviation (%)
Bank debt	74.3	31.4
Senior secured notes	47.2	36.9
Senior unsecured notes	31.8	33.7
Senior subordinated notes	16.1	26.7
Subordinated notes	15.0	24.7
Junior subordinated notes	2.5	4.1

Source: Standard & Poor's (*Special Report: Ratings Performance 2001* 2002)

(1995) analyzed defaulted loans of Citibank from 1970 until 1993 and found an average recovery rate of 65%. The recovery rates as reported by S&P are shown in tables 2.28 and 2.29. The average recovery rates over the time interval $[1998, 3rd\ quarter\ 2002]$ are much smaller than the average recovery rates over the time interval $[1988, 3rd\ quarter\ 2002]$. The ability to repay debt in default has decreased over the last years. Moody's compares recovery rates in Europe and the US (see table 2.30) and finds that the recovery rates in the US are on average higher than in Europe (with the exception of bank loans).

2.4.3 The Impact of the Industry

There are different studies which come to different conclusions: some of them find that industry matters some of them disagree. Altman & Kishore (1996) using corporate bond data from 1971 – 1995 and Grossman, O'Shea & Bonelli (2001) using Fitch rated bonds and loans from 1997 – 2000 find strong evidence that there is an important impact of industry on recovery rates (see tables 2.31 and 2.32). Brennan, McGirt, Roche & Verde (1998) using Fitch rated loans support these findings. However, Gupton et al. (Moody's Special Comment) using Moody's rated loans don't find that industry matters.

Table 2.30. Average recovery rates based on the time frame [1995, 2001] Europe vs. US as reported by Moody's.

Type	Recovery rate/ Europe (%)	Recovery rate/ US (%)
Bank loan - senior secured	71.8	66.8
Senior secured notes	55.0	56.9
Senior unsecured notes	20.8	50.1
Senior subordinated notes	24.0	32.9
Subordinated notes	13.0	31.3
All bonds	22.0	42.8

Source: Moody's (Hamilton, Cantor, West & Fowlie 2002)

Table 2.31. Industry specific recovery rates based on Fitch rated bonds and loans from 1997-2000.

Industry	Asset type	Avg. Recovery Rate (%)
Asset rich	Loans	95
Asset rich	Bonds	60
Service oriented	Loans	42
Service oriented	Bonds	3
Supermarkets and Drug Stores	Loans	89

Source: Grossman et al. (2001) and Schuermann (2003).

Table 2.32. Industry specific recovery rates based on corporate bond data from 1971-1995.

Industry	Avg. Recovery Rate (%)
Utilities	70
Services	46
Food	45
Trade	44
Manufacturing	42
Building	39
Transportation	38
Communication	37
Financial Institutions	36
Construction, Real Estate	35
General Stores	33
Textile	32
Paper	30
Lodging, Hospitals	26

Source: Altman & Kishore (1996) and Schuermann (2003).

2.4.4 The Impact of the Business Cycle

There is strong empirical evidence that recovery rates are dependent on the business cycle. In recessions they are much lower than they are during times of expansion. Speculative grade debt is even more sensitive to changes in the business cycle than investment grade debt. In fact, recovery and default rates are inversely related (see, e.g., Altman (2001)). Hence, for modeling purposes one can postulate that the recovery rate is a function of the default rate.

Example 2.4.1.
Bakshi, Madan & Zhang (2001) assume that the recovery rate is related to the underlying intensity as follows:

$$w_t = \varpi_0 + \varpi_1 e^{-\lambda_t},$$

where $\varpi_0 \geq 0$, $\varpi_1 \geq 0$, and $0 \leq \varpi_0 + \varpi_1 \leq 1$. There are some interesting properties of the model:

- Well defined: As $\lambda \to 0$, $w \to \varpi_0 + \varpi_1$, as $\lambda \to \infty$, $w \to \varpi_0$. The recovery rate is always between ϖ_0 and $\varpi_0 + \varpi_1$.
- Economically appealing: The recovery rate is negatively related to the default rate.
- Technically tractable.

Figure 2.20 shows the dependence of w_t on λ_t for $\varpi_0 = 0.2$ and $\varpi_1 = 0.8$. If $\lambda_t = 0$ (i.e. there is no default risk over the next infinitesimal time interval) then the recovery rate equals 1. As the default risk increases the recovery rate decreases and tends to 0.2 as λ_t tends to infinity. Especially, the recovery rate is never below 0.2. In the following, we consider the dynamics of w dependent on the dynamics of λ :

1. Constant intensity model: Assume $\lambda_t = 0.1$ for all t, $\varpi_0 = 0.2$, $\varpi_1 = 0.8$, and no default in the observed time frame. Then $w_t = 0.92387$ for all t.
2. Deterministically time-varying intensity model: Assume

$$\lambda_t = \begin{cases} 0.1 \text{ for } t \in [0,1) \\ 0.2 \text{ for } t \in [1,2) \\ 0.3 \text{ for } t \in [2,3) \\ \dots \quad \dots \end{cases},$$

   $\varpi_0 = 0.2$, $\varpi_1 = 0.8$, and no default in the observed time frame (see figure 2.21).
3. Stochastic intensity model: Assume the intensity λ follows an Ornstein-Uhlenbeck process, i.e.:

$$d\lambda_t = a\,(b - \lambda_t)\,dt + \sigma dW_t,$$

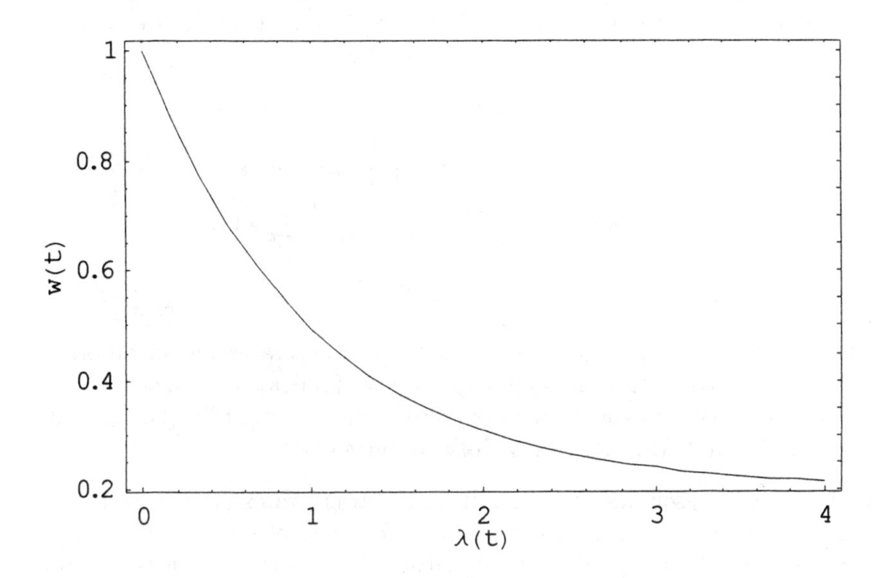

Fig. 2.20. Recovery rate at a fixed time $t: w_t = 0.2 + 0.8 \cdot e^{-\lambda_t}$.

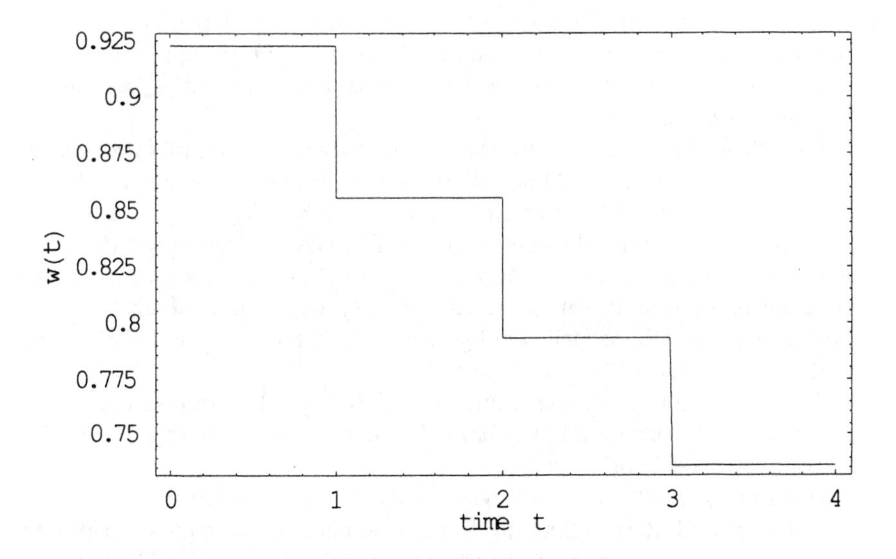

Fig. 2.21. Deterministically time-varying intensity model. $\varpi_0 = 0.2$, $\varpi_1 = 0.8$.

where W is a one-dimensional standard Brownian motion and the initial value λ_0, and the parameter values $a > 0$, b, and $\sigma > 0$ are constants. Then the recovery rate follows the dynamics

$$\begin{aligned} dw_t &= \varpi_1 d\left(e^{-\lambda_t}\right) \\ &= -\varpi_1 e^{-\lambda_t} d\lambda_t + \frac{1}{2}\varpi_1 e^{-\lambda_t}\sigma^2 dt \\ &= -\varpi_1 e^{-\lambda_t}\left(a\left(b-\lambda_t\right) - \frac{1}{2}\sigma^2\right) dt \\ &\quad -\varpi_1 e^{-\lambda_t}\sigma dW_t. \end{aligned}$$

The link between aggregate default and recovery rates on an empirically sound basis have been examined, e.g., by Frye (2000a), Frye (2000b), Jokivuolle & Peura (2000), Jarrow (2001), Hu & Perraudin (2002), Bakshi et al. (2001) and Altman, Brady, Resti & Sironi (2003):

- Frye (2000a) assumes that default and recovery rates are driven by a single common systematic factor, namely the state of the economy. In good times the default rates decrease whereas the recovery rates increase, in bad times vice versa. The (negative) correlation between the two variables is only explained by their common dependence on the state of the economy. Frye finds that in a severe downturn, recovery rates might decline $20 - 25$ percentage points from their normal year average.
- Jarrow (2001) assumes that default and recovery rates depend on the state of the economy, too. His methodology differs from Frey's approach in two ways: Jarrow's approach incorporates equity and a liquidity premium in the estimation procedure.
- Jokivuolle & Peura (2000) present a model in which collateral is correlated with default probabilities. The collateral value is in turn assumed to be the only stochastic variable determining recovery rates. As other authors they find an inverse relationship between default rates and recovery rates.
- Hu & Perraudin (2002) use Moody's historical bond market data to find that correlations between quarterly recovery rates and default rates for bonds issued by US-domiciled obligors are -0.22 for the period $1983-2000$ and -0.19 for the $1971 - 2000$ period.
- Bakshi et al. (2001) analyze a sample of BBB-rated corporate bonds and find that a 4% worsening in the hazard rate is associated with a 1% decline in recovery rates (both risk-neutral).
- Altman et al. (2003) consider recovery rates on corporate bond defaults over the period $1982 - 2002$ and try to explain these recovery rates by specifying linear, logarithmic and logistic regression models. They are the first ones to analyze specific determinants of recovery rates. Especially, they claim that aggregate recovery rates are a function of supply and demand for the defaultable securities. The universe of explanatory variables they consider are as follows:

3. Pricing Corporate and Sovereign Bonds

"Prediction is very difficult, especially of the future."
- Niels Bohr -

"When I was an undergraduate 30 or so years ago, banks were originate-and-hold institutions focused primarily on traditional loans; that is no longer the case."
- Charles Smithson, 2001 -

3.1 Introduction

3.1.1 Defaultable Bond Markets

Credit risk is one of the oldest forms of risk in the financial markets, and still revolutionary changes and developments are taking place in the credit markets today. To emphasize the growing importance of this market segment and its risks which shouldn't be ignored, we first give a brief overview of the latest credit market and credit risk developments.

Corporate Defaults. Corporate default risk[1] can't be neglected: in the mid-1980's, there was a huge number of defaults on bank loans and corporate bonds in the United States, and in the early-1990's about 10% of the junk bonds defaulted. Now, in the end-1990's and beginning-2000's, according to Standard & Poor's, corporate defaults are running at their highest level ever and are still on the rise[2]. In 1999, 2000, 2001, 2002 S&P registered defaults by 101, 117, 220, 234 rated or formerly rated companies, which is equivalent to a default ratio of 2.06%, 2.27%, 3.49%, 3.63% on a total of US$ 37.8 billion, US$ 42.3 billion, US$ 119.0 billion, and US$ 178.0 billion of debt, respectively. These were the highest one year overall default ratios and US$ values ever.

[1] For a detailed definition of corporate default see appendix A.2.

[2] For all information based on S&P data see *Ratings Performance 1996: Stability & Transition* (1997), *Ratings Performance 1997: Stability & Transition* (1998), *Ratings Performance 1998: Stability & Transition* (1999), *Ratings Performance 1999: Stability & Transition* (2000), *Ratings Performance 2000: Default, Transition, Recovery, and Spreads* (2001), *Special Report: Ratings Performance 2001* (2002), and *Special Report: Ratings Performance 2002* (2003).

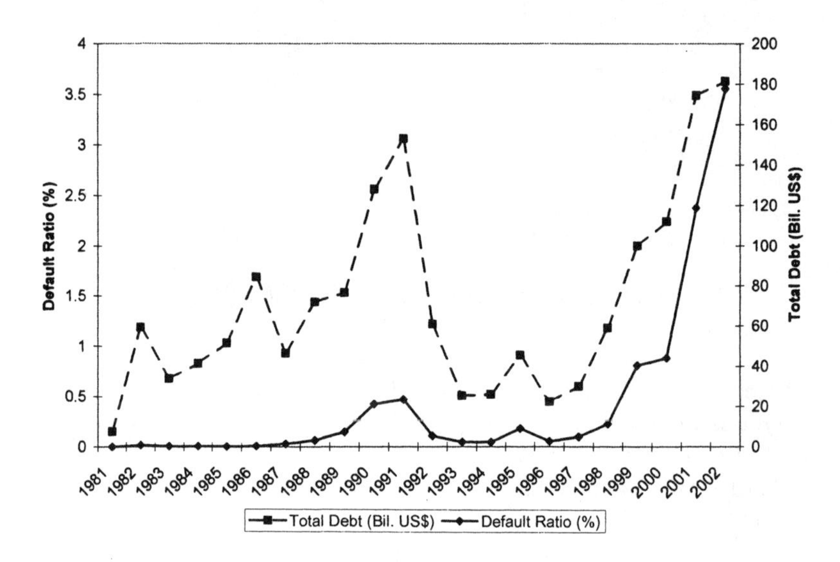

Fig. 3.1. S&P corporate default ratios and total corporate debt in default, 1981 – 2002. Source: (*Special Report: Ratings Performance 2002* 2003).

The median corporate rating assigned by S&P was "A"[3] in 1981, "A-" in 1991 and "BBB" by year-end 1999. In 1999, downgrades exceeded upgrades almost three to one, and the number of newly rated obligors with a speculative-grade rating[4] outnumbered those with an investment-grade rating[5]. This fact pushed the speculative-grade component from 36% in 1999 to 37.5% in 2000. 4.09 % of investment-grade companies were downgraded to speculative-grade during 2002, the highest rate ever. The downgrade to upgrade ratio also peaked in 2002 at a value of 3.41. However, one of the most disturbing trends in 2002 was the very high level of investment-grade defaults which was around 0.50%.

According to Moody's Investors Services (Moody's)[6], 1999 147 corporate and sovereign issuers defaulted on US$ 44.6 billion of debt, which is more than double as much as in 1998. In 2001 there were already 212 defaults on US$ 135 billion of debt. With 3.7% of all Moody's rated corporate bond issuers default rates reached decade highs in 2001. For speculative-grade issuers, the default rate even reached 10.2%. During 1999, there were four downgrades

[3] For the definition of letter ratings see, e.g., section 2.

[4] Rating of "BB+" or lower by S&P and "Ba1" or lower by Moody's. For more details see, e.g., section 2.

[5] A rating of "BBB-" or higher by S&P or a rating higher than "Ba1" by Moody's.

[6] See *Historical Default Rates of Corporate Bond Issuers, 1920-1999* (2000a) and Hamilton, Cantor & Ou (2002).

of corporate ratings for every upgrade. In 2001 rating downgrades exceeded rating upgrades almost by 2 to 1.

Sovereign Defaults. Default risk is not only concentrated on corporate issuers. Although sovereign nations have the right to print money to repay their debts, for political and economical reasons, such as inflationary pressures or social welfare costs, they can decide not to do so. Sovereign obligors have come to the international financial markets in ever greater numbers. This expansion of credit is for example supported by the joint BIS (Bank for International Settlements) - IMF (International Monetary Fund) - OECD (Organization for Economic Cooperation and Development) - World Bank statistics on external debt[7]. While in the mid-1970's we heard Walter Wriston, a former Chairman of Citibank, say "Countries don't go bankrupt", the crises of the 1980's and 1990's in Russia, Asia and Latinamerica taught us better. Past sovereign defaults[8] were caused by such a different variety of factors as, e.g., wars, lax fiscal and monetary policies, or external economic shocks. For example, on September 3rd, 1999, repackaged bonds based on Ecuadorian Brady bonds[9] jointly issued by Dresdner Bank / J.P. Morgan and Merrill Lynch / Credit Suisse First Boston, respectively, defaulted with zero recovery. Ecuador was the first country not to pay its interest rates on Brady bonds ever: in August 1999, Ecuador defaulted on about US$ 6 billion in Brady bonds. In addition, in October 1999 the government defaulted on about US$ 500 million in Eurobonds. This decision followed a serious deterioration in the country's economic conditions:

- Ecuador was hit by "El Niño" which caused a damage to nearly 20% of GDP (Gross Domestic Product).
- Decline in prices of Ecuador's two largest export commodities.
- Increase in the cost of external finance.
- Deeply stressed banking system.
- Internal political uncertainties.

Given these political and economic circumstances, despite the risk of partly loosing access to the international capital markets - after all, at the end of 1998 Brady bonds represented nearly 40% of external liabilities - servicing

[7] Can be downloaded from http://www.bis.org/publ/index.html.

[8] For a detailed definition of sovereign default see appendix A.2.

[9] A program established in 1990 to restructure the heavy debt burdens of emerging markets countries. Named after U.S. Treasury Secretary Nicholas Brady, it uses multilateral agencies such as the International Monetary Fund and the World Bank to provide funding. Under the program, a country's outstanding debt is converted to new long-term fixed-rate and/or floating-rate bonds whose principal is backed by U.S. Treasury bonds. Debtor nations are responsible for long-term interest payments. Bradys or Brady-style issues have been set up for Mexico, Venezuela, Brazil, Argentina, Panama, Philippines, Poland, Bulgaria, Morocco, Nigeria and Russia. Some of those nations are in the process of eliminating much of that form of debt.

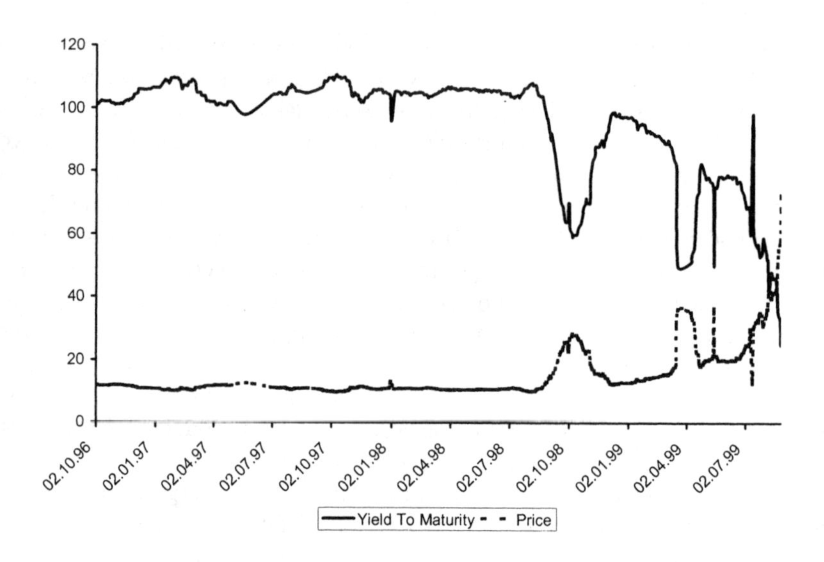

Fig. 3.2. Dresdner Bank's Synthetic Ecuador Bond LTD 12.25% 02/10/03 DEM - yields to maturity and prices. Source: Bloomberg.

the country's debt became impossible. Ecuador was the only new sovereign defaulting during 1999. Defaults by other sovereigns had originated earlier. Some of the most serious defaults are those of the Russian Federation, Indonesia, Ukraine, Mongolia, Pakistan and former Yugoslavia. Looking back in history, according to figure 3.3 the sovereign default rate started rising in the end 1970's. The Latin American debt crises in 1982 accelerated and pushed this trend further peaking in 1990 with Brazil's default on US$ 62 billion of debt. At the same time more and more sovereigns from other regions like Eastern Europe became involved. More conservative economic policy-making in the last decade changed this trend and made default ratios decrease constantly. However, in 2002 the default rate and the total debt in default sharply increased to 13.4% and US$ 132.6 billion. Let us finally mention that an increasing number of speculative-grade ratings has been assigned since 1991 (see figure 3.4), which supports this opinion.

Credit Losses from Derivatives[10]**.** Debt instruments are not the only financial instruments with embedded credit risk. For example, all kinds of

[10] A derivative is a financial contract whose value is derived from the performance of assets, interest rates, currency exchange rates, or indexes. Derivative transactions include a wide assortment of financial contracts including structured debt obligations and deposits, swaps, futures, options, caps, floors, collars, forwards and various combinations thereof.

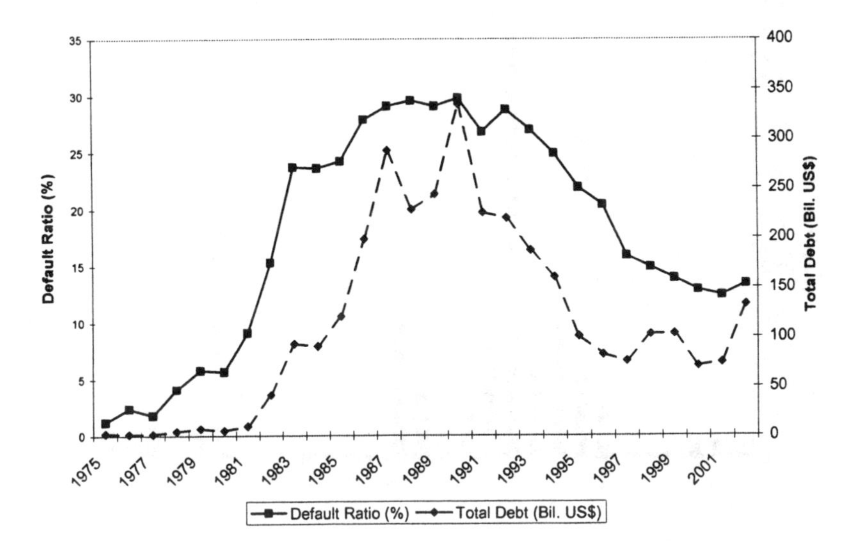

Fig. 3.3. S&P sovereign default ratios and sovereign debt in default, 1975-2002. Source: (*Special Report: Ratings Performance 2002* 2003).

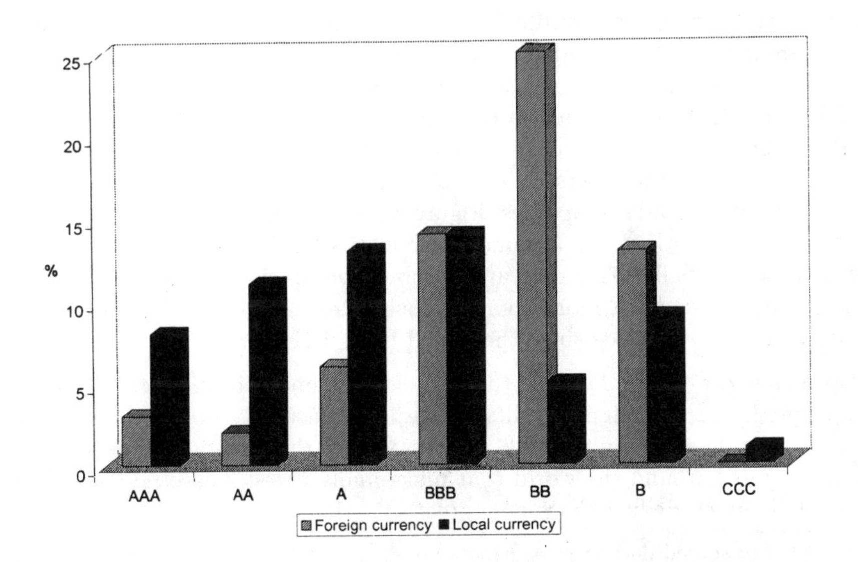

Fig. 3.4. New sovereign ratings, 01/91 – 09/02. Source: (*Special Report: Ratings Performance 2002* 2003).

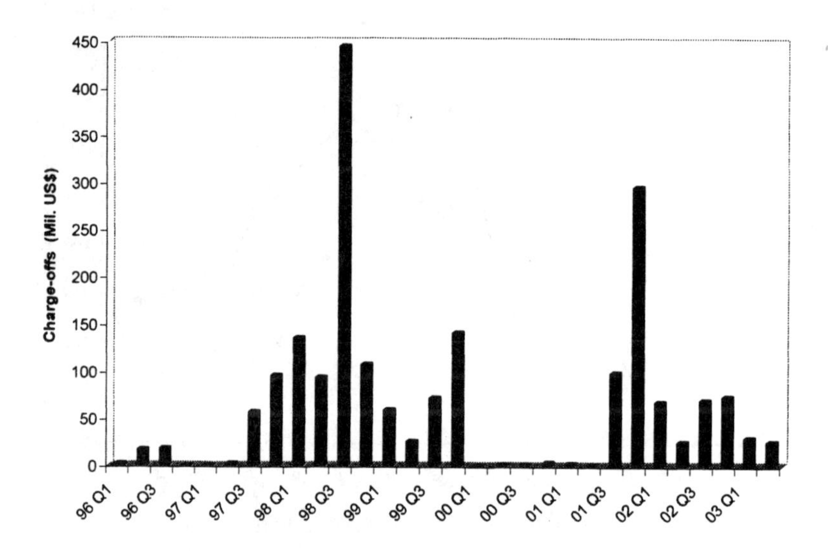

Fig. 3.5. Quarterly charge-offs (credit losses) from derivatives of all US commercial banks with derivatives. Source: Call report of the Office of the Comptroler, 2nd Quarter 2003.

derivative contracts are subject to counterparty default risk, especially if they are not exchange traded[11]. Over-the-counter (OTC) derivatives[12] usually don't have any kind of guarantees and, in many cases, are unsecured and without any collateral posted. In a study of financial reports, Basin (1996) finds out that not only highly rated firms serve as derivative counterparties. Quite the reverse is true. A significant part of the counterparties are speculative-grade companies. Figure 3.5 shows the quarterly credit losses from derivatives of all US commercial banks with derivatives. It seems evident that especially OTC derivatives should be priced with regard to credit risk. This is usually a more complex task than in the case of pricing debt such as fixed rate bonds since the payoff from derivatives is stochastic.

Appetite for Risk. Despite the increasing number of corporate defaults (and predicted sovereign defaults) there is a constantly growing number of investors willing to take this risk. One reason are the historically low nominal interest rates around the world that make many investors seek higher yields by taking more credit risk. Such a great appetite for risk has attracted large

[11] Exchange traded derivative contracts are standardized derivative contracts that are transacted on an organized exchange and that usually have margin requirements.

[12] Over-the-counter derivative contracts are privately negotiated derivative contracts that are transacted off organized exchanges.

numbers of companies to the bond markets, especially many issuers of high-yield[13] and emerging market[14] bonds from very risky industries and countries. Often these companies are brand-new, poorly capitalized and new in business. Granted high-yield debt has existed for decades - but whereas junk bonds used to come from formerly higher rated firms whose quality had declined for reasons like economic recessions, today, issuing junk bonds is a quite natural strategy for firms which don't have access to other forms of credit. Routinely they sell their debt in the capital markets and very often can't pay the interest on it. According to Caouette et al. (1998), in 1997, the high-yield public bond issuance was US$ 120 billion, and new leverage loan volume reached US$ 197 billion. A report from Moody's (*Historical Default Rates of Corporate Bond Issuers, 1920-1999* 2000b) notes that such a surge in low-rated, first-time issuers that entered the market in late 1997 and 1998 was a key factor in pushing the default rate to a new high. This group of issuers contributed 40% of 1999's rated bond defaults. S&P reports (*Ratings Performance 1999: Stability & Transition* 2000) that out of the 101 companies, that defaulted in 1999, 92 were originally rated speculative-grade. 77% of these companies were first rated after the year 1994. 60% of 1999's defaults were not related to worse economic conditions caused by last years' Asian and Russian crises.

Growth of the Corporate Bond Market. Credit markets have undergone a dramatic change over the past few years. The growth rates of traditional loan products have stagnated relative to other corporate financing markets[15]. Although the European corporate bond market is still small in relation to the US market, according to McLeish (2000) it has grown rapidly over the past years. And this gap between the European and US bond market is expected to narrow more over time. Some of the main reasons are the following:

- Migration away from bank loans and similar products towards the fixed-rate bond market due to pressure from stockholders to generate higher returns[16].

[13] High-yield bonds are speculative-grade rated. Technically they are referred to as non-investment grade credits.

[14] Broadly defined, an emerging market is a country making an effort to change and improve its economy with the goal of raising its performance to that of the world's more advanced nations. The World Bank classifies economies with a Gross National Product per capita of US$ 9656 and above as high-income countries. Antoine W. van Agtmael, an employee of the World Bank's International Finance Corporation, is credited with coining the term "emerging markets" in 1981. But the concept of investing in less developed countries with potential for economic expansion has been a part of individual and institutional investment strategies since the 19th century.

[15] See, e.g., Wilson (1998), page 220.

[16] See, e.g., exhibit 2 on page 222 in Wilson (1998) which shows the dramatically falling margins which banks earn from the traditional loan product.

- Explosion in the European Mergers & Acquisitions activities (partially) funded by bond issuance.
- Growth in pension and mutual fund assets as a demand driver.
- Existing institutional assets are shifted out of government bonds into corporate bonds due to the low European yield environment.
- The European monetary union replaced the segmentation of the European corporate bond market into national markets by one Euro-zone market. This increased competition and liquidity[17].

In euro, outstanding industrial bonds expanded by 115% in 1998 and by 271% in 1999, outstanding financial bonds by 83% in 1998 and by 29% in 1999. In sterling, outstanding industrial bonds expanded by 25% in 1998 and by 44% in 1999, outstanding financial bonds by 16% in 1998 and by 13% in 1999[18]. And Neil McLeish from Morgan Stanley Dean Witter explains (McLeish 2000):

> "Although we do not expect the same percentage rates of growth for euro and sterling industrial bonds to be maintained going forward, we do expect solid growth in the overall size of these markets, which over time should converge with the US in terms of size. Our view is based on several factors that support both increased supply of and demand for European industrial bonds."

3.1.2 Pricing Defaultable Bonds

Corporate bonds are bonds issued by corporations. Sovereign or government bonds are bonds issued by a government or government agency, examples of which are the Kingdom of Thailand bonds, Malaysia bonds and U.S. Government bonds (also known as Treasuries as they are issued by the U.S. Treasury). Within the sovereign category are quasi-sovereign bonds, which are securities issued by entities supported by the sovereign. An example of a quasi-sovereign bond issuer is Petroliam Nasional Berhad, the government owned national oil and gas company of Malaysia. By issuing bonds a corporation or government commits itself to making specific future payments to the bondholders. In some cases the debtors may not be able to meet these payment obligations. This means that the bondholders will not receive the promised payments in full or even worse will lose all money invested in the bonds. Therefore, investors willing to invest in defaultable bonds should be compensated for the risk they take. But what is a fair risk premium ?

Traditional methods for pricing defaultable bonds use historical default probabilities (see, e.g., section 2.3.1) to determine the expected loss on the specific

[17] European monetary union didn't affect only the corporate bond markets by making investors comparing credit risk over the entire Euro-zone, but rather made credit risk play the key role in the pricing of European government bonds.
[18] See McLeish (2000).

bond. The price is then determined such that it compensates the investor for this expected loss under the empirical probability measure, i.e. there is no risk premium. One example for this approach is the research of Fons (1994).

There are three other groups of defaultable bond pricing models that particularly account for risk premiums:

- structural/asset based models,
- reduced-form/intensity based models,
- hybrid models: they present a middle ground between structural and reduced-form models to ideally combine the strengths of both approaches.

Before we go into details of the different approaches, let us discuss the general framework and give a mathematical definition of non-defaultable and defaultable bonds first.

Definition 3.1.1.
A non-defaultable $T-$maturity zero coupon bond[19] *(discount bond) is a contract that guarantees its holder the payment of one unit of currency at time T, with no intermediate payments. The contract value at time $t < T$ is denoted by $P(t,T)$.*
A defaultable $T-$maturity zero coupon bond (discount bond) is a contract that promises its holder the payment of one unit of currency at time T, with no intermediate payments. If default occurs up to maturity of the bond (for what reasons ever) the payoff may be only a fraction of the promised payoff. The contract value at time $t < T$ is denoted by $P^d(t,T)$.

In the following we show how to price non-defaultable and defaultable bonds. Therefore we fix a finite horizon date $T^* > 0$ and a filtered probability space $\left(\Omega, \mathcal{F}, \mathbf{F} = (\mathcal{F}_t)_{0 \le t \le T^*}, \mathbf{P}\right)$. As usual the filtration $\mathbf{F}$ represents the arrival of information over time and $\mathbf{P}$ denotes the real-world measure. The theory of pricing non-defaultable bonds was originally based on the assumption of specific one-dimensional dynamics for the instantaneous short rate process r. Modeling directly such dynamics is very convenient as bonds are readily defined, by no-arbitrage arguments, as the expectation of a functional of the process r. If we assume that there is no arbitrage in our financial market model, absence of arbitrage is guaranteed by the existence of an equivalent martingale measure $\mathbf{Q}$. By definition, an equivalent martingale measure has to satisfy $\mathbf{Q} \sim \mathbf{P}$[20] and the discounted security price processes (for any tradeable

[19] Note, that in this text we do not treat the theory of pricing non defaultable bonds at length. Readers not familiar with this theory at all are refered to the excellent textbook of Brigo & Mercurio (2001).

[20] A probability measure $\mathbf{Q}$ on $(\Omega, \mathcal{F})$ is absolutely continuous relative to $\mathbf{P}$ if

$$\forall\, A \in \mathcal{F} \quad \mathbf{P}(A) = 0 \Rightarrow \mathbf{Q}(A) = 0.$$

The probabilities $\mathbf{P}$ and $\mathbf{Q}$ are equivalent ($\mathbf{P} \sim \mathbf{Q}$) if each one is absolutely continuous relative to the other.

securities which pay no coupons or dividends) have to be $\mathbf{Q}-$martingales with respect to a suitable numéraire[21]. As numéraire we choose the non-defaultable money market account[22] as given in the following definition.

Definition 3.1.2.
We define B_t to be the value of a non-defaultable money market account at time $t \geq 0$. We assume $B_0 = 1$ and that the money market account evolves according to the following differential equation

$$dB_t = r_t B_t dt.$$

Obviously,

$$B_t = e^{\int_0^t r_l dl}.$$

The existence of a risk-neutral measure $\mathbf{Q}$ implies that the arbitrage-free price of a non-defaultable discount bond at time t is given by

$$P(t,T) = E^{\mathbf{Q}}\left[e^{-\int_t^T r(l)dl} \;\middle|\; \mathcal{F}_t\right], \; 0 \leq t < T.$$

Hence, we can compute $P(t,T)$, whenever we can characterize the distribution of $e^{-\int_t^T r(l)dl}$ in terms of chosen dynamics for r and with regards to the measure $\mathbf{Q}$, conditional on the information available at time t.

Now, let us consider defaultable bonds. The various approaches to pricing defaultable bonds vary with regard to the way they model the default time T^d (see also chapter 2). From a mathematical point of view T^d is a $\mathbf{F}$-stopping time. In most structural models $\mathbf{F}$ is generated by a standard Brownian motion and therefore T^d is even a predictable $\mathbf{F}$-stopping time. This implies that the default time can be forecasted with some degree of certainty. In contrast to structural models, in the intensity based approach default occurs as a surprise.

In the remaining chapter we discuss structural/asset based and reduced-form/intensity based models. Hybrid models are introduced in chapter 6.

As already mentioned, we denote the time t ($t \leq T$) price of a non-defaultable discount bond by $P(t,T)$ and the time t price of a T-maturity defaultable discount bond by $P^d(t,T)$. Obviously, $P(T,T) = 1$ and $P^d(T,T) = 1$ if $T^d > T$. In general, using the default indicator function $H_t = 1_{\{T^d \leq t\}}$,

- in case of fractional recovery of par we can write

$$P^d(T,T) = (1 - H_T) + w_{T^d} P^{-1}\left(T^d, T\right) H_T,$$

[21] A numéraire is a price process X_t almost surely strictly positive for each $t \in [0,T]$.

[22] A money market account represents a riskless investment, where profit is accrued continuously at the risk-free rate prevailing in the market at every instant.

- in case of fractional recovery of a default-free but otherwise equivalent bond

$$P^d(T,T) = (1 - H_T) + w_{T^d} H_T,$$

- and finally, in case of fractional recovery of market value

$$P^d(T,T) = (1 - H_T) + w_{T^d} P^d(T^d-, T) P^{-1}(T^d, T) H_T.$$

In terms of the default indicator function, the cumulative dividend process of the defaultable discount bond is given by

$$\int_0^{t \wedge T} Z_u dH_u + 1_{\{T^d > T, t \geq T\}}, \ t \geq 0,$$

where Z is the process describing the payoff upon default (e.g., in case of fractional recovery of market value $Z_t = w_t P^d(t-, T)$). We let $\tau = \min(T, T^d)$. The arbitrage free price at time t, $P^d(t,T)$, of the defaultable discount bond can be obtained as the discounted expected value of the future cash flows[23]. This expectation has to be taken with respect to $\mathbf{Q}$, that is for all $0 \leq t < \tau$,

$$P^d(t,T) = E^{\mathbf{Q}}\left[\int_t^T e^{-\int_t^u r(l)dl} Z_u dH_u + e^{-\int_t^T r(l)dl} (1 - H_T) \middle| \mathcal{F}_t\right], \quad (3.1)$$

and $P^d(t,T) = 0$ for $t \geq \tau$ since there are no dividends after time τ.

Based on equation (3.1) we find that each model has to specify the following details and that each specification may yield a new model:

- the short rate interest rate process r,
- the recovery process Z,
- the default indicator process H and therefore the default time T^d,
- the correlation structure between the processes.

Note, that for arbitrary specifications of the processes equation (3.1) may be violated.

Let us finally consider a special case of equation (3.1). Assume that default only happens at maturity of the bond and the money market process is stochastically independent of the other processes. Then

$$P^d(t,T) = P(t,T) E^{\mathbf{Q}}[Z_T H_T + (1 - H_T) | \mathcal{F}_t], \ 0 \leq t < \tau.$$

Assume further that Z_T is some constant, then

$$\begin{aligned} P^d(t,T) &= P(t,T) E^{\mathbf{Q}}[1 - (1 - Z_T) H_T | \mathcal{F}_t] \\ &= P(t,T)\left(1 - (1 - Z_t) \mathbf{Q}(T^d \leq T | \mathcal{F}_t)\right), \ 0 \leq t < \tau. \quad (3.2) \end{aligned}$$

To price a defaultable discount bond in this model, we only need three pieces of information:

[23] See, e.g., Harrison & Pliska (1981) and Duffie, Schroder & Skiadas (1996).

- the price of an equivalent non-defaultable discount bond,
- the constant recovery rate,
- and the default probability under the measure $\mathbf{Q}$.

Let us finally mention, that using equation (3.2) we can easily deduce risk neutral default probabilities implicitly given by defaultable bond prices:

$$\mathbf{Q}\left(T^d \leq T \middle| \mathcal{F}_t\right) = \frac{1 - \frac{P^d(t,T)}{P(t,T)}}{1 - Z_t}, \; 0 \leq t < \tau,$$

given that $0 \leq Z_t < 1$.

3.2 Asset Based Models

3.2.1 Merton's Approach and Extensions

The pricing of defaultable bonds in the asset based setting goes back to the initial proposal of Merton (1974), where he assumes that the dynamics for the value of the assets of a firm across time can be described by a diffusion-type stochastic process and the defaultable security can be regarded as a contingent claim on the value of the assets of the firm. Merton's approach is based on the standard assumptions of continuous time no-arbitrage models[24] and some further rather strict assumptions such as the short-term interest rate is constant and equals $r > 0$. In section 2.3.3 we have already shown how to apply Merton's approach to the modeling of default probabilities under the real-world measure $\mathbf{P}$. As we are concerned with pricing issues, we model the relevant processes under the martingale measure $\mathbf{Q}$ now. We assume that a firm is only financed by equity and one type of zero coupon debt with promised terminal constant payoff $L > 0$ at maturity time T. The firm is able to redeem its debt if the total market value of its assets V_T at time T is worth more than the amount L owed to the creditor. Otherwise the obligor defaults on its obligation and the bondholders receive V_T. In other words the payoff to the bondholders equals the minimum of L and V_T. As

$$\min\left(L, V_T\right) = L - \max\left(0, L - V_T\right),$$

the price of a defaultable discount bond must be equal to the difference of the value of an equivalent non-defaultable discount bond with face value L and the value of a European put option on the firm value V with strike L and maturity T:

$$P^d\left(t, T\right) = LP\left(t, T\right) - Put_t, \; 0 \leq t \leq T.$$

[24] No taxes and transaction costs, perfectly divisible assets, no borrowing-lending spread, continuous costless trading and short-selling, equal access to information for all investors.

If we specify V under the measure $\mathbf{Q}$ as

$$\frac{dV_t}{V_t} = rdt + \sigma_V d\hat{W}_V(t), \; 0 \leq t \leq T,$$

where σ_V is the constant volatility parameter, and $\hat{W}_V$ is a standardized Brownian motion, then Put_t is simply given by the classic Black-Scholes formula for the price of a European put option (see, e.g. Lamberton & Lapeyre (1996), page 70). Therefore, we can formulate the following proposition:

Proposition 3.2.1. *Given the assumptions of Merton (1974) the price of a defaultable discount bond with face value L and maturity T is given by*

$$\begin{aligned} P^d(t,T) &= LP(t,T) - Put_t \\ &= LP(t,T) - LP(t,T)\Phi(-d_2(t)) + V_t\Phi(-d_1(t)) \\ &= LP(t,T)\Phi(d_2(t)) + V_t\Phi(-d_1(t)), \end{aligned} \tag{3.3}$$

where

$$d_1(t) = \frac{l_t + \left(r + \frac{1}{2}\sigma_V^2\right)(T-t)}{\sigma_V\sqrt{T-t}}, \; d_2(t) = d_1(t) - \sigma_V\sqrt{T-t},$$

and $l_t = \ln\left(\frac{V(t)}{L}\right)$ *for all* $0 \leq t \leq T$.

Remark 3.2.1. Sometimes it makes sense to include a constant payout ratio κ in the process V :

$$\frac{dV_t}{V_t} = (r - \kappa)\, dt + \sigma_V d\hat{W}_V(t).$$

Equation (3.3) can easily be adjusted to this more general case.

Apart from the standard assumptions of continuous time no-arbitrage models there are many shortcomings of this model: the liabilities of the firm are supposed to consist only of a single class of debt, the debt has a zero coupon, bankruptcy is triggered only at the maturity of the debt, bankruptcy is costless and interest rates are assumed to be constant over time. In addition, it is assumed that the absolute priority rule[25] is always enforced. But Franks & Torous (1994) find that the strict absolute priority rule was violated in 78% of the bankruptcies they considered. Thus, the assumptions of the Merton model are highly stylized versions of reality. As Kim et al. (1992), Kim, Ramaswamy & Sundaresan (Autumn 1993), and Jones et al. (1984) point out in their papers, Merton's model doesn't generate the levels of yield spreads which can be observed in the market. Rather they show that this model is

[25] If the strict priority rule is enforced, in case of bankruptcy, bondholders are paid off as good as possible while shareholders receive nothing.

unable to generate yield spreads in excess of 120 basis points whereas over the 1926 – 1986 period, the yield spreads of AAA-rated corporates ranged from 15 to 215 basis points. This inability to account for the magnitude of yield spreads was the motivation for a huge number of papers written during the last two decades to generalize Merton's risky debt pricing model: Black & Cox (1976) incorporate classes of senior and junior debt, safety covenants, dividends, and restrictions on cash distributions to shareholders, Geske (1977) considers coupon bonds by using a compound options approach and provides a formula for subordinate debt within this compound option framework, Ho & Singer (1982) allow for different maturities of debt and examine the effect of alternative bond indenture provisions such as financing restriction of the firm, priority rules and payment schedules. Leland (1994) extends the model further to incorporate bankruptcy costs and taxes which makes it possible to work with optimal capital structure. Because the empirical literature shows that credit spread curves can be flat or even downward sloping, that short-term debt often doesn't have zero credit spreads, and that sudden drops in the value of firms are possible, Mason & Bhattacharya (1981) consider a pure jump process, and Zhou (1997) incorporates a jump diffusion for the underlying asset value.

Shimko et al. (1993) allow for stochastic non-defaultable short rates as in Vasicek (1997) with interest rates that follow a mean-reverting process with constant volatility σ_r, i.e.

$$dr_t = a\left(b - r_t\right)dt + \sigma_r d\hat{W}_r\left(t\right), \tag{3.4}$$

where $\hat{W}_r$ is a one-dimensional standard Brownian motion under the measure $\mathbf{Q}$ and the initial value r_0, $a > 0$, b, and the volatility parameter $\sigma_r > 0$ are constants. In addition, they model V under the measure $\mathbf{Q}$ as

$$\frac{dV_t}{V_t} = r_t dt + \sigma_V d\hat{W}_V\left(t\right), \tag{3.5}$$

where σ_V is the constant volatility parameter, $\hat{W}_V$ is a standardized Brownian motion, and the Brownian motions $\hat{W}_r$ and $\hat{W}_V$ are correlated, with constant instantaneous correlation coefficient ρ_{Vr}. As we already know, pricing a defaultable discount bond in our setting more or less reduces to pricing an European put option. Jamshidian (1989) has derived a closed-form solution for the value of a European put option on a stock under stochastic interest rates. So Shimko et al. (1993) simply apply this result to the pricing of defaultable discount bonds within Merton's generalized model. The pricing formula is shown in the following proposition:

Proposition 3.2.2.
Given the assumptions of Merton (1974) slightly generalized

- *to allow for Vasicek type stochastic interest rates as given in equation (3.4)*
- *and a firm-value process as specified in equation (3.5) that is instantaneously correlated with the stochastic interest rates through $\hat{W}_r$ and $\hat{W}_V$*

$$Cov\left(d\hat{W}_r(t), d\hat{W}_V(t)\right) = \rho_{Vr}dt,$$

in analogy to Jamshidian (1989), the price of a defaultable discount bond can be calculated according to

$$\begin{aligned} P^d(t,T) &= LP(t,T) - Put_t \\ &= LP(t,T) - LP(t,T)\,\Phi(-d_2^*(t)) + V_t\Phi(-d_1^*(t)) \\ &= LP(t,T)\,\Phi(d_2^*(t)) + V_t\Phi(-d_1^*(t)), \end{aligned}$$

where

$$d_1^*(t) = \frac{l_t^* + \frac{1}{2}\sigma^2(t,T)}{\sigma(t,T)},\ d_2^*(t) = d_1^*(t) - \sigma(t,T),$$

with

$$l_t^* = \ln\left(\frac{V_t}{LP(t,T)}\right),$$

$$\sigma^2(t,T) = \int_t^T \left(\sigma_V^2 - 2\rho_{Vr}\sigma_V b(u,T) + b^2(u,T)\right)du,$$

and

$$b(t,T) = \frac{\sigma_r}{a}\left(1 - e^{-a(T-t)}\right),$$

for all $0 \le t \le T$. $P(t,T)$ is Vasicek's non-defaultable bond price:

$$P(t,T) = e^{\alpha(t,T)+\beta(t,T)r_t}$$

where

$$\begin{aligned} \beta(t,T) &= -\frac{1}{a}\left(1 - e^{-a(T-t)}\right), \\ \alpha(t,T) &= \left(b - \frac{\sigma^2}{2a^2}\right)(\beta(t,T) - T + t) - \frac{\sigma^2}{4a}\beta^2(t,T), \end{aligned}$$

for all $0 \le t \le T$.

There are quite a lot of further examples for asset based models with stochastic non-defaultable interest rates such as the papers of Brennan & Schwartz (1980) and Cox, Ingersoll & Ross (1980). In addition, Kim et al. (1992) and Kim et al. (Autumn 1993) study the role of call features in corporate and treasury bonds and find that the call feature is relatively more valuable in treasury than it is in corporate issues. Wang (1999) uses a Cox-Ingersoll-Ross type model for the stochastic non-defaultable short rates. Non-defaultable short rates are assumed to be stochastically independent of the firm-value process. Szatzschneider (2000) generalizes this result by dropping the independence assumption.

3.2.2 First Passage Time Models

First passage time models generalize Merton's model to account for defaults that may happen prior to maturity. Black & Cox (1976) introduce so called safety covenants which provide the bondholders the right to force the firm into bankruptcy as soon as the firm value hits some deterministic time-dependent threshold. Other types of default can only happen at the maturity of the bond. The recovery process is assumed to be proportional to the firm value process. Kim et al. (Autumn 1993) assume a Cox-Ingersoll-Ross type model for the stochastic non-defaultable short rates, i.e.

$$dr_t = a\left(b - r_t\right)dt + \sigma_r\sqrt{r_t}d\hat{W}_r\left(t\right),$$

with initial value $r_0 > 0$, and where $\hat{W}_r$ is a one-dimensional standard Brownian motion under the measure $\mathbf{Q}$ and $a, b, \sigma_r > 0$ are constants. The firm value process V is assumed to evolve according to

$$\frac{dV_t}{V_t} = \left(r_t - \kappa\right)dt + \sigma_V d\hat{W}_V\left(t\right), \ V_0 > \frac{c}{\kappa},$$

where κ is the constant payout ratio, $\sigma_V > 0$ the constant volatility parameter of V, $\hat{W}_V$ a standardized Brownian motion, and the Brownian motions $\hat{W}_r$ and $\hat{W}_V$ are correlated, with constant instantaneous correlation coefficient ρ_{Vr}. Kim et al. (Autumn 1993) assume that the bondholders continuously receive a coupon at a rate c (i.e. c units of currency) and that these payments are higher prioritized than possible dividend payments to the stockholders. The firm only defaults prior to maturity if the firm is not able to make the regular coupon payments, in other words if the firm value falls below $\frac{c}{\kappa}$. What is the intuitive idea behind this ? If at some time t $c \leq \kappa V_t$, the dividends still cover the coupon payments. If $c > \kappa V_t$ the dividend payments are not large enough to cover the coupon payments. Hence, before maturity of the bond Kim et al. (Autumn 1993) model the flow-based insolvency. Other types of default can only happen at maturity of the bond. Here the default condition is as usual. Default happens if the firm value falls below the notional of the bond.

Longstaff & Schwartz (1995b) develop a model similar to Shimko et al. (1993) except for the bankruptcy procedure. While in most papers using option pricing frameworks bankruptcy is triggered at the moment when the value of the firm reaches the value of the debt, they model default as the time when the value of the debt reaches some constant threshold value D which serves as a distress boundary, i.e. the default time T^d can then be expressed formally as $T^d = \inf\left\{t \geq 0 : V_t \leq D\right\}$, the first passage time for V_t (the value of the firm's assets at time t) to cross the lower bound D. If the value of the assets breaches this level, default is triggered, some form of restructuring occurs and the remaining assets of the firm are allocated among the firm's claimants. Implicit in this formulation is the assumption that once this level is reached, default occurs on all outstanding liabilities at the same time. Thus, contrary to Merton's and the other models discussed so far, default can occur prior to maturity - and not only for indenture provisions or coupon payment reasons. The idea is that as long as the firm value stays above D the firm is solvent. The recovery payment received by the bondholders in case of default is paid at maturity and is proportional to the face value of the bond. For each bond the recovery rate w is assumed to be constant. Then the bond's payoff at maturity T is given by

$$P^d(T,T) = (1-w)\,LH_T + L(1-H_T),$$

and therefore taking the money market account as numéraire

$$P^d(t,T) = E^{\mathbf{Q}}\left[e^{-\int_t^T r(l)dl}\left((1-w)\,LH_T + L(1-H_T)\right)\Big|\, \mathcal{F}_t\right]. \tag{3.6}$$

By taking $P(t,T)$ as numéraire and applying the change of numéraire theorem[26] to equation (3.6) we can express $P^d(t,T)$ according to

$$P^d(t,T) = LP(t,T)\left(1 - w\mathbf{Q}^T\left(T^d \le T\big|\, \mathcal{F}_t\right)\right),$$

where $\mathbf{Q}^T$ is the so called T-forward measure. Longstaff & Schwartz (1995b) assume that the interest rate uncertainty is driven by a Vasicek (1977) representation, i.e.

$$dr_t = a(b - r_t)\,dt + \sigma_r d\hat{W}_r(t),$$

with initial value $r_0 > 0$, and where $\hat{W}_r$ is a one-dimensional standard Brownian motion under the measure $\mathbf{Q}$ and $a, b, \sigma_r > 0$ are constants. The firm value process V is assumed to evolve according to

$$\frac{dV_t}{V_t} = r_t dt + \sigma_V d\hat{W}_V(t), \tag{3.7}$$

where $\sigma_V > 0$ is the constant volatility parameter of V, $\hat{W}_V$ a standardized Brownian motion, and the Brownian motions $\hat{W}_r$ and $\hat{W}_V$ are correlated, with constant instantaneous correlation coefficient ρ_{Vr}. Longstaff & Schwartz (1995b) show that the value of the defaultable discount bond at time $t = 0$ is given according to the following proposition.

Proposition 3.2.3.
Given the assumptions of Longstaff & Schwartz (1995b) the price of a defaultable discount bond can be numerically calculated according to

$$P^d(0,T) = LP(0,T)\left(1 - w\mathbf{Q}^T\left(T^d \le T\big|\, \mathcal{F}_0\right)\right),$$

where

---

[26] See, e.g., Briys, Bellalah, Mai & de Varenne (1998), p. 78.

$$P(0,T) = e^{\alpha(0,T)+\beta(0,T)r_0},$$

$$\beta(0,T) = -\frac{1}{a}\left(1 - e^{-aT}\right),$$
$$\alpha(0,T) = \left(b - \frac{\sigma^2}{2a^2}\right)\left(\beta(0,T) - T\right) - \frac{\sigma^2}{4a}\beta^2(0,T),$$

$$\mathbf{Q}^T\left(T^d \leq T \middle| \mathcal{F}_0\right) = \lim_{n\to\infty} \mathbf{Q}_n^T\left(T^d \leq T \middle| \mathcal{F}_0\right),$$

with

$$\mathbf{Q}_n^T\left(T^d \leq T \middle| \mathcal{F}_0\right) = \sum_{i=1}^{n} q_i,$$
$$q_1 = \Phi(a_1),$$
$$q_i = \Phi(a_i) - \sum_{j=1}^{i-1} q_j \Phi(b_{ij}) \text{ for } i = 1, ..., n,$$

$$a_i = \frac{-\ln\frac{V_0}{D} - \mu\left(\frac{iT}{n}, T\right)}{\sqrt{\sigma\left(\frac{iT}{n}\right)}} \text{ for } i = 1, ..., n,$$
$$b_{ij} = \frac{\mu\left(\frac{jT}{n}, T\right) - \mu\left(\frac{iT}{n}, T\right)}{\sqrt{\sigma\left(\frac{iT}{n}\right) - \sigma\left(\frac{jT}{n}\right)}} \text{ for } i = 1, ..., n \text{ and } j = 1, ..., i-1,$$

and

$$\mu(t,T) = \left(b - \frac{\rho_{Vr}\sigma_V\sigma_r}{a} - \frac{\sigma_r^2}{a^2} - \frac{\sigma_V^2}{2}\right)t$$
$$+ \left(\frac{\rho_{Vr}\sigma_V\sigma_r}{a^2} + \frac{\sigma_r^2}{a^3}\right)e^{-aT}\left(e^{at} - 1\right)$$
$$+ \left(\frac{r_0}{a} - \frac{b}{a} + \frac{\sigma_r^2}{a^3} - \frac{\sigma_r^2}{2a^3}e^{-aT}\right)\left(1 - e^{-at}\right),$$

$$\sigma(t) = \left(\frac{\rho_{Vr}\sigma_V\sigma_r}{a} + \frac{\sigma_r^2}{a^2} + \sigma_V^2\right)t$$
$$-\left(\frac{\rho_{Vr}\sigma_V\sigma_r}{a^2} + \frac{2\sigma_r^2}{a^3}\right)\left(1 - e^{-at}\right)$$
$$+\frac{\sigma_r^2}{2a^3}\left(1 - e^{-2at}\right).$$

Longstaff & Schwartz (1995b) show that the convergence is very rapid and that setting n to 200 is sufficient. However, as Briys et al. (1998) mention, the model suffers from the defect that pricing equations do not ensure that the payment to bondholders is no greater than the firm's value upon default when the defaultable bond reaches maturity. Indeed, the firm can find itself in a solvent position at maturity according to the threshold but with assets insufficient to match the face value of the bond.

Briys & de Varenne (1997) try to correct these problems. They assure that the payment to bondholders is not greater than the firm value upon default, and whenever the firm is "threshold-solvent", it is able to repay the face value of the bond. They define a stochastic default barrier. To be more precise, they define the default barrier as a fixed quantity discounted at the default free interest rate up to maturity of the defaultable bond. As soon as this barrier is reached the obligors receive an exogenously specified fraction of the firm's remaining assets. Briys & de Varenne (1997) assume that the interest rate uncertainty is driven by the following Gaussian representation:

$$dr_t = a_t \left(b_t - r_t\right) dt + \sigma_r \left(t\right) d\hat{W}_r \left(t\right),$$

with initial value $r_0 > 0$, and where $\hat{W}_r$ is a one-dimensional standard Brownian motion under the measure $\mathbf{Q}$ and for some deterministic functions $a_t, b_t, \sigma_r\left(t\right)$. The firm value process V is assumed to evolve according to

$$\frac{dV_t}{V_t} = r_t dt + \sigma_V d\hat{W}_V \left(t\right),$$

where $\sigma_V > 0$ is the constant volatility parameter of V, $\hat{W}_V$ a standardized Brownian motion, and the Brownian motions $\hat{W}_r$ and $\hat{W}_V$ are correlated, with constant instantaneous correlation coefficient ρ_{Vr}. Briys & de Varenne (1997) include safety covenants that allow the bondholders to trigger early bankruptcy. As soon as V_t hits the exogenously specified safety covenant

$$\alpha L P \left(t, T\right),$$

where $0 \leq \alpha \leq 1$ is some fixed value, bankruptcy is enforced. The closer α is to 0, the less protective is the safety covenant. There is default at maturity if the value of the firm's asset at time T is less than L. Hence, we have T^d implicitly redefined as

$$T^d = \min\left\{T^{d,1}, T^{d,2}\right\},$$

where

$$T^{d,1} = \begin{cases} \inf\left\{0 \leq t < T : V_t \leq \alpha L P\left(t,T\right)\right\}, & \text{if } V \text{ hits } \alpha L P \text{ before } T, \\ \infty, & \text{else,} \end{cases}$$

and

$$T^{d,2} = \begin{cases} T, & \text{if } V_T < L, \\ \infty, & \text{else.} \end{cases}$$

The recovery that is received by the bondholders at default equals $0 \leq \beta_1 \leq 1$ if default occurs prior to maturity and $0 \leq \beta_2 \leq 1$ if default occurs at maturity. If we assume that there is no default before maturity then V_t always remains above $\alpha LP(t,T)$ and bankruptcy can only happen at maturity. Then the payoff to the bondholders at maturity T can be expressed according to

$$L(1 - H_T) + \beta_1 \alpha L H_{T-} + \beta_2 V_T (H_T - H_{T-}).$$

Therefore, the price of a defaultable discount bond with face value L is given by

$$\begin{aligned} &P^d(t,T) \\ &= E^{\mathbf{Q}}\left[e^{-\int_t^T r(l)dl} \left(L(1 - H_T) + \beta_1 \alpha L H_{T-} + \beta_2 V_T (H_T - H_{T-}) \right) \Big| \mathcal{F}_t \right]. \end{aligned}$$

- $L(1 - H_T)$ is the payoff in case of solvency i.e. no default.
- $\beta_1 \alpha L H_{T-}$ captures the pay-off in case of forced bankruptcy before maturity.
- $\beta_2 V_T (H_T - H_{T-})$ is the pay-off in case of no premature forced bankruptcy but final default, i.e. the asset value of the firm at maturity is higher than the threshold value but less than the face value of the bond.

By applying change of numéraire and change of time[27] techniques one can show the following proposition.

Proposition 3.2.4. *Given the assumptions of Briys & de Varenne (1997) the price of a defaultable discount bond is given by the following closed-form solution:*

$$\begin{aligned} P^d(t,T) = LP(t,T) \cdot \Big[& 1 - Put_1(t) + \alpha Put_2(t) \\ & - (1-\beta_1) \frac{V_t}{LP(t,T)} \left(\Phi(-d_3(t)) + \frac{\alpha LP(t,T)}{V_t} \Phi(-d_4(t)) \right) \\ & - (1-\beta_2) \frac{V_t}{LP(t,T)} \Big(\Phi(d_3(t)) - \Phi(d_1(t)) \\ & + \frac{\alpha LP(t,T)}{V_t} \left(\Phi(d_4(t)) - \Phi(d_6(t)) \right) \Big) \Big], \end{aligned}$$

where

[27] For an introduction to change of numéraire and change of time techniques see, e.g., Briys et al. (1998), chapter 4.

$$d_1(t) = \frac{-\ln \frac{LP(t,T)}{V_t} + \frac{\sigma^2(t,T)}{2}}{\sigma(t,T)} = d_2(t) + \sigma(t,T),$$

$$d_3(t) = \frac{-\ln \frac{\alpha LP(t,T)}{V_t} + \frac{\sigma^2(t,T)}{2}}{\sigma(t,T)} = d_4(t) + \sigma(t,T),$$

$$d_5(t) = \frac{-\ln \frac{\alpha^2 LP(t,T)}{V_t} + \frac{\sigma^2(t,T)}{2}}{\sigma(t,T)} = d_6(t) + \sigma(t,T),$$

$$\sigma^2(t,T) = \int_t^T \left[(\rho_{Vr}\sigma_V + \sigma_P(u,T))^2 + \left(1 - \rho_{Vr}^2\right)\sigma_V^2 \right] du,$$

$$\sigma_P(t,T) = \sigma_r(t) \int_t^T e^{-\int_t^u a(s)ds} du,$$

and

$$Put_1(t) = -\frac{V_t}{LP(t,T)} \Phi(-d_1(t)) + \Phi(-d_2(t)),$$

$$Put_2(t) = -\frac{V_t}{\alpha^2 LP(t,T)} \Phi(-d_5(t)) + \Phi(-d_6(t)).$$

Nielsen, Saà-Requejo & Santa-Clara (1993), and Saá-Requejo & Santa-Clara (1997) allow for stochastic default boundaries and deviation from the absolute priority rule, too. The approach of Saá-Requejo & Santa-Clara (1997) is somehow more general than the approach of Nielsen et al. (1993). Therefore, we concentrate on the former. The short-term interest rate under the spot martingale measure $\mathbf{Q}$ is assumed to follow an Itô process

$$dr_t = \mu_r dt + \sigma_r d\hat{W}_r(t).$$

The firm value process V is assumed to evolve according to

$$\frac{dV_t}{V_t} = (r_t - \kappa) dt + \sigma_V d\hat{W}_V(t),$$

where κ is the constant payout ratio, $\sigma_V > 0$ the constant volatility parameter of V, $\hat{W}_V$ a standardized Brownian motion, and the Brownian motions $\hat{W}_r$ and $\hat{W}_V$ are correlated, with constant instantaneous correlation coefficient ρ_{Vr}. Note, that we can express $\hat{W}_V(t)$ as

$$\hat{W}_V(t) = \rho_{Vr}\hat{W}_r(t) + \sqrt{1 - \rho_{Vr}^2}\,\hat{W}(t),$$

where $\hat{W}(t)$ is a standard Brownian motion independent of $\hat{W}_r$. In addition, there is a boundary process D_t modeled according to the following SDE

$$\frac{dD_t}{D_t} = (r_t - \zeta)\,dt + \sigma_{D,1} d\hat{W}_r(t) + \sigma_{D,2} d\hat{W}(t)\,,\ D_0 < V_0,$$

where ζ, $\sigma_{D,1} > 0$ and $\sigma_{D,2} > 0$ are constants . It is assumed that default occurs as soon as V_t falls below D_t, i.e.

$$\begin{aligned} T^d &= \inf\{t \in [0,T] : V_t < D_t\} \\ &= \inf\left\{t \in [0,T] : l(t) = \ln\left(\frac{V_t}{D_t}\right) < 0\right\}. \end{aligned}$$

The recovery rate w is either assumed to be constant or a $\mathcal{F}_T$–measurable random variable. Therefore the payoff at maturity T equals

$$(1 - w)\,LH_T + L\,(1 - H_T)\,.$$

If we assume a constant recovery rate w and take $P(t,T)$ as numéraire we can express $P^d(t,T)$ according to

$$P^d(t,T) = LP(t,T)\left(1 - w\mathbf{Q}^T\left(T^d \leq T \middle| \mathcal{F}_t\right)\right),$$

where $\mathbf{Q}^T$ is the T-forward measure. Then in some special cases a closed form solution for $P^d(t,T)$ is easily available.

Proposition 3.2.5.
In addition to the general assumptions of Saá-Requejo & Santa-Clara (1997) we assume that

$$\sigma_V \rho_{Vr} - \sigma_{D,1} = 0,$$

then the price of a defaultable discount bond is given by the following closed-form solution:

$$\begin{aligned} &P^d(t,T) \\ &= LP(t,T)\left(1 - w\Phi\left(\frac{-l_t - \mu_l(T-t)}{\sigma_l\sqrt{T-t}}\right) - we^{-2\frac{\mu_l l_t}{\sigma_l^2}}\Phi\frac{-l_t + \mu_l(T-t)}{\sigma_l\sqrt{T-t}}\right), \end{aligned}$$

where

$$\mu_l = \varsigma - \kappa + \frac{1}{2}\left(\sigma_{D,1}^2 + \sigma_{D,2}^2 - \sigma_V^2\right)$$

and

$$\sigma_l^2 = \left(\sigma_V\sqrt{1 - \rho_{Vr}^2} - \sigma_{D,2}\right)^2.$$

Of course, there are a lot of more works devoted to generalizing structural models. However, because of the complexities of all these extensions, often a closed-form solution can't be obtained any more and numerical procedures must be used. Although the structural approach is conceptually important, gathering and analyzing corporate balance sheet data issuer by issuer is impractical. Complicated balance sheets can't be captured by the structural approach, anyway. And default is usually determined by much more factors than assets and liabilities.

3.3 Intensity Based Models

Despite the immense efforts devoted to generalize Merton's methodology, all of these structural models have only limited success in explaining the behavior of prices of debt instruments and credit spreads. This has led to attempts to use models that make more direct assumptions about the default process. These alternative approaches, called reduced-form models, don't consider the relation between default and asset value in an explicit way but model default as a stopping time of some given hazard rate process, i.e. the default process is specified exogenously. This achieves two effects. The first is that the model can be applied to situations where the underlying asset value is not observable. Secondly, the default time is unpredictable and therefore the behavior of credit spreads for short maturities can be captured more realistically. In addition, the approach is very tractable and flexible to fit the observed credit spreads. A long list of papers has appeared recently that follows this approach. Some of the most important ones are the following:

- Jarrow & Turnbull (1995) present a model where the bankruptcy process is compared to a spot exchange rate process. Default is driven by a Poisson process with a constant intensity parameter[28] and a given payoff at default.
- In Lando (1994), Lando (1996), and Lando (1998) default is driven by Cox processes, which can be thought of as Poisson processes with random intensity parameters.
- Duffie & Singleton (1997) and Duffie & Singleton (1998b) show that valuation under the risk-adjusted probability measure can be executed by discounting the non-defaultable payoff on the debt by a discount rate that is adjusted for the parameters of the default process.
- Schönbucher (1996) presents a generalization in a Heath-Jarrow-Morton framework[29] that allows for restructuring of defaulted debt and multiple defaults.

Despite of all the attractive properties of the reduced-form models, their main draw-back is the missing link between firm value and corporate default.

Finally, it should be mentioned, that Duffie & Lando (1997) show that there is a close connection between structural and reduced-form models.

3.3.1 Short Rate Type Model

As always we fix a finite horizon date $T^* > 0$ and a filtered probability space $\left(\Omega, \mathcal{F}, \mathbf{F} = (\mathcal{F}_t)_{0 \le t \le T^*}, \mathbf{P}\right)$ where $\mathbf{P}$ denotes the real-world measure. We assume that $\mathcal{F}_0$ is trivial, i.e. $\mathcal{F}_0 = \{\emptyset, \Omega\}$, that $\mathcal{F}_{T^*} = \mathcal{F}$ and that the filtration

[28] For a definition see section 2.

[29] For an introduction to the Heath-Jarrow-Morton framework see, e.g., Heath, Jarrow & Morton (1992).

$\mathbf{F} = (\mathcal{F}_t)_{0 \leq t \leq T^*}$ satisfies the usual conditions[30]. The filtration $\mathbf{F}$ represents the arrival of information over time. A predictable short rate process r is also fixed such that the non-defaultable money market account (given by $B_t = e^{\int_0^t r_l dl}$) is well defined. Let us recall some conventions from section 2.3.4, where we have already introduced the concept of intensity modeling. With the stopping time T^d we associate the first jump time of the point or counting process $\{A_t\}_{0 \leq t \leq T^*}$ defined by $A_t = \sum_{n \geq 1} 1_{\{T_n \leq t\}}$. The sequence T_n is a sequence of jump times (in our context of default events) with T_n defined as the time of occurrence of the nth jump. If we define H_t as the process A_t stopped at time T^d, then H_t can be interpreted as the default indicator function which equals 0 before default and 1 if default has already occurred, i.e. $H_t = 1_{\{T^d \leq t\}}$.We assume that the default indicator function H_t admits a $(\mathbf{P},\mathcal{F}_t)$ −intensity λ_t, more specific we assume a Cox process model (see remark 2.3.4), and allow for correlation between the processes r and λ. In addition, we assume that there exists an equivalent martingale measure $\mathbf{Q}$ relative to the short rate process r and that under the risk-neutral probabilities default is also a Cox process. We denote the risk-neutral intensity process by $\widehat{\lambda}$.

Pricing With Zero Recovery. First we assume that the recovery rate of the defaultable discount bond equals zero. We denote the price of such a bond by $P^{d,zero}$.

Proposition 3.3.1.
As shown by Lando (1998) under technical conditions, provided that default has not already occurred by time t, under the zero recovery assumption, the price of the defaultable discount bond is given by

$$\begin{aligned} P^{d,zero}(t,T) &= E^{\mathbf{Q}}\left[e^{-\int_t^T r_l dl} \left(1 - H_T\right) \Big| \mathcal{F}_t \right] \\ &= E^{\mathbf{Q}}\left[e^{-\int_t^T (r_l + \widehat{\lambda}_l) dl} \Big| \mathcal{F}_t \right], \ 0 \leq t < \tau, \end{aligned} \tag{3.8}$$

and $P^{d,zero}(t,T) = 0$ for $t \geq \tau$ since there are no dividends after time τ.

The intuition behind this pricing formula is as follows: The price of a defaultable instrument can either be calculated as the expected value of promised payoffs discounted by the non-defaultable discount factor $e^{-\int_t^T r_l dl}$ or as the expected value of certain payoffs but discounted with a default adjusted discount factor $e^{-\int_t^T (r_l + \widehat{\lambda}_l) dl}$. From a technical point of view, pricing of defaultable discount bonds is the same as pricing of non-defaultable discount bonds.

[30] Jacod & Shiryaev (1987), p.2: The stochastic basis $\left(\Omega, \mathcal{F}, (\mathcal{F}_t)_{t \geq 0}, \mathbf{P}\right)$ is called complete, or equivalently is said to satisfy the usual conditions if the σ-field $\mathcal{F}$ is $\mathbf{P}$−complete and if every $\mathcal{F}_t$ contains all $\mathbf{P}$−null sets of $\mathcal{F}$. It is always possible to complete a given stochastic basis.

Example 3.3.1. Under the measure $\mathbf{Q}$, $\widehat{\lambda}_t$ and r_t are affine functions of the factors X, i.e. $\widehat{\lambda}_t = l_{\widehat{\lambda},0}(t) + l_{\widehat{\lambda},1}(t) \cdot X_t$ and $r_t = l_{r,0}(t) + l_{r,1}(t) \cdot X_t$ for bounded continuous functions $l_{\lambda,0}$, $l_{r,0}$ and $l_{\lambda,1}$, $l_{r,1}$ on $[0,\infty)$, where X is a strong $n-$dimensional Markov process valued in $\mathbb{R}^n$ uniquely solving the stochastic differential equation

$$dX_t = \mu(X_t, t)\, dt + \sigma(X_t, t)\, d\hat{W}_t.$$

$\hat{W}$ is an $\mathbf{F}-$adapted $n-$dimensional standard Brownian motion in $\mathbb{R}^n$, $\mu : \Theta \to \mathbb{R}^n$, $\sigma : \Theta \to \mathbb{R}^{n\times n}$, $\Theta \subset \mathbb{R}^n \times [0,\infty)$, and for each t, $\{x : (x,t) \in \Theta\}$ containing an open subset of $\mathbb{R}^n$. Then, under certain restrictions[31] on the parameters of the process X and a specific choice of Θ the non-defaultable and defaultable discount bond pricing formulas are of the form

$$P(t,T) = e^{\alpha(t,T)+\beta^T(t,T)\cdot X_t} \text{ and}$$
$$P^{d,zero}(t,T) = e^{\alpha^{d,zero}(t,T)+\left(\beta^{d,zero}(t,T)\right)^T\cdot X_t},$$

where $\alpha(t,T)$, $\alpha^{d,zero}(t,T) \in \mathbb{R}$ and $\beta(t,T)$, $\beta^{d,zero}(t,T) \in \mathbb{R}^n$ are coefficients, depending on t and T that are given explicitly.

Pricing With Positive Recovery. In section 2.4 we have introduced different concepts of recovery rate modeling. The approach of fractional recovery of market value can be easily included in intensity based models. Therefore, we suppose that recovery payments are made at the time of default. Then, prior to default

$$P^d(t,T) = E^{\mathbf{Q}}\left[\int_t^T e^{-\int_t^u r_l dl} Z_u dH_u + e^{-\int_t^T r_l dl}\,(1-H_T)\Big|\,\mathcal{F}_t\right],\ 0 \le t < \tau. \tag{3.9}$$

Z_t is a risk-neutral fraction w_t of the market value of another bond, yet to default, but otherwise of the same quality, i.e.

$$Z_t = w_t P^d(t-,T),\ 0 \le w_t \le 1.$$

Proposition 3.3.2.
Under a lot of technical conditions Duffie & Singleton (1997) show that equation (3.9) can be transformed into

$$P^d(t,T) = E^{\mathbf{Q}}\left[e^{-\int_t^T\left(r_l+(1-w_l)\widehat{\lambda}_l\right)dl}\Big|\,\mathcal{F}_t\right],\ 0 \le t < \tau,$$

which is a straightforward generalization of equation (3.8).

---
[31] See, e.g., Duffie & Kan (1996).

4. Correlated Defaults

"Incorporating default correlation in any portfolio credit risk analysis is difficult because of the lack of good data on default correlation, and the complexity of developing realistic models of default correlations that capture its dependence on credit quality, region, industry and time horizon."
- Krishan Nagpal and Reza Bahar -

"We are developing toward the Titanic solution. We only build bigger and bigger ships, but not remembering there are still icebergs."
- Norbert Walter -

4.1 Introduction

Default dependencies among many different issuers play an important role in the quantification of a portfolio's credit risk exposure for many reasons. E.g.:

- Specifying an appropriate model for dependent defaults is the core problem in valuing CDOs, multi-name credit derivatives and other financial instruments where portfolios of other defaultable financial instruments are present. Different dependence structures produce different default distributions, which in turn affect the pricing of these instruments. They are actively traded which requires a methodology to measure default and market risks on a day-by-day basis. A consistent model for default correlations is essential to price and hedge these instruments.
- While the actual loss in a portfolio due to the default of a single obligor may be small (unless the risk exposure is very large) the effects of simultaneous defaults of several issuers can be catastrophic. However, little is known about the drivers of default risk at the portfolio level.

4.2 Correlated Asset Values

If we want to study default behavior in a portfolio context we have to consider and model dependent defaults[1]. Therefore, we consider a set of m obligors

[1] In this section we basically follow the work of Zhou (2001).

$1, ..., m$. $H_i(t)$ is the default indicator function of the i–th obligor. As a first approach we model dependent defaults in the context of structural models. We assume that the default behavior of the i–th firm is determined by the behavior of the firm value process V_i. Default occurs if the firm specific process V_i falls below a threshold process D_i, i.e. the default time of the i–th firm can be expressed as

$$\begin{aligned} T_i^d &= \inf \left\{ t \geq 0 : V_i(t) < D_t^i \right\} \\ &= \inf \left\{ t \geq 0 : V_i(t) \leq D_t^i \right\}. \end{aligned}$$

Let $T_{i,j}^d = \min\left(T_i^d, T_j^d\right)$ and $H_{i,j}(t) = 1_{\{T_{i,j}^d \leq t\}}$. The default correlation $\rho_{i,j}^d$ between firms i and j over the time interval $[0, t]$ is defined as the correlation between the random variables $H_i(t)$ and $H_j(t)$, $i, j \in \{1, ..., m\}$, $i \neq j$, i.e.

$$\begin{aligned} \rho_{i,j}^d(t) = &\left(E^{\mathbf{P}}\left[H_i(t) H_j(t) \middle| \mathcal{F}_0\right] - E^{\mathbf{P}}\left[H_i(t) \middle| \mathcal{F}_0\right] E^{\mathbf{P}}\left[H_j(t) \middle| \mathcal{F}_0\right]\right) / \\ &\left(\sqrt{E^{\mathbf{P}}\left[H_i(t) \middle| \mathcal{F}_0\right]\left(1 - E^{\mathbf{P}}\left[H_i(t) \middle| \mathcal{F}_0\right]\right)} \right. \\ &\left. \sqrt{E^{\mathbf{P}}\left[H_j(t) \middle| \mathcal{F}_0\right]\left(1 - E^{\mathbf{P}}\left[H_j(t) \middle| \mathcal{F}_0\right]\right)}\right). \end{aligned} \tag{4.1}$$

Note, that

$$E^{\mathbf{P}}\left[H_i(t) H_j(t) \middle| \mathcal{F}_0\right] = E^{\mathbf{P}}\left[H_i(t) \middle| \mathcal{F}_0\right] + E^{\mathbf{P}}\left[H_j(t) \middle| \mathcal{F}_0\right] - E^{\mathbf{P}}\left[H_{i,j}(t) \middle| \mathcal{F}_0\right]. \tag{4.2}$$

Therefore, if we can calculate the expected values of $H_i(t)$, $H_j(t)$ and $H_{i,j}(t)$, we can easily determine $\rho_{i,j}^d(t)$. To calculate these values we have to specify the assumptions on the processes V_i and D_i.

Assumption 4.2.1
The dynamics of any two-dimensional process (V_i, V_j), $i, j = 1, ..., m$, $i \neq j$, are given by the following vector stochastic process

$$\begin{pmatrix} d\ln V_i(t) \\ d\ln V_j(t) \end{pmatrix} = \begin{pmatrix} \mu_i \\ \mu_j \end{pmatrix} + \Sigma_{i,j} \begin{pmatrix} dW_i(t) \\ dW_j(t) \end{pmatrix},$$

where μ_i, μ_j are constant drift terms, W_i, W_j are independent standard Brownian motions, and

$$\Sigma_{i,j} = \begin{pmatrix} \sigma_{i,i}\ \sigma_{i,j} \\ \sigma_{j,i}\ \sigma_{j,j} \end{pmatrix}$$

is a constant 2×2 matrix such that

$$\Sigma_{i,j} \cdot \Sigma_{i,j}^T = \begin{pmatrix} \sigma_i^2 & \rho_{i,j}\sigma_i\sigma_j \\ \rho_{i,j}\sigma_i\sigma_j & \sigma_j^2 \end{pmatrix}.$$

The coefficient $\rho_{i,j}$ reflects the correlation between the movements in the asset values of firms i and j.

Assumption 4.2.2
For $i = 1, ..., m$ the threshold processes D_i are given by

$$D_i = e^{\mu_i t} \kappa_i,$$

where $\kappa_i \geq 0$ are constants and μ_i the constant drift terms.

Proposition 4.2.1.
Given assumptions 4.2.1 and 4.2.2, for $i = 1..., m$, we have

$$E^{\mathbf{P}} \left[H_i(t) \middle| \mathcal{F}_0 \right] = \mathbf{P} \left(T_i^d \leq t \middle| \mathcal{F}_0 \right) = 2\Phi \left(-\frac{Z_i}{\sqrt{t}} \right),$$

where

$$Z_i = \frac{\ln \frac{V_i(0)}{\kappa_i}}{\sigma_i}.$$

Proof. See, e.g., Harrison (1990).

According to equations (4.1) and (4.2), to calculate $\rho_{i,j}^d(t)$ it remains to determine $E^{\mathbf{P}} \left[H_{i,j}(t) \middle| \mathcal{F}_0 \right]$.

Proposition 4.2.2.
Given assumptions 4.2.1 and 4.2.2, for $i, j = 1..., m$, $i \neq j$, we have

$$E^{\mathbf{P}} \left[H_{i,j}(t) \middle| \mathcal{F}_0 \right] = \mathbf{P} \left(T_{i,j}^d \leq t \middle| \mathcal{F}_0 \right) = 1 - \frac{2\varrho_{i,j}}{\sqrt{2\pi t}} e^{-\frac{\varrho_{i,j}^2}{4t}} \sum_{k=1,3,5,\dots} \frac{1}{k} \sin \left(\frac{k\pi\theta_{i,j}}{\alpha_{i,j}} \right) \cdot \left[I_{\frac{1}{2}\left(\frac{k\pi}{\alpha_{i,j}}+1\right)} \left(\frac{\varrho_{i,j}^2}{4t} \right) + I_{\frac{1}{2}\left(\frac{k\pi}{\alpha_{i,j}}-1\right)} \left(\frac{\varrho_{i,j}^2}{4t} \right) \right],$$

where I_v is the modified Bessel function I with order v,

$$Z_i = \frac{\ln \frac{V_i(0)}{\kappa_i}}{\sigma_i}, \quad i = 1, ..., m,$$

$$\theta_{i,j} = \begin{cases} \tan^{-1} \left(\frac{Z_j \sqrt{1-\rho_{i,j}^2}}{Z_i - \rho_{i,j} Z_j} \right), & \text{if } (\cdot) > 0, \\ \pi + \tan^{-1} \left(\frac{Z_j \sqrt{1-\rho_{i,j}^2}}{Z_i - \rho_{i,j} Z_j} \right), & \text{otherwise,} \end{cases}$$

$$\varrho_{i,j} = \frac{Z_j}{\sin \theta_{i,j}},$$

$$\alpha_{i,j} = \begin{cases} \tan^{-1} \left(-\frac{\sqrt{1-\rho_{i,j}^2}}{\rho_{i,j}} \right), & \text{if } \rho_{i,j} < 0, \\ \pi + \tan^{-1} \left(-\frac{\sqrt{1-\rho_{i,j}^2}}{\rho_{i,j}} \right), & \text{otherwise.} \end{cases}$$

Proof. If we define for $i = 1, ..., m$

$$Y_i(t) = -\ln \frac{e^{-\mu_i t} V_i(t)}{V_i(0)}$$

and

$$b_i = \ln \frac{V_i(0)}{\kappa_i},$$

then it is straightforward to show that $(Y_i(t), Y_j(t))^T$ follows a two dimensional Brownian motion:

$$\begin{pmatrix} dY_i(t) \\ dY_j(t) \end{pmatrix} = -\Sigma_{i,j} \cdot \begin{pmatrix} dW_i(t) \\ dW_j(t) \end{pmatrix}.$$

After this transformation, finding $E^{\mathbf{P}}[H_{i,j}(t)|\mathcal{F}_0]$ is equivalent to finding the first passage time of the two dimensional Brownian motion $(Y_i(t), Y_j(t))^T$ with initial condition $(Y_i(0), Y_j(0))^T = (0,0)^T$ to a boundary consisting of two intersecting lines $b_i > 0$ and $b_j > 0$. For notational convenience, we denote the barrier by $\partial(b_i, b_j)$.
As visualized in figure 4.1 the two dimensional Brownian motion process $(Y_i(t), Y_j(t))^T$ represents the position of a particle at time t and $\partial(b_i, b_j)$ is an absorbing barrier. Let $f(y_1, y_2, t)$ be the transition probability density of the particle in the region $\{(y_1, y_2)|\, y_1 < b_i \text{ and } y_2 < b_j\}$, i.e. the probability density that the particle does not reach the absorbing barrier in $[0, t)$. The corresponding distribution function is given by

$$\begin{aligned} &F(y_1, y_2, t) \\ &= \mathbf{P}(Y_i(s) < b_i,\ Y_j(s) < b_j,\ \text{for } 0 \le s < t,\ Y_i(t) < x_1,\ Y_j(t) < x_2) \\ &= \int_{-\infty}^{x_1} \int_{-\infty}^{x_2} f(y_1, y_2, t)\, dy_1 dy_2. \end{aligned}$$

Thus,

$$E^{\mathbf{P}}[H_{i,j}(t)|\mathcal{F}_0] = 1 - F(b_i, b_j, t).$$

According to Cox & Miller (1972) and Karatzas & Shreve (1988), the transition probability density $f(y_1, y_2, t)$ satisfies the following Kolmogorov forward equation

$$\frac{\sigma_i^2}{2} \frac{\partial^2 f}{\partial y_1^2} + \rho_{i,j} \sigma_i \sigma_j \frac{\partial^2 f}{\partial y_1 \partial y_2} + \frac{\sigma_j^2}{2} \frac{\partial^2 f}{\partial y_2^2} = \frac{\partial f}{\partial t}, \quad (y_1 < b_i,\ y_2 < b_j)$$

subject to certain boundary conditions. Solving for the density function f from this PDE and integrating, we obtain the desired result.

Bielecki & Rutkowski (2002) use a probabilistic approach instead of a PDE approach to find the same result. Results for other specifications of D_i (such as constant thresholds) can be determined in a similar way.

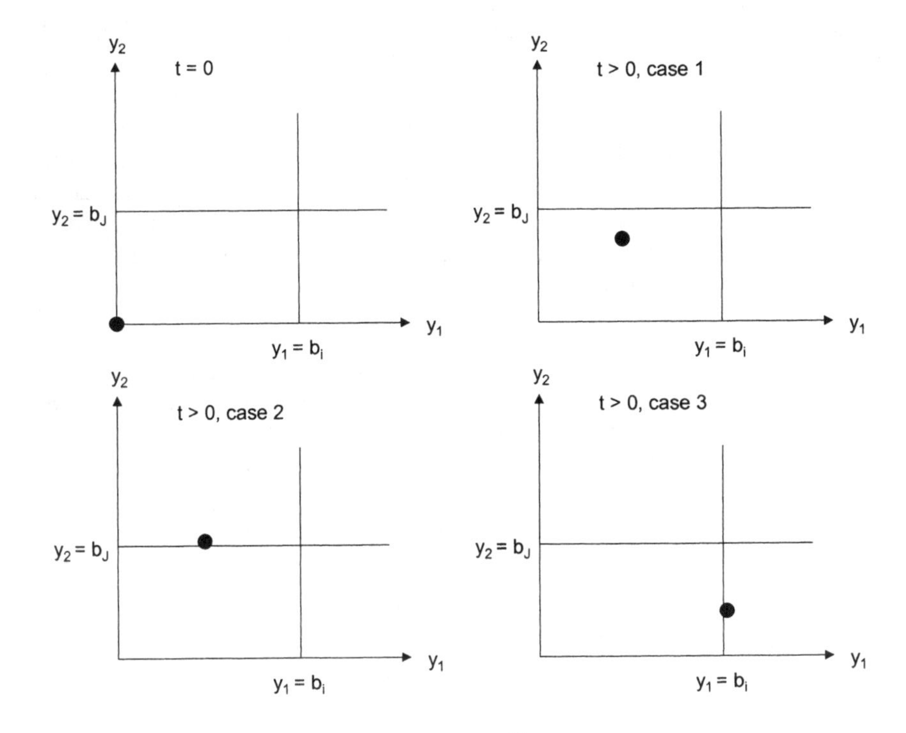

Fig. 4.1. Upper left: At time $t = 0$ $\left(Y_0^i, Y_0^j\right)^T = (0,0)^T$. Upper right: After time $t = 0$ the two dimensional process $\left(Y_i\left(t\right), Y_j\left(t\right)\right)^T$ moves through space. Lower left: As soon as the second component of the process reaches/crosses b_j default occurrs. Lower right: As soon as the first component of the process reaches/crosses b_i default occurrs.

In the framework we have discussed in this section, dependence between default events is derived from the dependence between latent variables, i.e. the firm's assets. It is common practice to assume that latent variables follow a multi-normal distribution, and consequently the default dependencies are completely captured by Pearson's correlations. However, it can be shown that joint default events generated from latent variables are very sensitive to the distributional assumptions and that models built on the assumption of multi-normal distributions may not adequately capture the true risk exposure in a portfolio.

4.3 Correlated Default Intensities

At the heart of intensity based models lies the assumption that defaults of different issuers are independent conditional on the sample paths of the default intensities. This is the assumption that facilitates the construction of Cox processes and directly implies that default correlation in such models is synonymous with the correlation of the default intensities. Theoretically, intensity based models provide a more flexible framework for modeling the dynamics and the term structure of credit risk based on market variables, such as credit spreads. However, calibration of the model to market variables is not trivial because of the sparsity of default data and the need to model a large number of variables simultaneously. There are some empirical studies that come to the conclusion that the intensity approach is problematic for modeling correlated defaults:

- Hull & White (2001) find that the range of default correlations that can be achieved is limited. Even when there is a perfect correlation between two intensities, the corresponding correlation between defaults in any chosen period of time is usually very low.
- Schoenbucher & Schubert (2001) believe that the default correlations that can be reached with the intensity approach are typically too low when compared with empirical default correlations, and furthermore it is hard to derive and analyze the resulting default dependency structure.

However, Yu (2003) argues that the default correlation in intensity based models can be quite sensitive to the common factor structure imposed on individual default intensities. On the one hand, he recognizes that using a model specification as in Duffee (1999) implies default correlations much lower than empirical observations. On the other hand, he shows that with a model specification as in Driessen (2002) the values are comparable or even higher than empirical observations.

The theory of modeling correlated defaults in an intensity based framework is a direct generalization of the single issuer intensity model as introduced in section 2.3.4. To start, we assume the existence of m default stopping times T_i^d, $i = 1, ..., m$, and default indicator functions $H_i(t)$, $i = 1, ..., m$, which admit intensities $\lambda_i(t)$, $i = 1, ..., m$. To construct the stopping times we use predictable nonnegative processes λ_i such that $\int_0^t \lambda_i(s)\,ds < \infty$ $\mathbf{P} - a.s.$ If we define

$$T_i^d = \inf\left\{t \geq 0 : \int_0^t \lambda_i(s)\,ds = \Lambda_i\right\},$$

where Λ_i is an exponentially distributed random variable with mean 1, independent of

$$Y_i(t) = E^{\mathbf{P}}\left[\exp\left(-\int_t^T \lambda_i(s)\,ds\right)\middle|\mathcal{F}_t\right],$$

then λ_i is the intensity of T_i^d and $\Delta Y_i(T_i^d) = 0$ a.s. Furthermore, we assume that Λ_i is independent of Λ_j for $i \neq j$. By equations (4.1) and (4.2) the default correlation over the time interval $[0, t]$ is given by

$$\begin{aligned}
&\rho_{i,j}^{d}(t) \\
&= \left(E^{\mathbf{P}}\left[H_i(t) \middle| \mathcal{F}_0 \right] + E^{\mathbf{P}}\left[H_j(t) \middle| \mathcal{F}_0 \right] - E^{\mathbf{P}}\left[H_{i,j}(t) \middle| \mathcal{F}_0 \right] \right. \\
&\quad \left. - E^{\mathbf{P}}\left[H_i(t) \middle| \mathcal{F}_0 \right] E^{\mathbf{P}}\left[H_j(t) \middle| \mathcal{F}_0 \right] \right) / \\
&\quad \sqrt{E^{\mathbf{P}}\left[H_i(t) \middle| \mathcal{F}_0 \right] \left(1 - E^{\mathbf{P}}\left[H_i(t) \middle| \mathcal{F}_0 \right]\right)} \\
&\quad \sqrt{E^{\mathbf{P}}\left[H_j(t) \middle| \mathcal{F}_0 \right] \left(1 - E^{\mathbf{P}}\left[H_j(t) \middle| \mathcal{F}_0 \right]\right)}.
\end{aligned} \tag{4.3}$$

To show how to calculate this default correlation we introduce a technical lemma first:

Lemma 4.3.1.
Suppose that the default time T_i^d admits an intensity process $\lambda_i(t)$, $i = 1, ..., m$, and that $\mathbf{P}\left(T_i^d = T_j^d\right) = 0$ for $i \neq j$. Then $\lambda_i(t) + \lambda_j(t)$ is an intensity process for $T_{i,j}^d = \min\left(T_i^d, T_j^d\right)$.

Proof. Because $\mathbf{P}\left(T_i^d = T_j^d\right) = 0$,

$$H_{i,j}(t) = H_i(t) + H_j(t), \ t \leq T_{i,j}^d.$$

According to section 2.3.4 there exist martingales defined by

$$M_i(t) = H_i(t) - \int_0^{T_i^d \wedge t} \lambda_i(s)\, ds = H_i(t) - \int_0^t \lambda_i(s)\, 1_{\{s \leq T_i^d\}} ds.$$

Then define a process M by

$$M_t = H_{i,j}(t) - \int_0^t \left(\lambda_i(s) + \lambda_j(s)\right) 1_{\{s \leq T_{i,j}^d\}} ds.$$

Note, that

$$M_t = M_i(t) + M_j(t) \ \text{ for } t \leq T_{i,j}^d$$

and

$$M_t = M_{T_{i,j}^d} \ \text{ for } t > T_{i,j}^d.$$

Hence, M is also a martingale and $\lambda_i(t) + \lambda_j(t)$ is the intensity process for $T_{i,j}^d$.

Proposition 4.3.1.
Given the assumptions specified above, we have for $i = 1, ..., m$,

$$\begin{aligned}
E^{\mathbf{P}}\left[H_i(t) \middle| \mathcal{F}_0 \right] &= \mathbf{P}\left(T_i^d \leq t \middle| \mathcal{F}_0 \right) \\
&= 1 - E^{\mathbf{P}}\left[\exp\left(- \int_0^t \lambda_i(s)\, ds \right) \middle| \mathcal{F}_0 \right],
\end{aligned} \tag{4.4}$$

and for $i \neq j$, $i, j = 1, ..., m$,

$$\begin{aligned}
&E^{\mathbf{P}}\left[H_{i,j}(t)\middle|\mathcal{F}_0\right] \\
&=\mathbf{P}\left(T_{i,j}^d \le t\middle|\mathcal{F}_0\right) \\
&=1-E^{\mathbf{P}}\left[\exp\left(-\int_0^t \left(\lambda_i(s)+\lambda_j(s)\right)ds\right)\middle|\mathcal{F}_0\right].
\end{aligned} \tag{4.5}$$

Therefore, default correlation (see equation (4.3)) is completely determined by the individual default intensities.

Proof. For the proof of equation (4.4) see section 2.3.4. For the proof of equation (4.5) apply lemma 4.3.1 to equation (4.4).

Various intensity based models can be distinguished for their specific choice of the state variables and the processes they follow. In case of the most important class of intensity based models, namely the affine jump diffusion intensity models, the generalization is as follows. Suppose we want to model the joint default behavior of m issuers. Then we denote the issuer specific intensity by $\lambda_i(t)$, $i = 1, ..., m$, where $\lambda_i(t)$ is an affine function of some factors X i.e.

$$\lambda_i(t) = k_i(t) + l_i(t) \cdot X_t, \; i = 1, ..., m,$$

for bounded continuous functions $k_i : [0, \infty) \to \mathbb{R}$, $l_i : [0, \infty) \to \mathbb{R}^n$, where X are strong $n-$dimensional Markov process valued in $\mathbb{R}^n$ uniquely solving the stochastic differential equation

$$dX_t = \mu(X_t, t)\,dt + \sigma(X_t, t)\,dW_t + dZ_t.$$

W is an $\mathbf{F}-$adapted $n-$dimensional standard Brownian motion in $\mathbb{R}^n$, $\mu : \Theta \to \mathbb{R}^n$, $\sigma : \Theta \to \mathbb{R}^{n \times n}$, $\Theta \subset \mathbb{R}^n \times [0, \infty)$, Z is a pure jump process whose jump-counting process A_t admits a stochastic intensity $\gamma : \Theta \to [0, \infty)$ depending on X with a jump-size distribution ν only depending on t and for each t, $\{x : (x, t) \in \Theta\}$ containing an open subset of $\mathbb{R}^n$. Note that the components of X are independent processes that consist of three different types: components governing the common aspects of performance (e.g., components that are directly related to the state of the business cycle), components that are sector specific and components that are obligor specific (e.g., book to market and leverage ratios). E.g., suppose there are three issuers $i = 1, 2, 3$ and six state variables, i.e. $n = 6$. All of the issuers belong to the same region, issuer 1 and 2 belong to the financial sector, issuer 3 belongs to the chemical sector. Then one possible model would be

$$\begin{aligned}
\lambda_1(t) &= k_1(t) + l_{1,1}(t)\,X_1(t) + l_{1,4}(t)\,X_4(t) + l_{1,6}(t)\,X_6(t), \\
\lambda_2(t) &= k_2(t) + l_{2,2}(t)\,X_2(t) + l_{2,4}(t)\,X_4(t) + l_{2,6}(t)\,X_6(t), \\
\lambda_3(t) &= k_3(t) + l_{3,3}(t)\,X_3(t) + l_{3,5}(t)\,X_5(t) + l_{3,6}(t)\,X_6(t),
\end{aligned}$$

where $X_i(t)$, $i = 1, 2, 3$, is the issuer specific state variable, $X_4(t)$ is the financial sector specific state variable, $X_5(t)$ is the chemical sector specific state variable and $X_6(t)$ is the common state variable.

Finally, we want to show how to simulate the default times T_i^d, $i = 1, ..., m$ given the intensity processes $\lambda_i(t)$, $i = 1, ..., m$. For simplicity, we assume $\mathbf{P}(T_i^d = T_j^d) = 0$ for $i \neq j$. The so called multi compensator simulation[2] is as follows. Assume that the compensator $\int_0^t \lambda_i(s)\, ds$ can be simulated for each $t \geq 0$ and $i = 1, ..., m$.

1. Simulate m independent unit-mean exponentially distributed random variables $\Lambda_1, ..., \Lambda_m$.
2. For each i, if $\int_0^T \lambda_i(s)\, ds < \Lambda_i$, then $T_i^d > T$.
3. Otherwise, let $T_i^d = \min\left\{t \geq 0 : \int_0^t \lambda_i(s)\, ds = \Lambda_i\right\}$.

Two approaches try to extend the intensity-based approach explained above in order to reach stronger default dependencies. In the infectious defaults model by Davis & Lo (1999) and Jarrow & Yu (2000) default intensities jointly jump upwards by a discrete amount at the onset of a credit crisis. Duffie (1998c) and Kijima (2000) introduce separate point processes that trigger joint default events at which several obligors default at the same time.

4.4 Correlation and Copula Functions

As in the previous sections we consider a set of m obligors $1, ..., m$. We denote the distribution function of T_i^d (which is as usual the default time of obligor i) by F_i. Then we have

$$F_i(t) = \mathbf{P}\left(T_i^d \leq t\right), \text{ for all } t \geq 0.$$

To make sure that a quantile function $F_i^{-1}(x)$ exists on $[0,1]$ we assume that $F_i(t)$ satisfies the following conditions:

$$F_i(0) = 0 \text{ and } \lim_{t \to \infty} F_i(t) = 1.$$

If U is a standard uniformly distributed random variable, i.e. $U \sim U(0,1)$, then $F_i^{-1}(U) \sim F_i$. We denote the joint distribution of $T_1^d, ..., T_m^d$ by F, i.e.

$$F(t_1, ..., t_m) = \mathbf{P}\left(T_1^d \leq t_1, ..., T_m^d \leq t_m\right), \text{ for all } t_1, ..., t_m \geq 0.$$

In addition, we consider dependent standard uniformly distributed random variables $U_1, ..., U_m$ and denote their joint distribution by C, i.e.

$$C(u_1, ..., u_m) = \mathbf{P}(U_1 \leq u_1, ..., U_m \leq u_m).$$

[2] For alternative default time algorithms see, e.g., Duffie & Singleton (1998a).

Definition 4.4.1.
A copula function C is a multivariate distribution function such that its marginal distributions are standard uniform. If considered in $\mathbb{R}^{\mathrm{m}}$

$$C : [0,1]^m \rightarrow [0,1] .$$

Given this so called copula function C and the marginal distribution function C it can be shown[3] that the joint distribution of $T_1^d, ..., T_m^d$ can be expressed as

$$F(t_1, ..., t_m) = C(F_1(t_1), ..., F_m(t_m)) .$$

The converse is also true[4]: If F is a m-dimensional multivariate distribution function with marginal continuous distributions $F_1, ..., F_m$. Then there exists a unique copula function C such that

$$F(t_1, ..., t_m) = C(F_1(t_1), ..., F_m(t_m)) .$$

The copula links marginal and joint distribution functions and separates the dependence between random variables and the marginal distributions $F_1, ..., F_m$. This greatly simplifies the estimation problem of a joint stochastic process for a portfolio with many issuers. Instead of estimating all the distributional parameters simultaneously, we can estimate the marginal distributions separately from the joint distribution. Given the marginal distributions, we use the appropriate copulas to construct the joint distribution with a desired correlation structure. The best copula can be found by examining the statistical fit of the different copulas to the data. There are basically two different approaches to using copulas in credit risk modeling:

- Given a copula, we can choose different marginal distributions for each issuer. By changing the types of the marginal distributions and their parameters, we can examine how the individual default affects the joint default behavior of many issuers in a credit portfolio.
- Given marginal distributions, we can vary the correlation structures by choosing different copulas, or the same copula with different parameter values. It therefore enables us to quantify the effects of default correlations on a credit portfolio.

There is a lively debate over the right choice of copula in the credit risk context. One of the simplest and widely used copulas is the normal copula.

Example 4.4.1.
Let the distribution function of the random vector $(X_1, ..., X_m)$ be given by the distribution function of the $m-$dimensional normal distribution $N_\rho^m(x_1, ..., x_m)$, then for $(U_1, ..., U_m)$ with $U_i = \Phi(X_i)$, $i = 1, ..., m$:

[3] See, e.g., Bluhm, Overbeck & Wagner (2003).
[4] See, e.g., Sklar (1959).

$$C_\rho(u_1,...,u_m) = N_\rho^m\left(\Phi^{-1}(u_1),...,\Phi^{-1}(u_m)\right)$$

and the random default times can be obtained by $T_i^d = F_i^{-1}(\Phi(X_i))$, $i = 1,...,m$. As shown, e.g., in Li (2000) to make the joint default probabilities of the Merton model coincide with those of the normal copula the asset correlations have to match the correlations of the normal copula. Then $\rho = (\rho_{i,j})_{i,j=1,...,m}$ denotes the matrix of asset correlations between credits i and j. To summarize, to generate correlated default times via the normal copula approach we have to proceed as follows:

- For each obligor $i = 1,...,m$ specify the cumulative default time distribution $F_i(t)$ (credit curve).
- Assign a standard normal random variable X_i, $i = 1,...,m$, to each obligor, where the correlation between X_i and X_j equals $\rho_{i,j}$ for each $i \neq j$.
- Obtain the default time T_i^d through $T_i^d = F_i^{-1}(\Phi(X_i))$, $i = 1,...,m$.

The popularity of the normal copula in credit risk portfolio modeling is due to this relation to the asset based model of Merton (1974) and the fact that it can be implemented easily.

Example 4.4.2.
Whereas the normal copula is directly linked to Merton's approach and is easy to implement, the Student's t copula better accounts for extreme event risk. Let $\Gamma_{\rho,\upsilon}$ be the standardized multivariate Student's t distribution with υ degrees of freedom and correlation matrix ρ. The multivariate Student's t copula is then defined as follows

$$C_{\rho,\upsilon}(u_1,...,u_m) = \Gamma_{\rho,\upsilon}\left(t_\upsilon^{-1}(u_1),...,t_\upsilon^{-1}(u_m)\right),$$

where t_υ^{-1} is the inverse of the univariate Student's t distribution.

Example 4.4.3.
The Gumbel copula can be expressed as follows

$$C(u_1,...,u_m) = \exp\left[-\left(\sum_{i=1}^m (-\ln u_i)^\alpha\right)^{\frac{1}{\alpha}}\right],$$

where α is the parameter determining the tail of the distribution.

Example 4.4.4.
The Clayton copula is defined as follows

$$C(u_1,...,u_m) = \exp\left[\sum_{i=1}^m u_i^{-\alpha} - 1\right]^{-\frac{1}{\alpha}},$$

where $\alpha > 1$ is the parameter determining the tail of the distribution.

For other copulas we refer to Nelsen (1999). Simulating $m-$dimensional random vectors from these copulas is discussed, e.g., by Wang (2000), Bouye, Durrelmann, Nikeghbali, Riboulet & Roncalli (2000), and Embrechts, Lindskog & McNeil (2001). How to apply the concept of copulas to standard credit risk problems is explained, e.g., by Frees & Valdez (1997), Li (2000), and Frey, McNeil & Nyfeler (2001).

5. Credit Derivatives

"We already know [...] that there has been an irreversible application of risk management technology without which banks would not be able to design, price, and manage many of the newer financial products, like credit derivatives. These same or similar technologies can and are beginning to be used to price and manage traditional banking products."
- Alan Greenspan -

"In my experience, securitization was born to a greater extent out of a need for capital than out of a desire for a lower price of capital "
- S.P. Baum -

5.1 Introduction to Credit Derivatives

Credit derivatives will lead to a revolution in banking. One is definitely true: the volume of the credit derivatives market has increased enormously (see, e.g., table 5.2) since credit derivatives have been first publicly introduced in 1992, at the International Swaps and Derivatives Association annual meeting in Paris, and the end of this tendency is not about to come. Much of the growth in the credit derivatives market has been aided by the growing use of the LIBOR swap curve as an interest rate benchmark. As it represents the rate at which AA-rated commercial banks can borrow in the capital markets, it reflects the credit quality of the banking sector and the cost at which they can hedge their credit risks. It is, therefore, a pricing benchmark. It is also devoid of the idiosyncratic structural and supply factors that have distorted the shapes of the government bond yield curves in a number of important markets. Also much of the growth of the credit derivatives market would not have been possible without the development of models for the pricing and management of credit risk. There is an increasing sophistication in the market as market participants have developed a more quantitative approach to analyzing credit. Electronic trading platforms such as CreditTrade (www.credittrade.com) and derivatives online CreditEx (www.creditex.com) appeared in the market. Both have proved successful and have had a significant impact in improving price discovery and liquidity in the credit derivative market.

Credit derivatives are probably one of the most important types of new financial products introduced during the last decade. Meanwhile, there is a great variety of products: The most important ones are credit options (section 6.4.2), credit spread products (section 6.4.3), default swaps/options (section 6.4.4), total rate of return swaps (section 5.3.6), and credit-linked notes (section 5.3.7). Credit derivatives can be based on a broad set of credit sensitive underlyings:

- Derivatives based on sovereign bonds: E.g., Brady bond derivatives allow investors to take a view on the narrowing of sovereign credit spreads.
- Derivatives based on corporate bonds: E.g., derivatives based on a single company's bond are often used for liability management.
- Derivatives based on bank loans: E.g., derivatives based on loans are used by banks to diversify or neutralize concentrations of issuer specific risk and offer investors favorable risk-return profiles.

Recent estimates show corporates accounting for just over 50% of the market, with the remainder split roughly equally between banks and sovereign credits. Traditionally, exposure to credit risk was managed by trading in the underlying asset itself. Or even worse: when a loan was made in the past, the credit risk remained on the lender's balance sheet till the obligation matured, was paid or charged off. Like other derivatives, credit derivatives facilitate the separate trading of individual attributes of an underlying asset in isolation from the asset itself. They have been developed for transferring, repackaging, replicating and hedging credit risk. They can change the credit risk profile of an underlying asset by isolating specific aspects of credit risk without selling the asset itself. This is even possible for highly illiquid instruments which couldn't have been traded regularly otherwise. Hence, they allow credit risk to be disaggregated and traded. Our definition of credit risk encompasses all credit-related events ranging from a spread widening, through a ratings downgrade, all the way to default. Banks in particular are using credit derivatives to hedge credit risk, reduce risk concentrations on their balance sheets, and free up regulatory capital in the process. Their payoff usually depends on the credit risk of a particular counterparty or a group of counterparties, or is based on the credit performance of some credit-sensitive instrument. Credit performance is measured by yield spreads relative to benchmarks (e.g., basis points over Treasuries or LIBOR), credit ratings or default status. And as Masters (1998), global head of credit derivatives marketing at J. P. Morgan, points out:

> "In bypassing barriers between different classes, maturities, rating categories, debt seniority levels and so on, credit derivatives are creating enormous opportunities to exploit and profit from associated discontinuities in the pricing of credit risk."

In addition, credit derivatives

- have a market-wide impact. They
  - have encouraged more participants in many markets such as high-yield debt, bank loans and credit enhancement products
  - have helped to make pure credit risk markets more liquid and well-developed
  - have enabled more efficient pricing across several market sectors
  - have led to a reduction in credit spreads
  - have improved institutional portfolio diversification
- allow banks maintaining banking relationships and diversification at the same time
- create new arbitrage opportunities
- allow investments based on credit views
- can be used for
  - risk protection, exchange of one set of risks for another, and speculation
  - offsetting over-the-counter counterparty exposure
  - loan portfolio hedging
  - extending lines of business
  - changing the maturity or currency of credit risks
  - getting access to classes of assets in which an institution may otherwise find it difficult to invest and trade

Example 5.1.1.
During Asia's debt crisis at the end of the 1990's many banks were offering short-term protection (by pitching short-term credit default swaps) on some Asian corporate bonds to entice buyers back into Asia's fixed-income markets. There was the view that if a corporate could survive the upcoming year, it would most probably survive in the long term. Short-term default swaps shortened the yield curve out to a year, while leaving the underlying corporate bond unhedged for later credit events. To get default protection investors paid a high premium for the first year of exposure to the Asian corporate bond, but in return they could lock in enormously high yields (up to 1200 basis points over US$ LIBOR) for the rest of the life of the bond.

The wide variety of applications of credit derivatives attracts a broad range of market participants. Historically, banks have dominated the market as the biggest hedgers, buyers, and traders of credit risk. Over time, new players have been entering the market. This observation is supported by the results of the BBA survey, which produces a breakdown of the market by the type of participant.

Not very surprisingly, the volumes traded in the global credit derivatives market have enormously risen during the last years. Based on an Canadian

Table 5.1. Participants in the credit derivatives market.

Counterparty	Protection Buyer (%)	Protection Seller (%)
Banks	63	47
Securities Firms	18	16
Insurance Companies	7	23
Corporations	6	3
Hedge Funds	3	5
Mutual Funds	1	2
Pension Funds	1	3
Government Credit Agencies	1	1

Source: (*BBA Credit Derivatives Report 1999/2000* 2000).

Imperial Bank of Commerce (CIBC) estimate[1], the credit derivatives market volume was about US$ 5 billion and US$ 40 billion in 1994 and 1995, respectively. According to a J.P. Morgan estimate[2] the outstanding volume doubled in 1996 again. A March 1998 RISK magazine's survey (Green, Locke & Paul-Choudhury 1998) estimates the global volumes ranging from US$ 140 billion to US$ 325 billion in 1997. The March 1999 RISK magazine's survey (Baldwin 1999) gives an average figure of all outstanding credit derivatives of US$ 477 billion, and in the RISK survey published in March 2000 (Hargreaves 2000) this figure has increased by 39% to US$ 661 billion. For the year 2000 RISK (Patel 2001) reports a total notional amount of outstanding contracts around US$ 810 billion. The British Bankers' Association's (BBA) credit derivatives surveys[3] of the years 1997/1998 and 1999/2000 show even higher numbers which are summarized in figure 5.1. The BBA numbers were derived by polling international member banks through their London office and asking about their global credit derivatives business. Given that almost all of the major market participants have a London presence, the overall numbers should, therefore, be representative of global volume. Nevertheless, as

[1] See, e.g., Wilson (1998), page 225, exhibit 5.

[2] See, e.g., Wilson (1998), page 225, exhibit 5.

[3] According to *BBA Credit Derivatives Report 1999/2000* (2000) over 30 institutions from Australia, Canada, France, Germany, Italy, Japan, Netherlands, Switzerland, United Kingdom and United States submitted detailed questionnaires to the British Bankers' Association. 85% of them considered themselves as Intermediary/market makers. The survey used the following definiton of credit derivatives:

> "Any instrument that enables the trading/management of credit risk in isolation from the other types of risk associated with an underlying asset. These instruments may include: credit default products, credit spread products, total return products, basket products, credit linked notes and asset swaps." (*BBA Credit Derivatives Report 1999/2000* (2000), page 6)

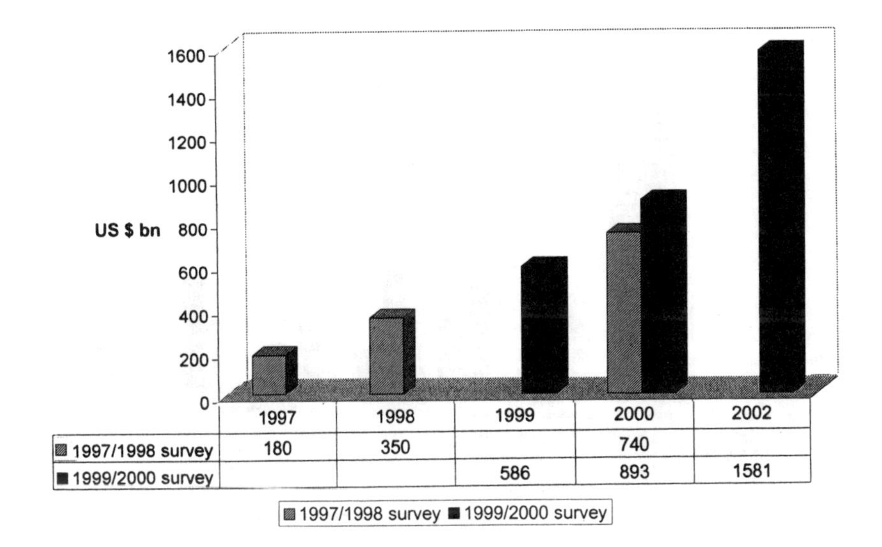

Fig. 5.1. Global credit derivatives market size. Source: (*BBA Credit Derivatives Report 1999/2000* 2000).

the numbers are based on interviews and estimations, they should be treated as indicative estimates rather than hard numbers.

Not only the market size is increasing but also the range of products used. Although standard credit default products still account for the largest market share by far, more and more hybrid products are being developed. Figure 5.2 shows the proportions of the global derivative market which is accounted for by the specific products.

According to figure 5.3, currently the biggest credit derivatives market is Europe, as the euro has helped to develop European debt markets. In less developed derivatives markets, due to the Asian and Russian crises, the use of credit derivatives has fallen from 34% in 1998 to 5% in 1999, but has recovered again to 18% in 2000. In emerging markets the fastest growing geographical sector is central and eastern Europe. The region accounts for 11% of the global market in 2000. Hungary, for example, is the region's most active market for credit derivatives and also has the greatest number of liquid bonds.

Besides such estimates, there are only a few official credit derivatives volumes figures reported yet. The US Office of the Comptroller of the Currency began publishing US commercial banks' credit derivatives volumes in 1997.

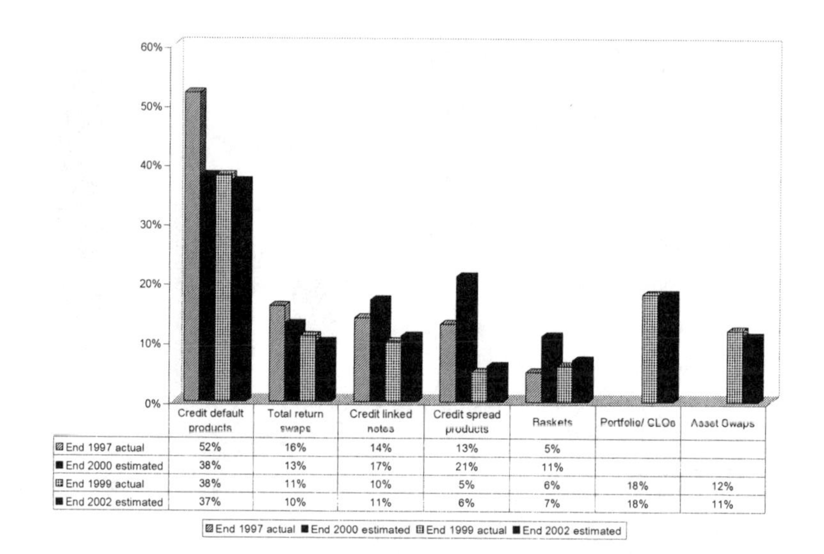

	Credit default products	Total return swaps	Credit linked notes	Credit spread products	Baskets	Portfolio/ CLOs	Asset Swaps
End 1997 actual	52%	16%	14%	13%	5%		
End 2000 estimated	38%	13%	17%	21%	11%		
End 1999 actual	38%	11%	10%	5%	6%	18%	12%
End 2002 estimated	37%	10%	11%	6%	7%	18%	11%

Fig. 5.2. Credit derivatives products market share. Source: (*BBA Credit Derivatives Report 1999/2000* 2000), page 13.

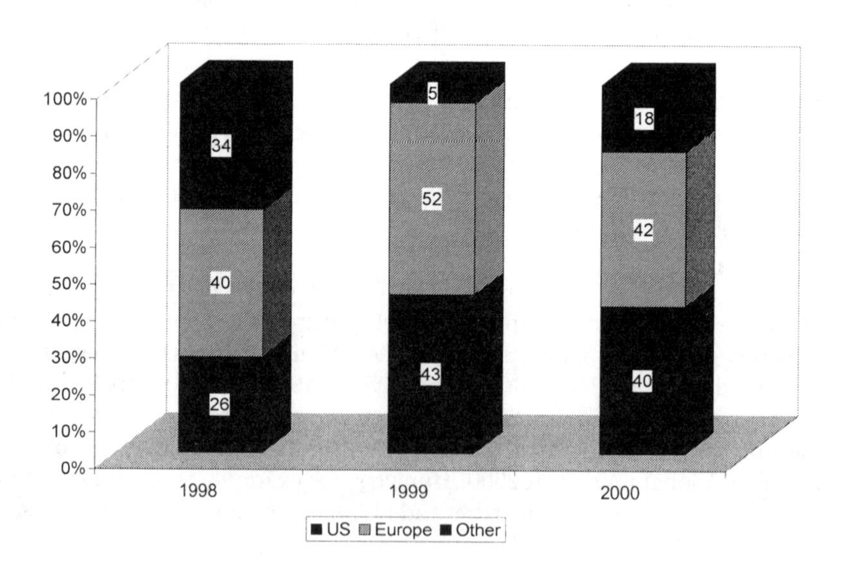

Fig. 5.3. Estimated geographical distribution of credit derivatives market 1998-2000. Source: (Hargreaves 2000).

The numbers are based on "call reports" filed by U.S.-insured banks and foreign branches and agencies in the United States. Unlike the BBA survey, it is based on hard figures. However it does not include investment banks, insurance companies or investors. Table 5.2 shows the absolute outstanding notional amounts in billion US$ and the number of US banks that had outstanding credit derivatives positions. The notional amounts have grown from less than US$ 20 billion in the first quarter of 1997 to about US$ 426 billion in the fourth quarter of 2000. The main reasons for the enormous credit derivatives' growth are a much greater awareness of these products, the better understanding and wider usage of risk management techniques and an increasing market liquidity. End-users of credit derivatives[4] are banks (in 2000: 64%), reinsurance companies (11%), corporates (8%), asset managers (7%), hedge funds (5%), insurance companies (3%), and special purpose vehicles (2%).

The main problem of credit derivatives is that they still suffer from price-discovery problems. James (1999), vice president and head of research, strategic risk management advisory, at First National Bank of Chicago in London, explains:

> "Perhaps the closest analogy with the credit derivatives market today is the options market some 25 years ago, shortly before the Black-Scholes pricing model was developed. The credit derivatives market seems to be waiting for a similar revolution. There is a feeling that just around the corner lies some "Black-Scholes model for credit derivatives" and, after it is found, pricing will become simple."

There are basically three approaches used by market participants to price credit derivatives today:

- Pricing based on guarantee product markets:
  The amount by which a guarantor should be compensated is determined by looking to other forms of credit enhancement as a reference. For example, if counterparty A is willing to pay counterparty B 100 basis points annually to guarantee the debt of counterparty C, a credit default swap should be priced similarly as it provides more or less the same form of protection. The insurance community, e.g., offers a variety of traditional insurance or guarantee products priced by using historical default tables. Of course, this approach can't be used for all types of credit derivatives, and doesn't take into account whatever specific market information about default may exist.
- Pricing based on replication and funding costs:
  This approach prices credit derivatives based on the price of the positions necessary to hedge the derivative contract and how much it costs to enter in these positions. The net hedging cost plus reserves and the dealer's profit

[4] See, e.g., Patel (2001).

Table 5.2. US credit derivatives market (notional amounts in US$ billions).

	Notional Amounts ($Millions)	Number of Banks
1st Qtr 97	19.139	504
2nd Qtr 97	25.649	463
3rd Qtr 97	38.880	475
4th Qtr 97	54.729	459
1st Qtr 98	91.419	451
2nd Qtr 98	129.202	461
3rd Qtr 98	161.678	464
4th Qtr 98	144.006	447
1st Qtr 99	190.990	439
2nd Qtr 99	210.351	435
3rd Qtr 99	234.131	426
4th Qtr 99	286.584	416
1st Qtr 00	301.913	389
2nd Qtr 00	361.685	416
3rd Qtr 00	378.585	421
4th Qtr 00	425.848	400
1st Qtr 01	352.405	395
2nd Qtr 01	351.235	367
3rd Qtr 01	359.548	359
4th Qtr 01	394.797	369
1st Qtr 02	437.532	379
2nd Qtr 02	492.263	391
3rd Qtr 02	572.784	408
4th Qtr 02	634.814	427
1st Qtr 03	709.928	488
2nd Qtr 03	802.306	530

Source: Office of the Comptroller of the Currency bank call reports. Credit derivatives data were being reported in the call reports in 1997 for the first time. Statistics include all US commercial banks holding derivatives. Call reports do not differentiate credit derivatives by contract type.

and loss is the price of the credit derivative. The strength of this approach is that it is easy to use and very intuitive. If a hedge can be constructed, the result is most useful for a dealer, but for many structures, a complete hedge is not available, or would be too costly.

- Pricing based on stochastic models such as structural, intensity based or hybrid models:
  Although the structural models have intuitive appeal, they are not widely used in trading rooms. Intensity based models are quite common in dealing rooms now. In the following we will apply our three-factor term structure model to the pricing of credit derivatives.

Most of the papers in the field of credit/default risk pricing are concerned with the pricing of defaultable bonds. There still have been only a few works especially on the pricing of credit derivatives. Das (1995) basically shows that in an asset based framework credit options are the expected forward values of put options on defaultable bonds with a credit level adjusted exercise price. Longstaff & Schwartz (1995a) develop a pricing formula for credit spread options in a setting where the logarithm of the credit spread and the non defaultable short rate follow Vasicek processes. Das & Tufano (1996) apply their model, which is an extension to stochastic recovery rates of the model of Jarrow et al. (1997), to the pricing of credit-sensitive notes. Das (1997) summarizes the pricing of credit derivatives in various credit risk models (e.g., the models of Jarrow et al. (1997), and Das & Tufano (1996)). All models are presented in a simplified discrete fashion. Duffie (1998a) uses simple no arbitrage arguments to determine approximate prices for default swaps. Hull & White (2000) provide a methodology for valuing credit default swaps when the payoff is contingent on default by a single reference entity and there is no counterparty default risk. Schönbucher (2000) develops various pricing formulas for credit derivatives in the intensity based framework.

5.2 Technical Definitions

Definition 5.2.1.

- *A credit derivative is a derivative security that has at least one payoff that is conditioned on the occurrence of a credit event (default payment). The credit event is defined with respect to a reference credit or a basket of reference credits and the reference credit assets issued by the reference credit. The maturity date of a credit derivative is deemed to be which ever is earlier - the originally agreed maturity date or the credit event related payment date. In some cases there is a pre-agreed time interval between the credit event and the payout calculation. Credit derivatives can be either quoted in terms of an up-front fee, that is payable at the beginning of the contract, or as a fee, that has to be paid in regular intervals until default or maturity.*
- *A credit event is a precisely defined default event, determined by negotiation between the parties at the outset of a credit derivative. Market standards typically specify the existence of publicly available information confirming the occurrence, with respect to the reference credit, of bankruptcy, insolvency, repudiation, restructuring, moratorium, failure-to-pay, cross-default or cross-acceleration. There is some risk of disagreement over whether the credit event has in fact occurred. For pricing issues we will ignore this enforceability risk.*
- *A reference credit is an issuer whose default triggers the credit event.*
- *The reference credit assets are a set of assets issued by the reference credit.*

- *Default payments are the payments which must be made in the case of a credit event.*

The key issues related to credit derivatives transactions are the exact definitions of the credit event and the credit event payout or settlement mechanism. Usually the payout mechanism is based on the market bid price of the reference credit asset at the declaration of the credit event, or it is a binary payout (i.e. there is a pre-determined amount of cash payout), or there is a physical sale or delivery of the defaulted reference credit asset at a pre-arranged price. If the reference credit asset is a loan, it is usually very difficult to obtain a market price, and binary payments are more common. In most cases these payouts are based on pre-agreed assessments of the potential changes in price of the reference loans. Naturally, this may expose the credit protection buyer to basis risk between the payout and the actual change in price until and including the credit event. In addition, binary payouts may expose the counterparties to additional tax payments. Although physical settlement mechanisms may overcome some of the problems of the other two forms of payout mechanisms, there may be other risks like settlement risk (the credit protection seller may not be long the reference credit asset at the time of the credit event) or there may be a ban on the delivery of the reference credit asset after the credit event. This may, e.g., happen if there is a court-imposed freeze on the reference credit assets of the defaulted reference credit. In the case of loans it may take a long time to transfer these, due to their administrative complexity, which often includes a requirement that the lender secures the consent of the borrower to reassign its loan. If a settlement period is not exactly specified, the protection seller may not know when it will receive the asset and therefore will be unable to estimate and hedge its risk. In addition, many credit derivatives bear credit risk by themselves. In the following we assume that we can neglect this risk.

5.3 Single Counterparty Credit Derivatives

For pricing single counterparty credit derivatives the models explained in chapters 2 and 3 can be used. In section 6.4 we develop specific pricing formulas for a couple of credit derivatives in a three-factor defaultable term structure framework.

5.3.1 Credit Options

Definition 5.3.1.
A credit option is a put or call option on the price of either

1. *a floating rate note, fixed rate bond, or loan,*

2. *an asset swap package, consisting of a credit-risky instrument with any payment characteristics and a corresponding derivative contract that exchanges the cash flows of that instrument for a floating rate cash flow stream, typically three or six month LIBOR plus a spread.*

Settlement may be on a cash or physical basis.

Definition 5.3.2.

1. *A European call option with maturity $T_O \leq T$ and exercise price K on a defaultable zero-coupon bond $P^d(t,T)$ pays off*

$$1_{\{T_O < T^d\}} \left(P^d(T_O, T) - K \right)_+,$$

if the call is knocked out at default.

2. *A European put option with maturity $T_O \leq T$ and exercise price K on a defaultable zero-coupon bond $P^d(t,T)$ pays off*

$$1_{\{T_O < T^d\}} \left(K - P^d(T_O, T) \right)_+,$$

if the put is knocked out at default.

For quite a big number of liquid bonds, such as some of the Latin American Brady bonds and bonds of large US corporates, there is a well-developed bond option market. Credit options offer protection against changing default free interest rates and changing credit spreads. Hence, they allow hedging of mark-to-market exposure to fluctuations in spreads and interest rates: hedging long positions with puts, and short positions with calls. For investors they are a good source of yield enhancement. E.g., a retail investor who is long the underlying bond can sell a bond option to enhance his yield. Options on fixed-rate bonds can be used to express a view on the credit spread of an issuer, interest rate movements in the currency of denomination of the bond, and interest rate and credit spread volatility. If investors take a view on volatility by using the bond to delta hedge a long option position, they are long volatility and can profit from increased uncertainty about the reference credit, as expressed through the volatility of the bond yield. Such a strategy is indifferent to the direction of movement of the yield. Bond options allow for correlation plays on movements between interest rates and credit spreads. For many investors, it is cheaper to buy a call option on the bond rather than fund the bond on balance sheet. Borrowers can lock in future borrowing costs without inflating their balance sheet by entering into credit options on their own name. They can buy the right to put their own paper to a dealer at a pre-agreed spread. Although physical or cash settlement is possible, most of the time bond options are physically settled.

For credit option pricing formulas see section 6.4.2.

5.3.2 Credit Spread Products

Definition 5.3.3.

1. *A credit spread call option on a defaultable bond $P^d(t,T)$ with maturity $T_O < T$ and strike spread K gives the holder the right to buy the defaultable bond at time T_O at a price that corresponds to a yield spread of K above the yield of an otherwise identical non defaultable bond $P(t,T)$.*
2. *A credit spread put option on a defaultable bond $P^d(t,T)$ with maturity $T_O < T$ and strike spread K gives the holder the right to sell the defaultable bond at time T_O at a price that corresponds to a yield spread of K above the yield of an otherwise identical non defaultable bond $P(t,T)$.*

Definition 5.3.4.

1. *A European credit spread call option with maturity $T_O \leq T$ and strike spread K on a defaultable zero-coupon bond $P^d(t,T)$ pays off*
$$1_{\{T_O<T_B^d\}}\left(P^d(T_O,T) - e^{-(T-T_O)K}P(T_O,T)\right)_+,$$
if the call is knocked out at default.
2. *A European credit spread put option with maturity $T_O \leq T$ and exercise price K on a defaultable zero-coupon bond $P^d(t,T)$ pays off*
$$1_{\{T_O<T_B^d\}}\left(e^{-(T-T_O)K}P(T_O,T) - P^d(T_O,T)\right)_+,$$
if the put is knocked out at default.

A Credit Spread Option is an option contract in which the decision to exercise is based on the credit spread of the reference credit relative to some strike spread. This spread may be the yield of a bond quoted relative to a Treasury or may be a LIBOR spread. In the latter case, exercising the credit spread option can involve the physical delivery of an asset swap, a floating-rate note, or a default swap. This reference asset may be either a floating rate note or a fixed rate bond via an asset swap. As with standard options, one must specify whether the option is a call or put, the expiry date of the option, the strike price or strike spread, and whether the option exercise is European (single exercise date), American (continuous exercise period), or Bermudan style (multiple exercise dates). The option premium is usually paid up front, but can be converted into a schedule of regular payments. A call on the spread (put on the bond price), expressing a negative view on the credit, will usually be exercisable in the event of a default. In this case, it would be expected to be at least as expensive as the corresponding default swap premium. For a put on the spread (call on the bond price), expressing a positive view on the credit, the option to exercise on default is worthless and, hence, irrelevant. The strike for a credit spread option is normally quoted in terms of a spread to LIBOR. Credit spread options offer investors whose portfolio values are highly sensitive to shifts in the spread between defaultable and non-defaultable yields an important tool for managing and hedging their exposure to this type of risk. Credit spread puts enable investors who are not allowed to invest in assets of too low quality to switch out of a defaultable bond investment as soon as the credit quality of the reference credit asset falls below some boundary. They can be viewed as an exchange option that give one counterparty the right to exchange one defaultable bond for a certain number of non defaultable bonds. Depending on the specification, the option can either survive a default or be knocked out by it.

Credit spread options present an unfunded way for investors to express a pure credit view. Unlike options on fixed rate bonds the decision to exercise has no dependency on interest rates. It simply depends on where the credit spread of the reference credit is relative to the strike spread. The more volatile the credit spread, the more time-value the option will have, and the more the option will be worth. And if the investors hedge the option by trading the underlying, they will be long volatility. As a result, credit spread options allow investors to express a view about spread volatility separate from a view about the direction of the credit spread. Buying an out-of-the money put option on a bond is similar to a buying protection with a default swap with one advantage—it can be exercised even when the credit deterioration is significant but formal default has not occurred. One extension of the credit-spread option is the exchangeable asset swap option. This gives the purchaser the right but not the obligation to swap one asset swap package for another asset swap package linked to a different credit. This makes it possible for the purchaser of the option to take a view on the difference between two asset swap spreads. As the hedge fund market for credit spread products grows, we expect to see more growth and development of the credit spread option market. Investors can use credit spread options to take a view about credit spread volatility.

Because of the optionality, pricing credit-spread options requires a model for the evolution of credit spreads. For European-style options, the simplest such model is a variation on Black's model for valuing interest rate caps and floors where instead of forward rates, we model the forward credit spread at option exercise as a random variable with a lognormal distribution. For American-style options, a tree-based approach must be used to take into account the early exercise decision. Other more sophisticated approaches exist that take into account other factors such as the correlation between interest rates and credit spreads. Because the seller of the option will typically have to hedge the short option position dynamically and because the reference credit may not be highly liquid, transaction costs will also have to be factored into the price of the option.

Example 5.3.1.
As an example how credit derivative pricing formulas can be developed by applying the theory explained in the previous chapters we show the approach of Longstaff & Schwartz (1995a) in the pricing of credit spread call options. Given the credit spread s_{T_O} at maturity T_O of the option and a strike K Longstaff & Schwartz (1995a) the payoff of the option at maturity is

$$\max\left(s_{T_O} - K, 0\right).$$

Suppose that the processes for the credit spread s under the measure $\mathbf{Q}$ are given by

$$d\ln s_t = a_s\left[b_s - \ln s_t\right]dt + \sigma_s d\hat{W}_s\left(t\right),$$

where a_s, b_s, and σ_s are parameters and $\hat{W}_s$ is a standard Brownian motion. In addition, assume that the non-defaultable short-rate process r under the measure $\mathbf{Q}$ is given by

$$dr_t = a_r\left[b_r - r_t\right]dt + \sigma_r d\hat{W}_r\left(t\right),$$

where a_r, b_r, and σ_r are parameters and $\hat{W}_r$ is a standard Brownian motion. The Brownian motions $\hat{W}_s$ and $\hat{W}_r$ are correlated with correlation $\rho_{r,s}$. The price of the credit spread call option can be calculated according to

$$F_t^{\text{csc}} = E^{\mathbf{Q}}\left[e^{-\int_t^{T_O} r_l dl}\max\left(s_{T_O} - K, 0\right)\Big|\,\mathcal{F}_t\right]. \tag{5.1}$$

Taking $P\left(t,T\right)$ as numéraire and applying the change of numéraire theorem to equation (5.1) we can express F^{csc} according to

$$F_t^{\text{csc}} = P\left(t,T\right)E^{\mathbf{Q}^T}\left[\max\left(s_{T_O} - K, 0\right)\middle|\,\mathcal{F}_t\right],$$

where $\mathbf{Q}^T$ is the so called T-forward measure. Under this measure the dynamics of s are given by

$$d\ln s_t = \left[a_s b_s - a_s\ln s_t - \frac{\rho_{r,s}\sigma_r\sigma_s}{a_r}\left(1 - e^{-a_r(T_O - t)}\right)\right]dt + \sigma_s d\overline{W}_s\left(t\right),$$

where $\overline{W}_s$ is a standard Brownian motion under the measure $\mathbf{Q}^T$. This SDE implies that s_{T_O} is conditionally normally distributed (as seen at pricing time 0, i.e. conditional on $\mathcal{F}_0$) with mean

$$\begin{aligned}\mu_0 = {} & e^{-a_s T}\ln s_0 + \frac{1}{a_s}\left(a_r b_r - \frac{\rho_{r,s}\sigma_r\sigma_s}{a_r}\right)\left(1 - e^{-a_s T}\right) \\ & + \frac{\rho_{r,s}\sigma_r\sigma_s}{a_r\left(a_s + a_r\right)}\left(1 - e^{-(a_s + a_r)T}\right),\end{aligned}$$

and variance

$$\sigma^2 = \frac{\sigma_s^2\left(1 - e^{-2a_s T}\right)}{2a_s}.$$

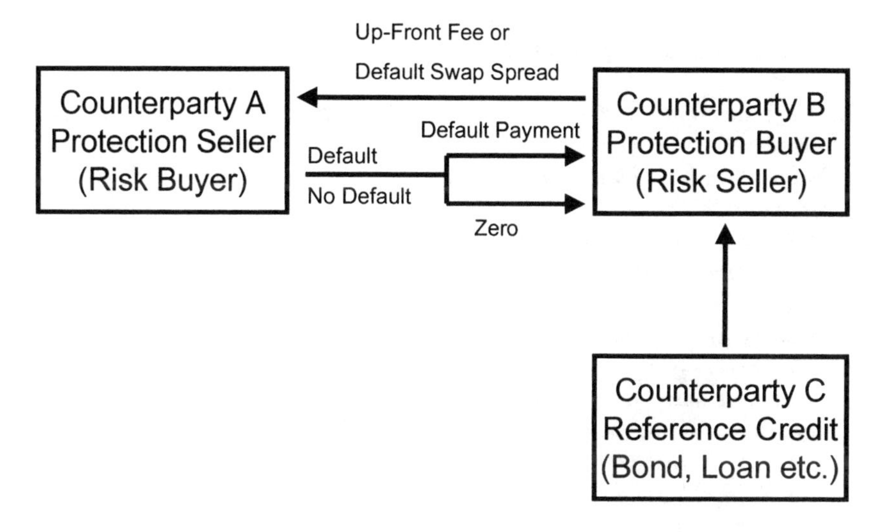

Fig. 5.4. Basic structure of a credit default swap and option.

With this framework it is easy to find that

$$F_0^{\mathrm{csc}} = P(0,T)\left(e^{\mu+\frac{\sigma^2}{2}}\Phi\left(\frac{-\log K+\mu+\sigma^2}{\sigma}\right) - K\Phi\left(\frac{-\log K+\mu-\sigma^2}{\sigma}\right)\right).$$

The value of an equivalent credit spread put option is given by

$$F_0^{csp} = F_0^{\mathrm{csc}} + P(0,T)\left(K - e^{\mu+\frac{\sigma^2}{2}}\right).$$

For further credit spread option pricing formulas see section 6.4.3.

5.3.3 Credit Default Products

The credit default option and swap have become the standard credit derivatives. Actually, they can be seen as the basic building blocks of the credit derivatives market. According to figure 5.2 they dominate the credit derivatives market. They are easy to understand and present to hedgers and investors a wide range of possibilities that did not previously exist in the cash market. Figure 5.4 shows the general structure of a default swap and option. The following definition gives an overview of the most important default swap and default option structures:

Definition 5.3.5.

1. *A default digital (put) option has a pre-agreed payoff at the time of default, e.g., expressed as a fixed percentage of the notional amount or as absolute value. The beneficiary pays the guarantor a lump-sum fee up-front.*
2. *A default digital swap has a pre-agreed payoff at the time of default, e.g., expressed as a fixed percentage of the notional amount or as absolute value. The beneficiary pays the guarantor regular fees instead of a single up-front payment.*
3. *A default option (put) is a credit derivative under which one party (the beneficiary) pays the other party (the guarantor) a fixed amount (lump-sum fee up-front). This is in exchange for the guarantor's promise to make a fixed or variable payment in the event of default in one or more reference assets to cover the full loss in default.*
4. *A default swap is a swap under which one party (the beneficiary) pays the other party (the guarantor) regular fees, amounts that are based on a generic interest rate, called the default swap spread or the default swap premium. This is in exchange for the guarantor's promise to make a fixed or variable payment in the event of default in one or more reference assets to cover the full loss in default.*
5. *A default swap with a dynamic notional is a default swap which for a fixed fee provides the protection buyer with a contingent payment that matches the mark-to-market value on any given day of a specified derivative.*
6. *A first loss default swap is a default swap whereby the protection seller commits to indemnify the protection buyer for a predefined amount of losses incurred following one or more credit events in a specified reference portfolio.*
7. *A first-to-default swap or basket swap is a default swap where the protection seller takes on exposure to the first entity suffering a credit event within a basket. The credit position in each name in the basket is typically equal to the notional of the first-to-default default swap. Losses are capped at the notional amount and the protection seller does not have exposure to subsequent credit events.*

There are still no standard default option and swap contracts, but they are very much individually negotiated. Before the trade can be executed some contract details such as the reference entity or the credit event must be clearly defined. Usually the reference entity is a corporate, a bank or a sovereign issuer. The credit event triggers the payment. It can either be bankruptcy, failure to pay, obligation acceleration, repudiation, restructuring and others[5]. The contract must clearly specify the payoff that is made following the credit event. Typically, the payoff will be the difference between par and the recovery value of the reference asset after the credit event occurred. Physical or cash

[5] These events are defined in the ISDA 1999 list of credit derivatives.

settlement is possible. In case of physical delivery the protection seller has to pay par in return for the physically delivered defaulted instrument. Note that the contract usually specifies a basket of obligations that are ranked pari passu that may be delivered in place of the reference asset. Actually, the protection buyer who has chosen physical delivery is effectively long a "cheapest to deliver" option. In case of cash settlement the protection buyer receives par minus the default price of the reference asset. The price of the defaulted asset is typically determined via a dealer poll conducted within 14-30 days of the credit event. The purpose of the delay is to let the recovery rate stabilize. In certain cases, it may not be possible to price the defaulted asset at all. For that case there may be provisions in the documentation to take the price of an equivalent asset of the same credit quality and similar maturity as substitute. Finally, in case of default digital options and swaps fixed cash settlement applies. The protection buyer in a default swap contract may stop paying the premium once the credit event has occurred. Of course, this property has to be incorporated into the cost of the default swap premium payments. It enables both counterparties to close out their positions right after the credit event and so eliminates the ongoing administrative costs that would otherwise occur. Current market standards for banks and corporates require that the protection buyer pays the accrued premium to the credit event; sovereign default swaps do not require a payment of accrued premium. The premium leg terminates at the earlier of the default swap maturity or the time of a credit event. A default swap is a par product: it does not completely hedge the loss on an asset that is currently trading away from par. If the asset is trading at a discount, a default swap overhedges the credit risk and vice-versa. This becomes especially important if the asset falls in price significantly without a credit event. To hedge this,the investor can purchase protection in a smaller face value or can use an amortizing default swap in which the size of the hedge amortizes to the face value of the bond as maturity is approached. For pricing formulas of credit default products see section 6.4.4.

5.3.4 Par and Market Asset Swaps

The asset swap market was born at the beginning of the 1990s. Asset swaps are widely used by banks to swap their long-term fixed rate assets to floating rate assets. In addition, investors can get exposure to the credit quality of bonds with minimal interest rate risk. Basically, an asset swap is a synthetic structure that allows an investor to buy a fixed rate bond and then hedge out the interest rate risk by swapping the fixed payments to floating. The investor keeps the credit exposure to the fixed rate bond. The most widely used structure is the so called par asset swap which consists of two trades:

- The asset swap buyer purchases a fixed rate bond from the asset swap seller in return for a dirty price of par (see figure 5.5).

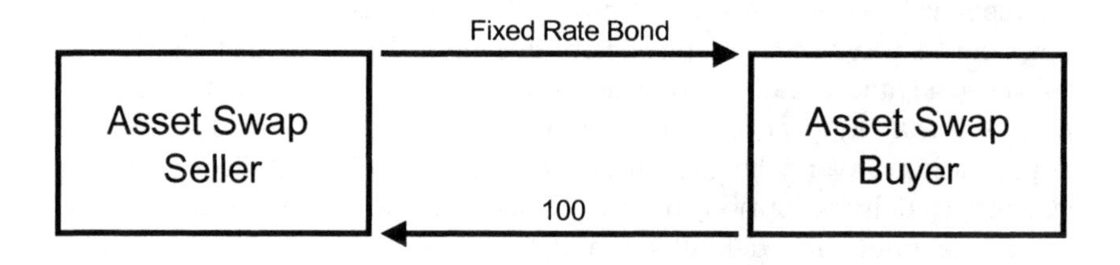

Fig. 5.5. First trade of a par asset swap.

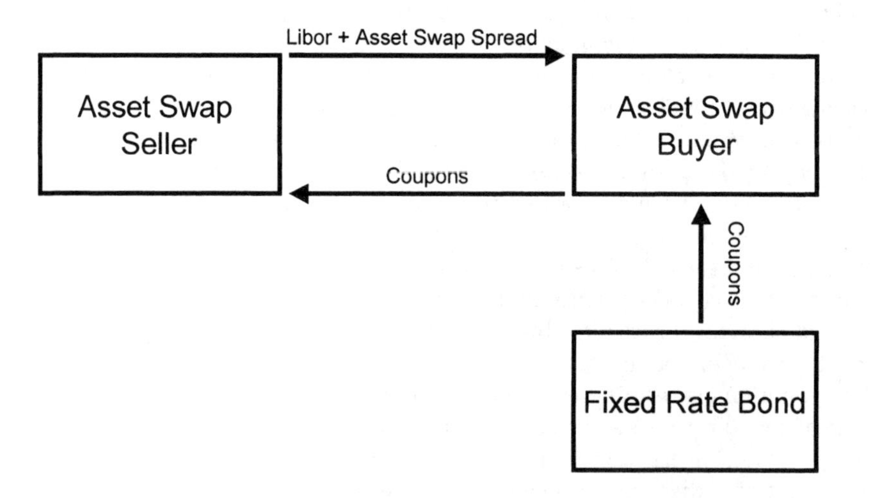

Fig. 5.6. Second trade of a par asset swap.

- The asset swap buyer enters into an interest rate swap contract to pay fixed coupons to the asset swap seller equal to the coupons from the fixed rate bond. Therefore, the asset swap seller pays Libor plus a fixed spread, the so called asset swap spread (see figure 5.6). The maturity of the asset swap equals the maturity of the fixed rate bond. The asset swap spread is calculated such that the net value of the sale of the fixed rate bond plus the interest rate swap is zero at inception.

If the fixed rate bond defaults prior to maturity the asset swap buyer defaults on the coupon payments and par redemption - only receives the recovery rate from the issuer of the fixed rate bond. The interest rate swap either continues until maturity or is closed out at the actual market value - although the fixed leg of the interest rate swap can't be funded with the coupon payments from the fixed rate bond any longer. Hence, the asset swap buyer takes the credit risk of the fixed rate bond. Therefore, the asset swap spread can be seen as the compensation of the asset swap buyer for taking the credit risk and is used as a bond specific measure for the expected loss given a default. For a

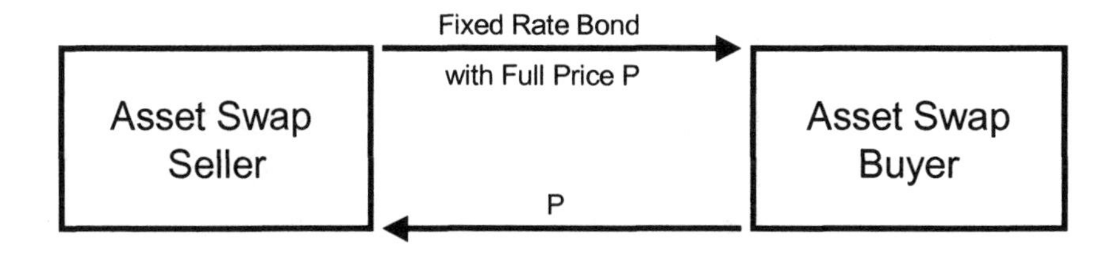

Fig. 5.7. First trade of a market asset swap.

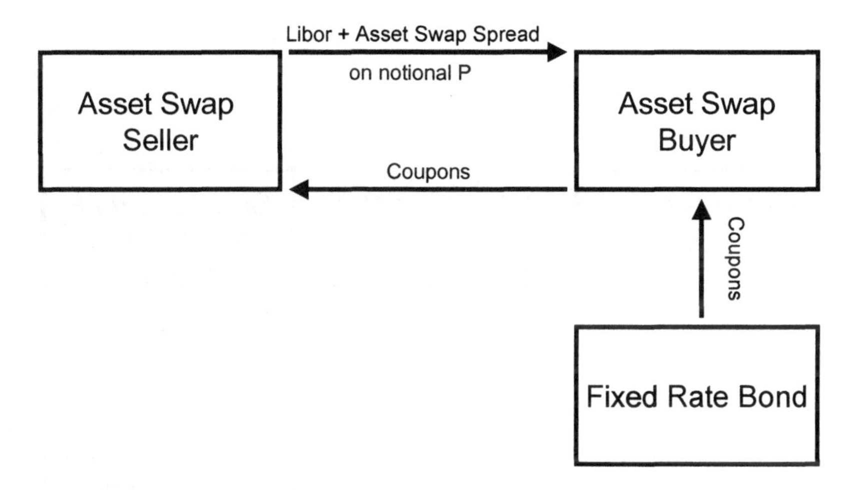

Fig. 5.8. Second trade of a market asset swap.

par asset swap the maximum loss the asset swap buyer can make is par minus the recovery rate. This makes an asset swap spread similar to a default swap spread since a default swap pays out par minus recovery to the protection buyer in case of a default. It should be also close to the spread of a par floating rate note since the loss on a floating rate note which trades at par is also par minus recovery. Obviously, in practice there can be great differences between these spreads because of differences in liquidity, market size and funding costs in the different markets.

In contrast to a par asset swap in the market asset swap, the full price for the fixed rate bond is paid. The notional of the Libor leg is scaled by the full price so that the resulting value of the asset swap spread is different. At maturity of the market asset swap there is an exchange of par for the original price of the fixed rate bond. The credit exposure of a market asset swap is the same as the credit exposure of a par asset swap. Figures 5.7 - 5.9 visualize a typical market asset swap contract.

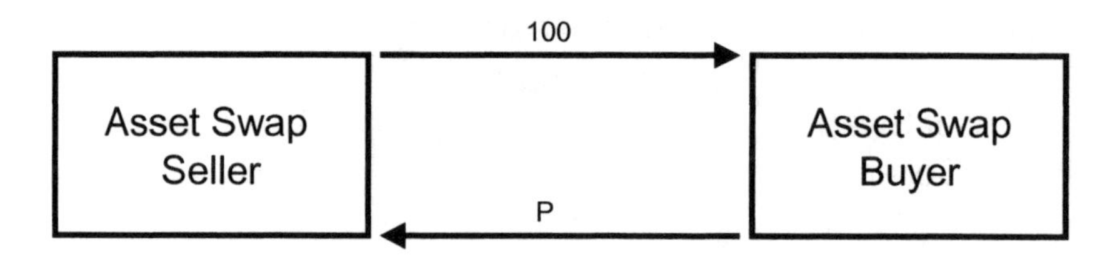

Fig. 5.9. Exchange of payments at maturity of a market asset swap.

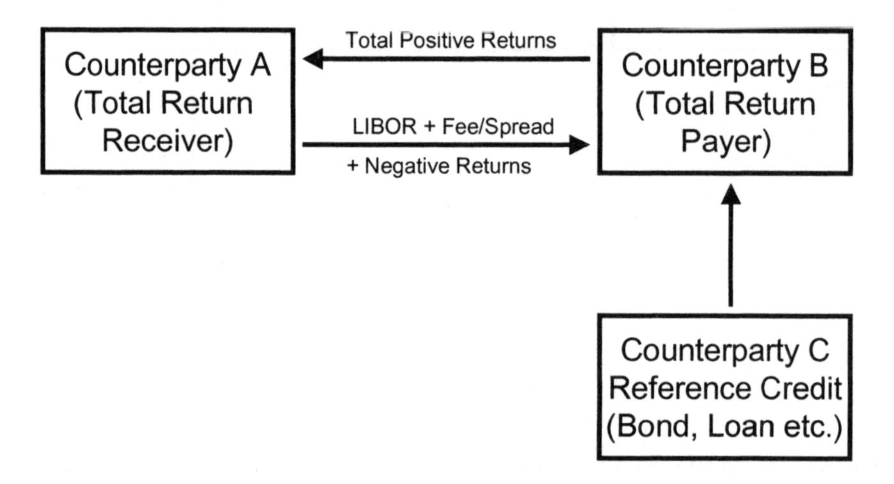

Fig. 5.10. Basic structure of a total return swap.

5.3.5 Other Credit Derivatives

Total Rate of Return Swaps (Total Return Swaps).

Definition 5.3.6.
A total rate of return swap is a swap under which two parties periodically pay each other the total return on one or two reference assets or baskets of assets which they do not necessarily hold. When at least one of the reference assets is a credit-sensitive instrument, the swap is a credit derivative. The payments of the two parties are netted.

A total return swap is a means of duplicating cash flows of either selling or buying an asset, without ever necessarily possessing the asset itself. As such, a total return swap is more a tool for balance sheet arbitrage than a credit derivative. However, as a derivative contract with a credit dimension (the asset can default) it usually falls within the remit of the credit derivatives trading desk of investment banks and so becomes classified as a credit derivative. It is distinct from the default swap in that it exchanges either the total economic performance of a specified asset for another cash flow or the total economic performances of two specified assets. Hence, payments between the counterparties to a total rate of return swap are based upon changes in the market valuation of one or more specific credit instruments, irrespective of whether a credit event has occurred. In contrast to the default swap, the total return swap doesn't only transfer credit risk but also market risk of the reference credit asset. Figure 5.10 shows the structure of a typical total return swap. Counterparty A pays counterparty B at regular intervals (usually weekly, monthly, or quarterly) a regular rate of LIBOR plus a fee, a possible price depreciation of the counterparty C bond since the last payment date, and in case of a default of the counterparty C bond the par value of the bond. Counterparty B pays counterparty A at the same regular intervals all coupon payments of the counterparty C bond, a possible price appreciation of the counterparty C bond since the last payment, the principal repayment of the counterparty C bond (in case of no default) and the recovery value of the bond (in case of a default). To determine the cash flows, the total rate of return swap has to be marked to market at the regular payment dates. As an alternative to cash settlement total rate of return swaps allow for physical delivery of the reference credit asset at maturity. Maturity of the total return swap is not required to match that of the reference credit asset, and, in practice, rarely does.

The static hedge for the payer in a total return swap is to buy the asset at trade inception, fund it on balance sheet, and then sell the asset at trade maturity. Indeed, one way the holder of an asset can hedge oneself against changes in the price of the asset is to become the payer in a total return swap. This means that the cost of the trade will depend mainly on the funding cost of the total return payer and any regulatory capital charge incurred. We can break out the total cost of a TRS into a number of components. First, there is the actual funding cost of the position. This depends on the credit rating of the total return payer that holds the bond on its balance sheet. If the asset can be repo'd, it depends on the corresponding repo rate. If the total return payer is a bank, it also depends on the BIS risk weight of the asset, with 20% for OECD bank debt and 100% for corporate debt. If the total return payer is holding the asset, then the total return receiver has very little counterparty exposure to the total return seller. However, the total return payer has a real and potentially significant counterparty exposure to the total return receiver. This can be reduced using collateral agreements or may be factored into the LIBOR spread coupon paid.

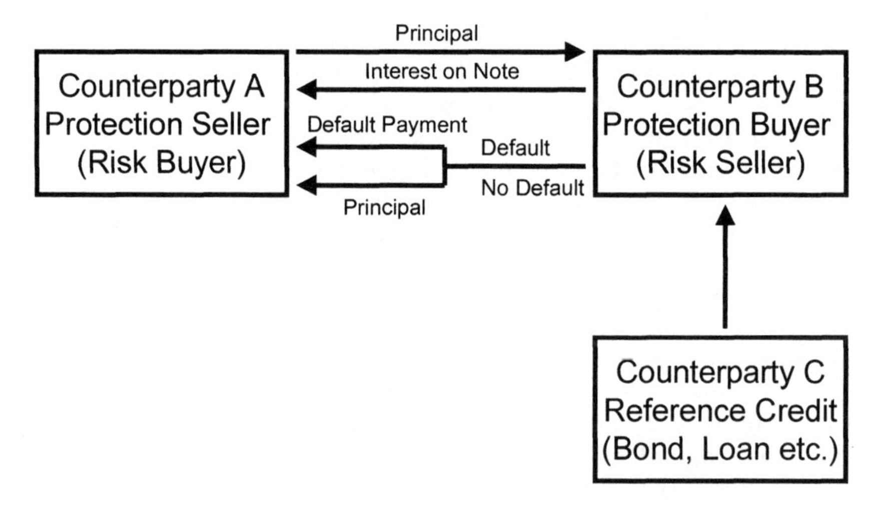

Fig. 5.11. Basic structure of a credit-linked note.

Credit-Linked Notes.

Definition 5.3.7.
A credit-linked note is a security that is typically issued from a collateralized special purpose vehicle, which may be a company or business trust, with redemption and/or coupon payments linked to the occurrence of a credit event. Credit-linked notes may also be issued on an unsecured basis directly by a corporation or financial institution. A capital-protected credit-linked note is a credit-linked note where the principal is partly or fully guaranteed to be repaid at maturity. In a 100% *principal-guaranteed credit-linked note, only the coupons paid under the note bear credit risk.*

For investors who wish to take exposure to the credit derivatives market and who require a cash instrument, one possibility is to buy it in a funded credit-linked note form. The note pays a fixed- or floating-rate coupon and has an embedded credit derivative. If the note issuer defaults, then the investors can lose some or all of their coupon and principal. The standard credit-linked note contains an embedded default swap. The investor pays par to buy the note, which then pays LIBOR plus a spread equal to the default swap spread of the reference asset plus a spread linked to the funding spread of the issuer. This issuer funding spread compensates the investors for their credit exposure to the note issuer. It will be less than the issuer spread to the note maturity to take into account the fact that the credit event may cause the note to terminate early. The issuer will also impose a certain cost for the administrative work. Credit linked notes can be used to embed credit derivatives in a fully funded note format. Like an asset swap, the credit-linked note is really a synthetic par floater. If the reference asset defaults, the credit-linked note accelerates, and the investor is delivered the defaulted asset. Unlike an asset swap, there is no default contingent interest rate risk.

5.4 Multi Counterparty Credit Derivatives

For pricing multi counterparty credit derivatives the models explained in chapters 2, 3, and 4 can be used.

5.4.1 Index Swaps

Sometimes total return swaps are not linked to a single security but to the total return of an index such as the Lehman Brothers Corporate Bond Index[6]. Such an instrument is called index swap. The advantage for the investor of such an instrument lies in the simultaneous exposure to a broad universe of corporate securities without exposing oneself to a possible default of one specific issuer. Although index swaps can be structured in a lot of different ways one popular way is that the buyer of the index swap receives the gain or loss in the value of the index plus any coupon accrual in return for floating-rate payments of LIBOR plus a fixed spread.

For pricing such an instrument the following points have to be considered:

- To hedge the instrument the total return payer has to buy the index. Hence, the funding costs of the assets in the index will affect the price.
- An index usually contains thousands of different bonds. To replicate it the total return payer only buys some subset of bonds. This causes a tracking error that must be included in the price.
- Buying the index involves transaction costs which have to be considered as well.

There are several reasons why index swaps are an efficient alternative to cash:

- Access to a diversified portfolio even without a significant amount of capital.
- Avoid the high bid-offer spreads faced in the smaller transactions if one buys the single positions of the index directly. Bid-offer spreads for index swaps will usually be much tighter as index swaps may be more liquid than the underlying assets.
- Exposure to a sector in which one might not have specialized knowledge.

[6] See, e.g., www.lehman.com.

- Clients can use the index swap to benchmark their portfolio to standard fixed income indices.
- Portfolio managers can replicate an index without incurring a tracking error.
- Investors can gain access to asset classes from which they might otherwise be precluded.

5.4.2 Basket Default Swaps

A basket default swap is a generalization of a default swap. The credit event is the default of some combination of the credits in a specified basket of credits. A well-known example is a first-to-default basket. The default payment is triggered by the first default of a credit in a basket of credits. Cash and physical settlements are possible. Somehow more common is physical delivery of the defaulted credit in return for a payment of the par amount in cash. For protection against the first-to-default, the protection buyer pays a basket spread to the protection seller as a set of regular accruing cash payments. These payments terminate following the first credit event. A simple extension of the first-to-default basket is the second-to-default basket. The default payment is triggered by two defaults in a basket of credits. Following this credit event, it is the second credit to default that is delivered in return for a payment of par. Second-to-default baskets belong to the so called second-loss products. They are attractive to investors that want to buy high quality products that return a higher yield than comparably high quality credits. In general, they are less risky than single credit products.

Investing in basket default swaps has a lot of advantages:

- Investors can leverage their credit risk and enhance their yield while at the same time being exposed to good-quality names.
- Basket default swaps do not increase the downside risk of investors compared to investing in the underlyings credits. The most they can lose is par minus the recovery of the first credit to default, which is the same as they would have lost had they simply purchased this credit in the first place. But the basket spread paid can be a multiple of the spread paid by the individual credits in the basket.
- More risk-averse investors can use default baskets to construct low risk assets: second-to-default baskets trigger a credit event after two or more assets have defaulted. As such, they are lower-risk second-loss exposure products that may pay a higher return than other similar risk assets.

As basket default swaps are essentially default correlation products the methods introduced in chapter 4 can be used to price these instruments. The basket spread highly depends on the probability of the reference credits in the basket to default together. Credits issued by companies within the same country and industrial sector usually have a higher default correlation than those within different countries and industrial sectors. In addition, there is a close link between default correlation and credit rating. Lower-rated companies are generally more leveraged, so in an economic downturn, they are be more likely to default together. The lack of default events makes it very difficult to determine default correlation from empirical analysis. If the credits in the basket have a low default correlation, the basket spread is close to the sum of the single spreads of the reference credits in the basket. If the default correlation is high, assets tend not to default alone, but together. Then the basket behaves like the worst credit in the basket, i.e. the basket spread is very close to the widest spread of the credits in the basket. This is a lower benchmark for the basket spread. There is no static but only a dynamic hedge for basket default swaps. Of course, hedging costs must be taken into account when pricing these instruments. For model based pricing of basket default swaps in an intensity based framework see, e.g., Duffie (1998c).

5.4.3 Collateralized Debt Obligations (CDOs)

Some Figures First. Even more complicated products than the instruments presented so far are structured finance transactions (SPs), such as collateralized debt obligations (CDOs), collateralized bond obligations (CBOs), collateralized loan obligations (CLOs), collateralized mortgage obligations (CMOs) and other asset-backed securities (ABSs). The key idea behind these instruments is to pool assets and transfer specific aspects of their overall credit risk to new investors and/or guarantors. According to J. P. Morgan (*Structured Products/ ABS Market Monthly* 1999), global SP/ABS supply from January 1999 through August 1999 rose by 24% versus the similar period in 1998 up to a record level of over US$ 333 billion. US$ public ABS issuance was close to US$ 150 billion, the European structured products market surpassed US$ 55 billion and the Japanese ABS market reached a volume of about US$ 13 billion. Compared to the corresponding 1998 period this was an increase of 25%, 50% and 230%, respectively. The amount of ABS contracts outstanding increased from US$ 316.3 billion in 1995 to US$ 1619.9 billion in the second quarter of 2003 (see figure 5.12).

Introduction to CDOs. CDOs are Special Purpose Vehicles (SPVs) that invest in a diversified pool of assets (collateral pool). These investments are done by an experienced portfolio manager, the so called collateral manager. He can only act within clearly defined rules and limits that are fixed in a contract and controlled by an independent trustee. The investments are financed by issuing several tranches of financial instruments some of them being rated some of them not. The repayment of the tranches depends on the performance of the underlying assets in the collateral pool. The rating of the single tranches is determined by the rank order they are paid off with the interest and nominal payments that are generated from the cash flows in the

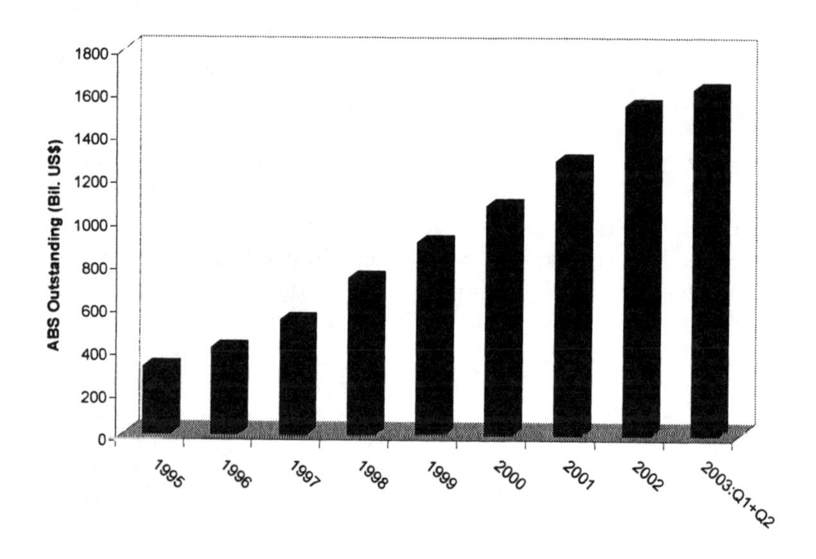

Fig. 5.12. Asset-backed securities outstanding 1995–2003 (in US$ billions). Source: The Bond Market Association.

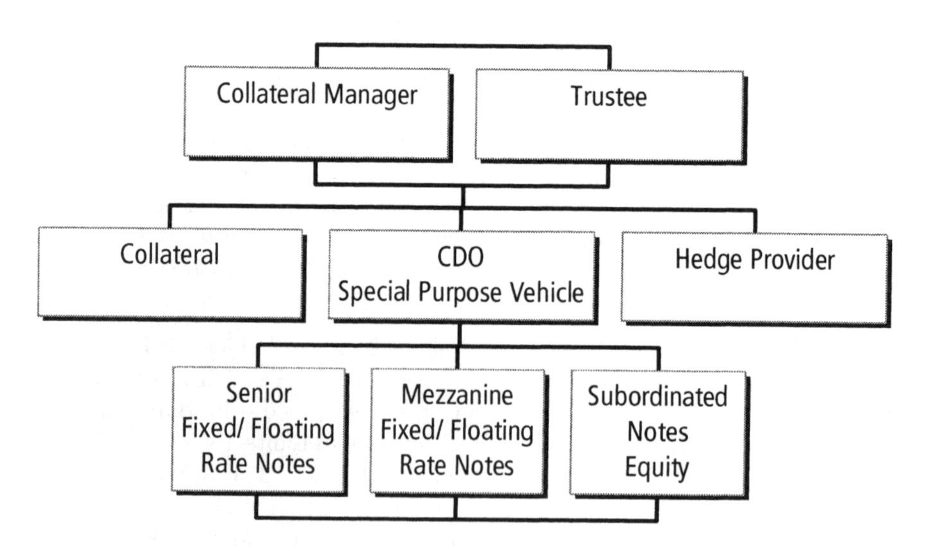

Fig. 5.13. Typical structure of a CDO.

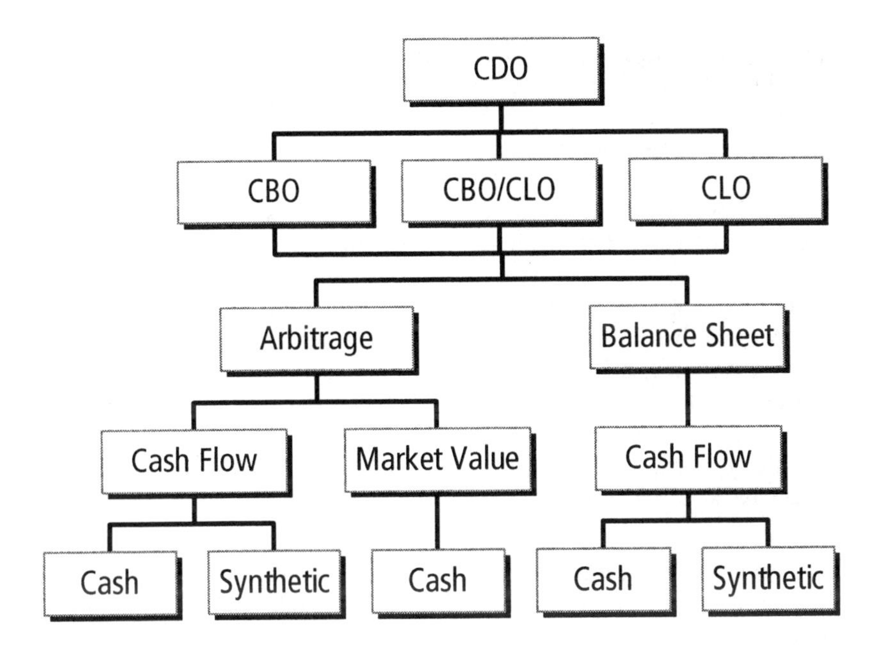

Fig. 5.14. Characterization of CDO contracts.

collateral pool. So called senior notes are usually rated between AAA and A and have the highest priority in interest and nominal payments, i.e. they are paid off first. Mezzanine notes are typically rated between BBB and B. They are subordinated to senior notes, i.e. they are only paid off if the senior notes have already been serviced. Finally, the subordinate notes do not have a rating. They are the residuum of the transaction. There are only interest rate and nominal payments to these tranches if the cash flows generated in the collateral pool are big enough to cover all fees and completely pay off the senior and mezzanine tranches. Interest rate payments may be postponed or completely cancelled. Equity tranches are so called first loss positions.
The different CDO contracts can be characterized the best way by:

- Assets in the collateral portfolio:
  - If the collateral pool of a CDO purely consists of bonds it is called a collateralized bond obligation (CBO).
  - If the collateral pool purely consists of loans it is called a collateralized loan obligation (CLO).

  The most common asset classes in CDO collateral pools are:
  - European and US high yield corporate bonds. The high coupons make them especially convenient for cash flow CDOs (see below).
  - Emerging market corporate and sovereign bonds. The high yield spread between emerging market bonds and senior notes of CDOs make them es-

pecially convenient for cash flow CDOs. Because of the high price volatility of these instrument they are less suitable for market value CDOs (see below). In most cases, exposure in emerging market bonds is combined with restrictions with respect to regions.
  - Bank loans of high quality. They are mostly used in balance sheet CDOs (see below). Because of the low interest rate payments they are less suitable for cash flow arbitrage CDOs (see below) and market value CLOs.
  - Bank loans of poor quality. Most of the time only loans are used that are collateralized. Because of their high interest rate payments they are suitable for cash flow arbitrage CLOs.
  - Distressed debt. It is especially used in market value arbitrage CDOs (see below).
  - Credit derivatives like credit swaps, credit linked notes and others.
  - Asset backed securities.
- Purposes:
  - Arbitrage transactions. The collateral pool is actively managed. There is a permanent monitoring and trading. The collateral manager earns a management fee. These transactions are especially interesting for asset managers or investors in equity tranches. The equity tranche investors take advantage of the spread difference between the credits in the collateral pool paying high coupons and the senior tranches. They hope to achieve a leveraged return between the after-default yield on assets and the financing cost due to debt tranches. This potential spread, or funding gap, is the arbitrage of the arbitrage CDO.
  - Balance sheet transactions. The collateral pool is not actively managed. Changes in the collateral pool only arise from instruments that have already matured. By using balance sheet transactions financial institutions can remove loans or bonds from their balance sheet in order to obtain capital, to increase liquidity or to earn higher yields. Therefore, these instruments transfer outstanding money of obligors into liquidity. This helps to reduce economic and regulatory capital. In addition, they are a good supplement to the classical instruments for asset liability management as they allow for active risk management on the one hand and are an alternative for financing and refinancing.
- Credit structures:
  We distinguish two ways CDOs protect debt tranches from default.
  - Cash flow transactions. Cash flow deals are based on collateral that generates enough cash flows to pay the regular interest payments and the nominal payments to the rated tranches.
  - Market value transactions. Market value deals are based on the mark-to-market of the instruments in the collateral pool. The collateral manager has to make sure (e.g., by actively trading) that the market value of the collateral is always high enough to assure that by selling collateral

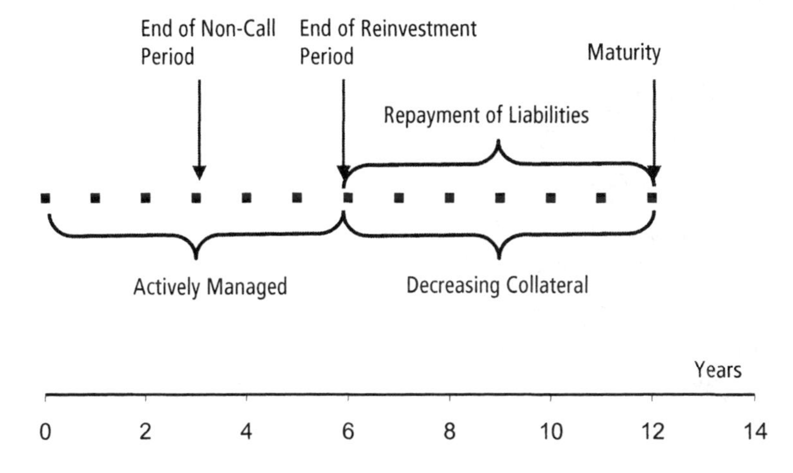

Fig. 5.15. Typical life cycle of a cash flow CDO.

instruments the interest rate and nominal payments to the tranches can always be made.

- Cash/synthetic property:
  In a cash sale the assets are sold to the CDO. This is not true for synthetic deals.
- Liabilities, i.e. the structure of the tranches:
  A variety of numbers of tranches is possible. The nominal can be repaid in a fast way, slowly or steady-going. The coupon payments can be either fixed or floating. The dimensions of the single tranches are chosen such that the financing costs are minimized. Usually, the biggest tranche is the senior tranche. The other debt tranches account for 5% to 15% and the equity tranche is between 2% and 15%.

The life cycle of a typical cash flow CDO consists of a reinvestment phase which lasts $5-6$ years. During this period, the collateral portfolio is actively managed. All cash flows from the interest payments in the collateral pool are used to pay fees and make interest payments to the debt and finally equity tranches. If some instruments in the pool mature the money is reinvested in new instruments. Most of the CDO managers stay away from concentrations on speculative assets but focus on generating steady cash flows. Not only investors in senior tranches benefit from this strategy but also equity investors - it is especially useful to optimize returns and volatility of equity tranches. After the first $3-5$ years (the non-call period) the investors can exercise a call option. The reinvestment period is followed by the repayment phase which

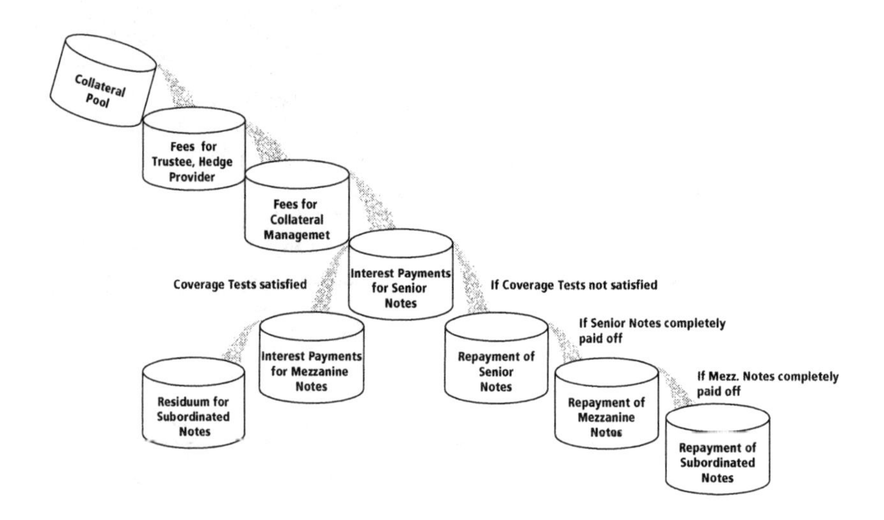

Fig. 5.16. Waterfall for Interest Rate Cash Flows.

typically lasts 6 – 7 years. During this period the liabilities (i.e. the tranches) are repaid according to their seniority from the cash flows, i.e. interest and principal of the instruments in the collateral pool. The amount of collateral is decreasing as the cash flows from instruments that mature is not reinvested. Consequently, the (interest) payments to the equity tranche are decreasing as well. After all senior and mezzanine tranches have been completely paid off all the cash flows from the collateral pool go to the equity tranche. Figures[7] 5.16 and 5.17 show how the interest cash flows and the principal cash flows from the collateral pool are used for interest payments and principal payments of the single tranches.

There are several reasons why investors should invest in CDO tranches.

- CDOs can be used as efficient diversification instruments. With the investment in a CDO tranche the investor gets exposure to different economies, rating classes, industries and so on.
- CDOs help to match specific risk/return profiles. One CDO offers investment in a broad spectrum of different tranches. Each tranche is able to generate a specific risk/return profile. Different tranches satisfy different risk appetites and return expectations of investors. E.g., investing in a AAA tranche of a CDO can give exposure to the high yield segment.
- CDOs pay high interest rates compared to other investments of the same rating classes.

[7] From Fabozzi & Goodman (2001).

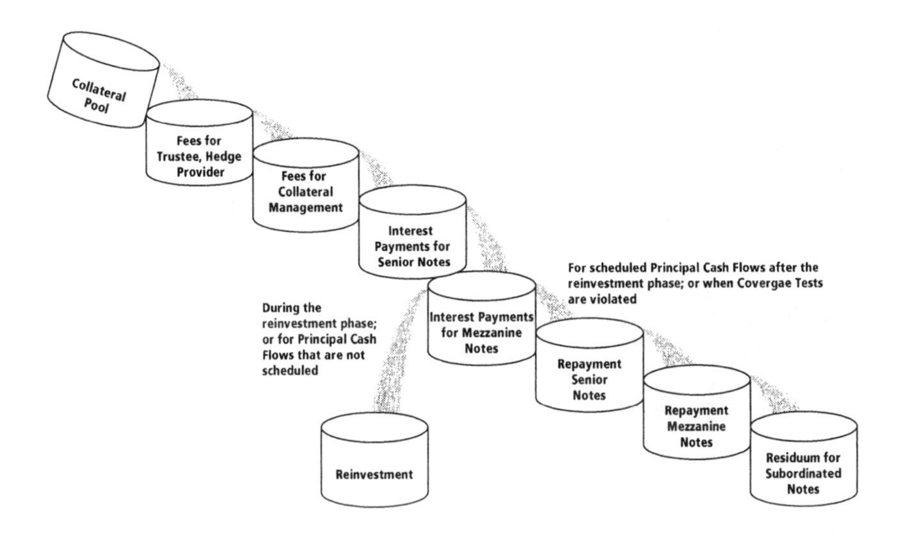

Fig. 5.17. Waterfall for principal cash flows.

- Mezzanine tranches are especially interesting for insurance companies. They provide the companies with steady long-term fixed interest rates that help them to cover their liabilities.

Example 5.4.1.
Figure 5.18 shows a real CDO structure. Using the Promise-A-2002-1 PLC-Structure the German based "Kreditanstalt fuer Wiederaufbau (KfW)" enables "HypoVereinsbank (HVB)" to reduce their economic and regulatory capital caused by credit exposure to German "Mittelstand" loans. Therefore, implicitly the structure is a kind of support for the German "Mittelstand" by KfW. Promise-A-2002-1 is structured as synthetic partially financed CLO and not as a true sale CLO. It is sponsored by KfW but arranged by HVB. The transaction allows a transfer of credit risk of about Euro 1.62 billions from the collateral pool to the investors. The pool consists of corporate loans of HVB and "Vereins- und Westbank (VuW)". Obviously, the risk of the investors is the probability that Promise-A-2002-1 can not pay off the tranches in full because of defaults in the reference pool. The credit risk transfer takes place in two steps:

- HVB buys a guarantee from KfW and therefore pays a guarantee fee. Are there any defaults in the collateral pool KfW compensates HVB for the losses. KfW itself hedges its risk by entering in a senior credit default swap contract and therefore pays a swap premium.

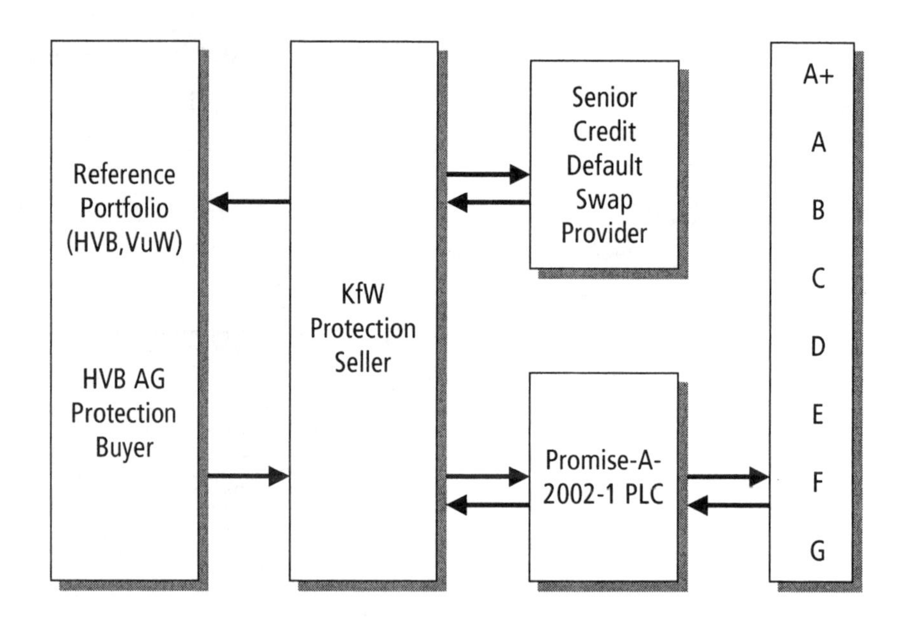

Fig. 5.18. Promise-A-2002-1 PLC - Structure.

- The second part of the hedge uses credit linked borrower's notes issued by KfW. The money from the investments in the tranches is partly used to finance these notes. For each tranche there exists a special series of borrower's notes synchronized with the interest and principal payments of the single tranches.

Figure 5.19 shows the tranches and coupons of the Promise-A-2002-1 PLC - structure. The biggest tranche is the senior AAA tranche with 36.3%, all investment grade tranches together amount roughly 48% , the speculative grade tranches 37% and the equity tranche 15%.

There are some additional features that make CDOs a somehow safer investment:

- There are rigorous collateral selection criteria and quality tests. E.g., there can be rules like the following:
  - It is not allowed that less than $x\%$ of the collateral pool belongs to a speculative rating class.
  - The weighted average rating of the collateral pool must be above a critical floor.
  - The exposure to specific obligors, regions, industries, financial instruments must be limited. Sometimes specific industry and diversity score tests or obligor concentration tests are applied.

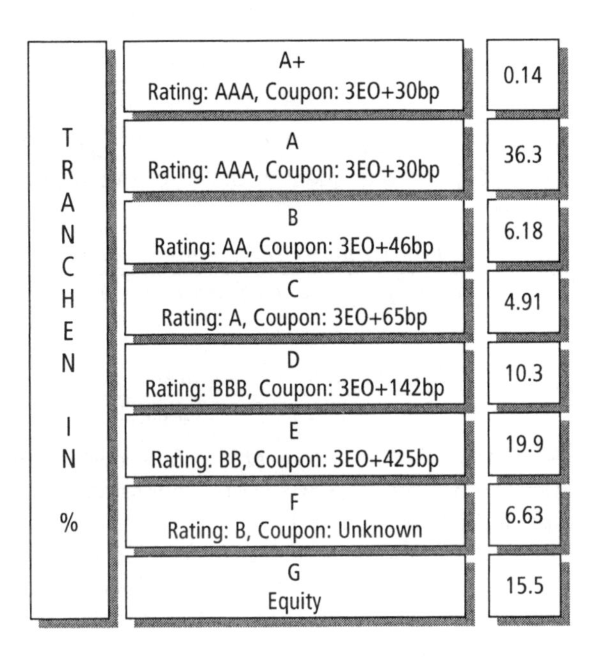

Fig. 5.19. Tranches of the Promise-A-2002-1 PLC - structure.

The idea behind all these rules is to assure that the quality of the pool is not too poor and that the level of diversification is high enough. The state of the collateral pool has to be reported to the trustee and to the investors.

- Sometimes the collateral pool is actively traded, i.e. the collateral manager can react if there are dangerous market developments. He is responsible for the performance of the collateral pool and has to make sure that the level of diversification, quality and structure are always comply with the given rules.
- Sometimes the management fees are dependent on the performance of the collateral manager. This is an additional incentive for the manager to perform well.
- There are two types of tests (so called coverage tests) that are intended as a warning if the cash flows from the collateral pool are running short to cover interest and/or principal payments to the tranches. If the coverage tests fail, the cash flows from the collateral are redirected. They are used to repay the outstandings on the tranches according to their seniority.
  - The overcollateralization test is done for every single tranche but the equity tranche. The test considers and measures if the principal coverage of collateral assets compared to the required amount for paying back the

notional of the tranche under consideration and the tranches senior to this tranche is satisfied. Suppose there are three debt tranches A, B, C. Then the O/C ratio for senior tranche A is defined as

$$O/C_A = \frac{Par\ value\ of\ the\ collateral\ pool}{Par\ value\ of\ tranche\ A}.$$

Typically there is a lower boundary $O/C_A^{\min}$ for O/C_A such that the test fails if $O/C_A < O/C_A^{\min}$. Consequently, the O/C ratios for tranches B and C are given by

$$O/C_B = \frac{Par\ value\ of\ the\ collateral\ pool}{Sum\ of\ par\ values\ of\ tranches\ A\ and\ B},$$

and

$$O/C_C = \frac{Par\ value\ of\ the\ collateral\ pool}{Sum\ of\ par\ values\ of\ tranches\ A,\ B\ and\ C}.$$

Again, the test fails if O/C_B and/or O/C_C falls below some specific lower boundaries $O/C_B^{\min}$ and $O/C_C^{\min}$, resp. The tests take into account that senior notes must be repaid first.

- The interest coverage test considers and measures for a specific payment period if the interest payments from the collateral assets can cover the fees and coupons that have to be paid to the tranches or not. If we assume again that there are three debt tranches A, B, C, then there are three interest coverage tests, one for each single tranche. Assuming an annual horizon the I/C ratio for tranche A is defined as

$$I/C_A = \frac{Exp.\ interest\ payments\ from\ the\ pool\ over\ 1\ year\ -\ Annual\ fees}{Exp.\ interest\ payments\ to\ tranche\ A\ over\ 1\ year}.$$

Typically there is a lower boundary $I/C_A^{\min}$ for I/C_A such that the test fails if $I/C_A < I/C_A^{\min}$. Similarly, the I/C ratios for tranches B and C are given by

$$I/C_B = \frac{Exp.\ interest\ payments\ from\ the\ pool\ over\ 1\ year\ -\ Annual\ fees}{Exp.\ interest\ payments\ to\ tranches\ A\ and\ B\ over\ 1\ year},$$

and

$$I/C_C = \frac{Exp.\ interest\ payments\ from\ the\ pool\ over\ 1\ year\ -\ Annual\ fees}{Exp.\ interest\ payments\ to\ tranches\ A,\ B\ and\ C\ over\ 1\ year}.$$

Again, the test fails if I/C_B and/or I/C_C falls below some specific lower boundaries $I/C_B^{\min}$ and $I/C_C^{\min}$, resp. The tests take into account that senior notes must be repaid first.

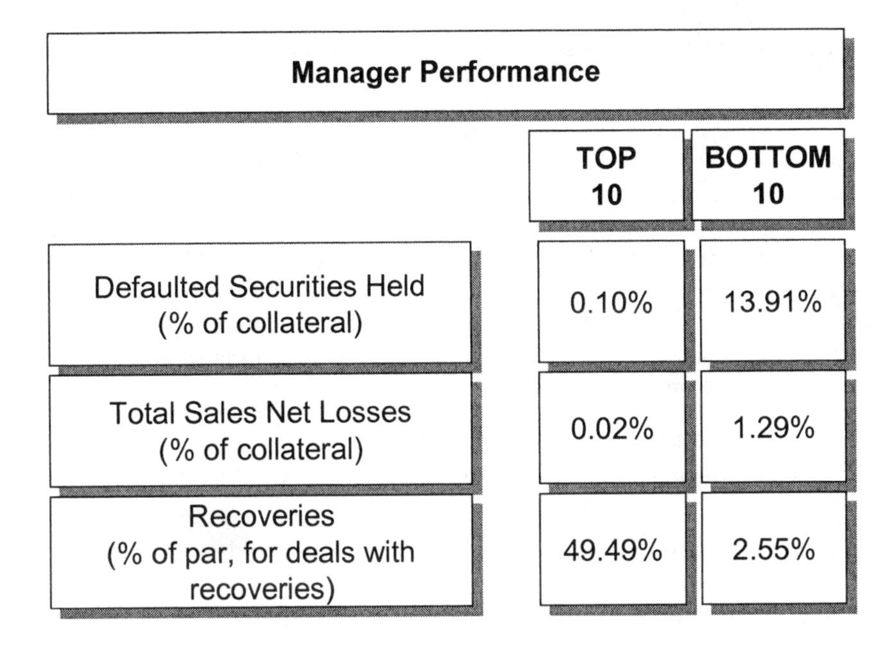

Fig. 5.20. Measures for the collateral manager performance.

The quality of the collateral manager is one of the most important but unfortunately only badly predictable factors of the performance of a CDO. The CDO performance within a specific class of CDOs can vary tremendously. See, e.g., figure[8] 5.20. A comprehensive analysis of the manager should cover qualitative and quantitative criteria as well. Qualitative factors are, e.g.,

- expertise in the relevant asset classes
- track record
- stability of the team
- quality of the investment process
- technical equipment of the team.

Important quantitative factors are, e.g., historical performance attribution, i.e. performance relative to Standard & Poor's CDO indices (asset-class and launch-year specific). There are two different indices:

- Standard & Poor's Arbitrage CBO Index. It aggregates monthly historical performance information across arbitrage CDO transactions collateralized primarily by high-yield bonds. The index was introduced in February of 2001 and contains monthly performance data starting May of 1999.
- Standard & Poor's Arbitrage CLO Index. It aggregates monthly historical performance information across arbitrage CDO transactions collateralized

[8] From *S&P CDO Surveillance* (2002).

primarily by leveraged loans. The index was introduced in August of 2001 and contains monthly performance data from April of 2000.

Each index tracks 22 categories of performance information, focused primarily on purchases, sales, and defaults of collateral debt securities, and on the over-collateralization ratios for the transactions. The information in the indexes is broken out by year of origination (i.e., by cohort or vintage) with separate tables showing the performance of transactions originated in 1997, 1998, 1999 and 2000. Each month, the most recent three months of performance information (along with one month of performance information from one year ago for comparison) is posted on RatingsDirect and on www.standardandpoors.com. The categories covers the following information.

- Information on total collateral debt securities (CDS) held:
  - average principal balance of CDS
- Information on defaults of collateral debt securities:
  - average monthly new defaults
  - average defaulted assets held
- Information on purchases of collateral debt securities:
  - average par value of CDS purchased
  - average market price of CDS purchased
- Information on total sales of collateral debt securities:
  - average par value of total CDS sales
  - average market price of total CDS sales
  - average par gain or loss from total CDS sales
- Information on sales of defaulted collateral debt securities:
  - average par value of defaulted CDS sales
  - average market price of defaulted CDS sales
  - average par gain or loss from defaulted CDS sales
- Information on sales of credit risk collateral debt securities:
  - average par value of credit risk CDS sales
  - average market price of credit risk CDS sales
  - average par gain or loss from credit risk CDS sales
- Information on sales of credit improved collateral debt securities:
  - average par value of credit improved CDS sales
  - average market price of credit improved CDS sales
  - average par gain or loss from credit improved CDS sales
- Information on discretionary sales of collateral debt securities:
  - average par value of discretionary CDS sales
  - average market price of discretionary CDS sales
  - average par gain or loss from discretionary CDS sales
- Information on the transactions' overcollateralization ratios:
  - average senior overcollateralization ratio test spread

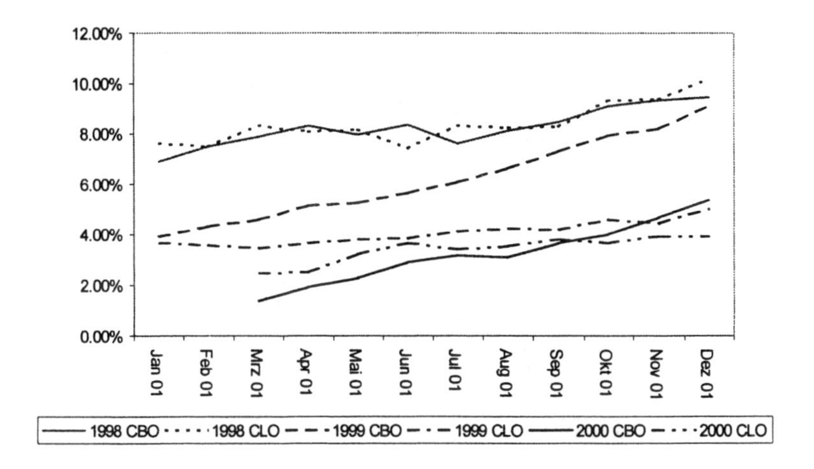

Fig. 5.21. Annualized rate of monthly defaults.
Source: www.standardandpoors.com.

– average subordinate overcollateralization ratio test spread

Figure 5.21 shows the annualized rates of monthly defaults of CBO and CLO transactions originated in 1998, 1999 and 2000 as observed in 2001. The monthly defaults are highest for the year 1998 transactions, followed by the year 1999 and finally year 2000 transactions. The year 1999 CBO transactions show much more defaults than the year 1999 CLO transactions. For the year 1998 and 2000 transactions the CBOs experience slightly more defaults than the CLOs.

Figure 5.22 shows the percentage of defaulted assets held within collateral pools. The defaulted assets are highest for the year 1998 transactions, followed by the year 1999 and finally the year 2000 transactions. In the collateral pools of CBOs are more defaulted assets than in the collateral pools of CLOs.

Figure 5.23 shows the par value of cumulative defaulted CDS sales during 2001. Most sales have been done in the year 1998 transactions. There have been almost no sales in the year 1999 and 2000 CBOs.
Figure 5.24 shows 12 month rolling averages of recovery rates on defaulted assets in the collateral pools. The recovery rates of loans are much higher than the recovery rates of bonds. The differences of the recovery rates within the group of bonds are smaller then the differences within the group of loans.

Figure 5.25 shows par losses from the sale of defaulted and credit risk assets. Obviously, the figure is very similar to figure 5.23.

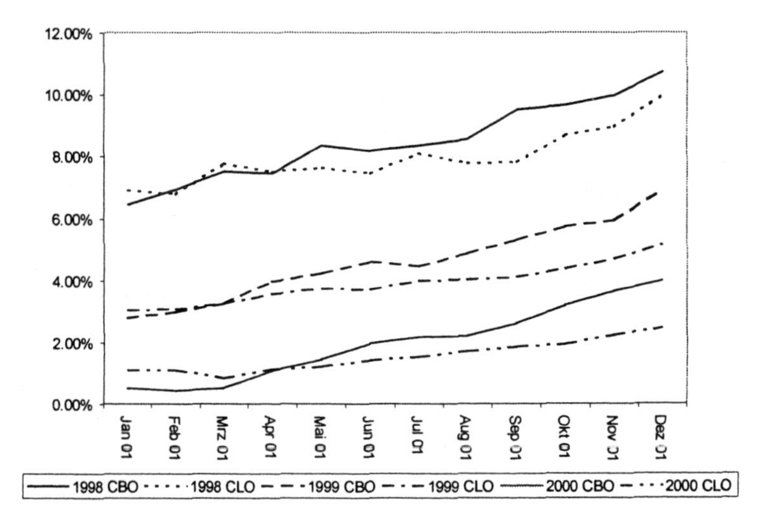

Fig. 5.22. Percentage of defaulted assets held within collateral pools. Source: www.standardandpoors.com.

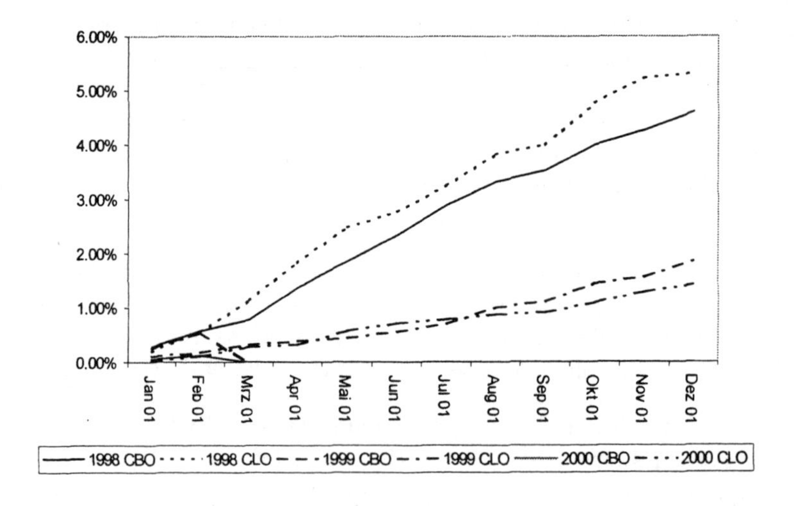

Fig. 5.23. Par value of cumulative defaulted CDS sales during 2001. Source: www.standardandpoors.com.

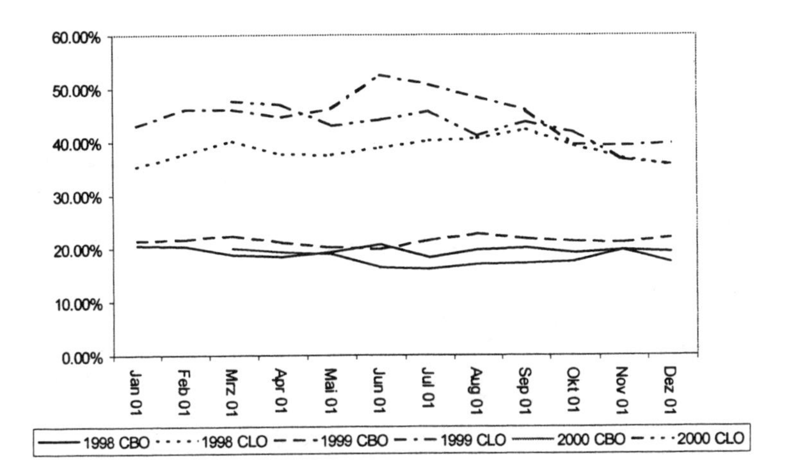

Fig. 5.24. Recovery rates on defaulted assets (12 month rolling averages). Source: www.standardandpoors.com.

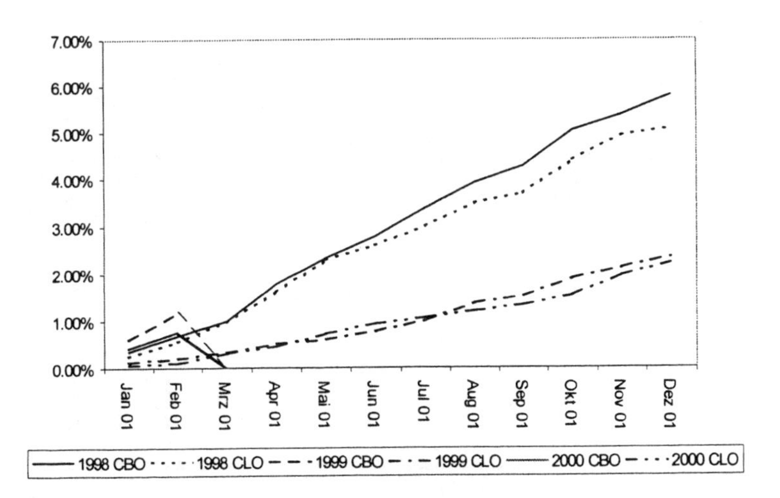

Fig. 5.25. Par losses from sale of defaulted and credit risk assets. Source: www.standardandpoors.com.

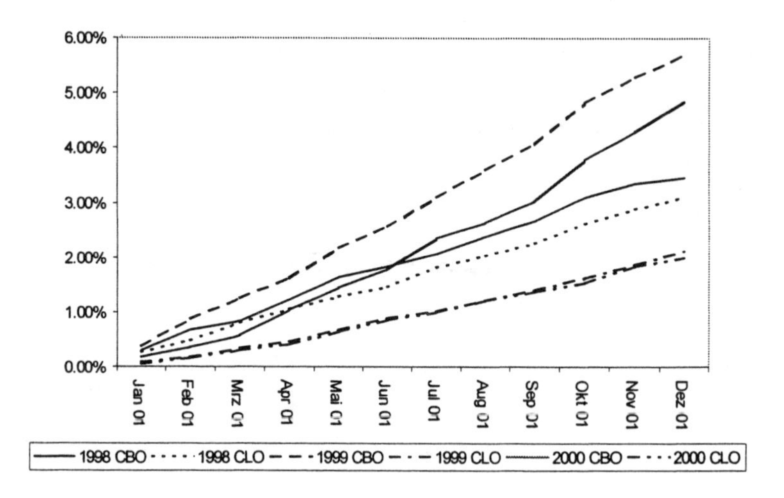

Fig. 5.26. Par gains from purchase of new assets.
Source: www.standardandpoors.com.

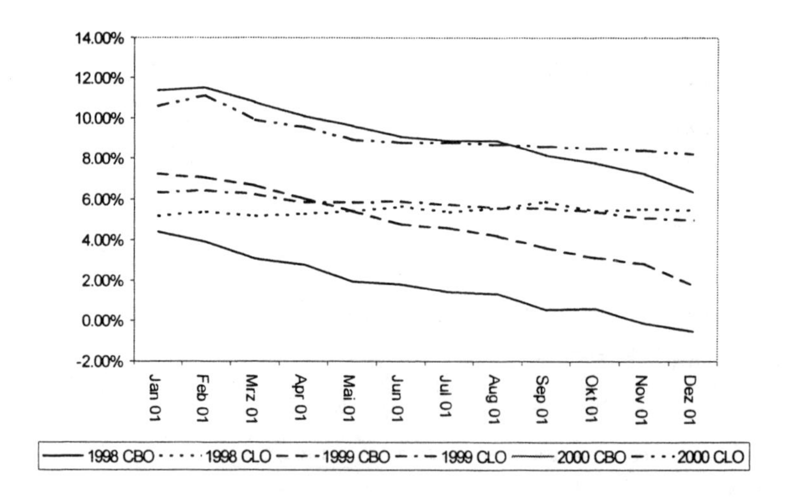

Fig. 5.27. Average senior O/C ratio spread.Source: www.standardandpoors.com.

Figure 5.26 shows the par gains from the purchase of new assets. The par gains are highest for the bonds.

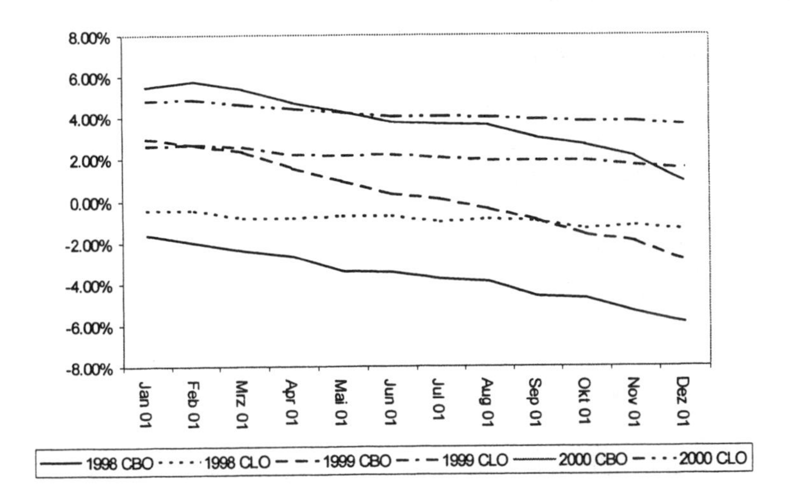

Fig. 5.28. Average subordinate O/C ratio spread.
Source: www.standardandpoors.com.

Figure 5.27 shows the results of the overcollateralization tests for the senior tranches. The result is best for the year 2000 transactions. Some of the 1998 CBO transactions must already have failed the test. All curves are decreasing.

Figure 5.28 shows the results of the overcollateralization tests for the subordinate tranches. The result is best for the year 2000 transactions. Some of the 1998 CBO and CLO as well as year 1999 CBO transactions must already have failed the test. All curves are flat or decreasing.

Moody's BET. In the following we give a short outline of the BET model. For detailed discussions see, e.g., Cifuentes, Efrat, Gluck & Murphy (1998) or Gluck & Remeza (2000). Suppose the collateral pool of a CDO consists of m different credits $1, ..., m$. Each of the credits has a notional amount of N_i. Actually, to model the cash flows in the pool we would have to model the joint behavior of the credits. Therefore we could apply methods as explained in chapter 4. But what if we could replace the pool of m correlated credits by a pool of $n < m$ uncorrelated credits that produce a similar loss distribution? Then modeling would be much easier. The loss distribution of uncorrelated credits is the easily tractable binomial distribution.

Therefore, choose n credits each with notional amount

$$N = \frac{\sum_{i=1}^{m} N_i}{m}$$

such that the first two moments of the loss distribution match those of the original pool. m is called diversity score of the collateral pool.

The expected loss and the variance of the original collateral pool at time 0 with respect to a time horizon T are given by

$$EL_T^{original} = \sum_{i=1}^{m} (1 - w_i) N_i \mathbf{P} \left(T_i^d \leq T \right), \tag{5.2}$$

$$VarL_T^{original} = \sum_{i=1}^{m} VarL_i (T) + \sum_{i,j,i \neq j} \rho_{i,j}^d (T) \sqrt{VarL_i (T) VarL_j (T)}, \tag{5.3}$$

where

$$VarL_i (T) = (1 - w_i)^2 N_i^2 \mathbf{P} \left(T_i^d \leq T \right) \left(1 - \mathbf{P} \left(T_i^d \leq T \right) \right).$$

As usual, w_i is the recovery rate and T_i^d the default time of credit i. $\rho_{i,j}^d (T)$ is the default correlation between credits i and j at the time horizon T.

Now we consider a collateral pool consisting of n independent and identically distributed credits with notional amounts N, recovery rate

$$w = \frac{\sum_{i=1}^m N_i w_i}{Nn}$$

and default probability $\mathbf{P} \left(T^d \leq T \right)$. We now want to choose n and $\mathbf{P} \left(T^d \leq T \right)$ such that the expected loss and the variance of the loss of both pools are the same. The expected loss and the variance of the loss of the modified pool are given by

$$EL_T^{\text{modified}} = (1 - w) Nn \mathbf{P} \left(T^d \leq T \right), \tag{5.4}$$

$$VarL_T^{\text{modified}} = (1 - w)^2 N^2 n \mathbf{P} \left(T^d \leq T \right) \left(1 - \mathbf{P} \left(T^d \leq T \right) \right). \tag{5.5}$$

If equations (5.2) and (5.3) match equations (5.4) and (5.5) then we can approximate the original collateral pool of m correlated credits by a collateral pool of n independent credits. Therefore, we have to choose n and $\mathbf{P} \left(T^d \leq T \right)$ as follows:

$$\mathbf{P} \left(T^d \leq T \right) = \frac{\sum_{i=1}^m (1 - w_i) N_i \mathbf{P} \left(T_i^d \leq T \right)}{(1 - w) Nn},$$

and

$$n = \frac{EL_T^{original} \left((1 - w) N - EL_T^{original} \right)}{VarL_T^{original}}.$$

6. A Three-Factor Defaultable Term Structure Model

"All models are wrong, but some are useful."
- George E. Box -

"The grand aim of all science is to cover the greatest number of empirical facts by logical deduction from the smallest number of hypotheses or axioms."
- Albert Einstein -

6.1 Introduction

6.1.1 A New Model For Pricing Defaultable Bonds

In the following, we develop a model for pricing defaultable debt that is capable of reflecting real world phenomena realistically. Before we actually introduce the model, we make some heuristic observations to motivate the theory. Throughout the following sections we examine and extend these considerations further:

- Since defaults actually happen, a bond pricing model must be able to deal with default events: For example, in August 1999 Ecuador defaulted on interest rate payments of Brady bonds with a total amount of about US$ 6 billions. In October 1999 there was a second default of US$ 500 millions. Figure 6.1 shows the yields to maturity and S&P ratings of an Ecuadorian sovereign bond that was downgraded to the rating D in September 1999.
- Rating agencies can't check the ratings on a daily basis or even continuously. Ratings of rating agencies can only be described by discrete time processes. But the quality of an obligor usually changes according to a continuous time process (see, e.g., figure 6.1). Thus, the quality or the distance to default of an obligor should be observed continuously.
- Very often the quality of obligors shows mean reversion properties. In case of a downgrade the management of a firm usually takes actions to improve the quality again. Figure 6.2 shows this phenomenon for the firm Hertz. The mean reversion level of the rating of Hertz is A- and A3 for S&P's and Moody's rating, respectively.

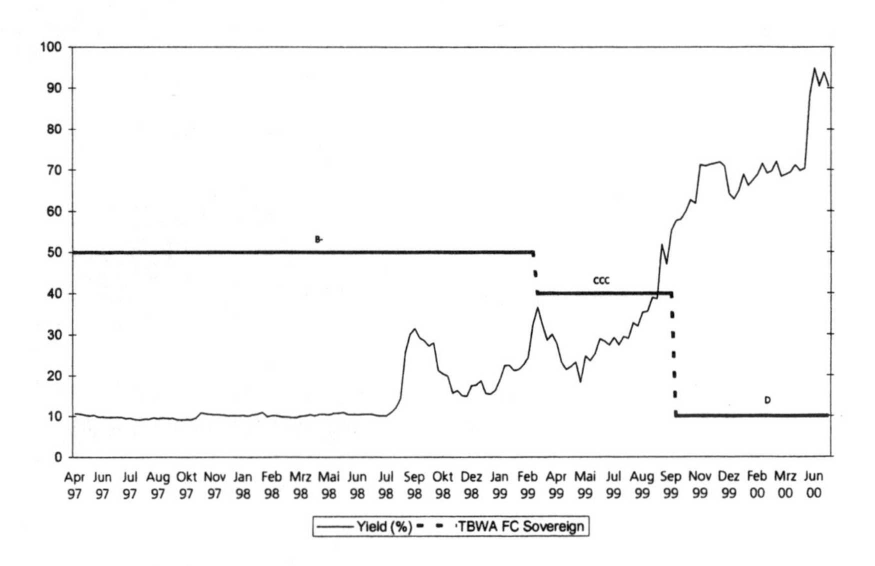

Fig. 6.1. Yields to maturity (in %) and Thomson Bankwatch Fitch IBCA (TBWA FC) ratings (B-, CCC, D) of the Ecuadorian sovereign bond BEEC 11.25% 25/04/2002 DEM from April 1997 until June 2000. Source: Reuters Information Services.

- Many models for pricing defaultable zero coupon bonds don't allow for positive short term credit spreads: As time to maturity tends to zero the model credit spreads tend to zero, too. Obviously, these models are far away from describing reality. In figure 6.3 we show observed corporate credit spreads for different maturities. Even bonds with a rating of A and time to maturity of seven days can have credit spreads which are higher than 500 basis points. Hence, a model must be capable of generating high credit spreads even for short term debt.
- Yields and credit spreads usually depend on the quality of the obligor. The better the quality the smaller the yields (compare to figure 6.4). The time series of the yields show high correlations between the rating classes A and AAA. Gains and losses are - at least to some extend - caused by changes in the quality of the obligor.
- Credit spread processes often show mean reversion properties. See, e.g., figure 6.5.
- Non-defaultable interest rates are usually mean reverting. See, e.g., figure 6.6.

In the following section we develop a three-factor defaultable term structure model for the pricing of a wide range of risky debt contracts and derivatives. It combines structural and reduced-form models. One of the factors that

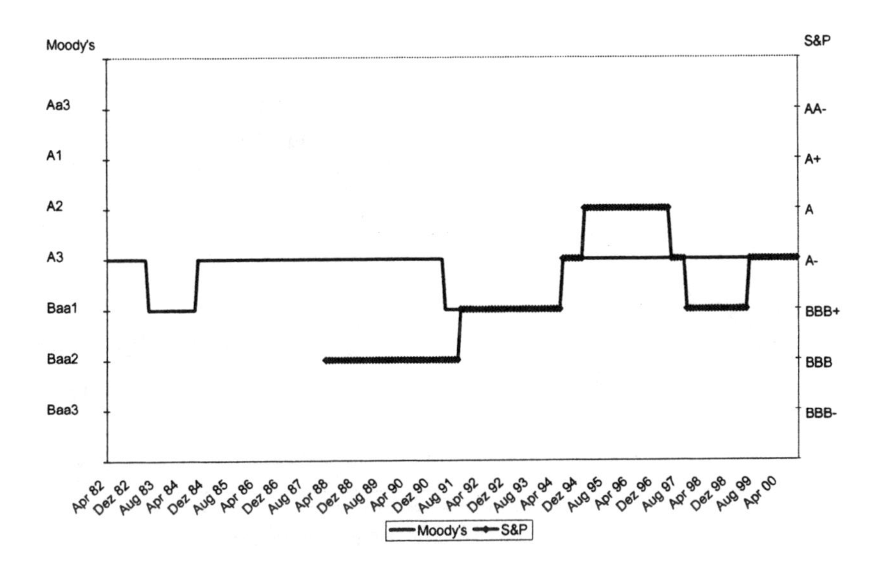

Fig. 6.2. History of the Moody's and S&P ratings of Hertz from April 1982 until August 2000. Source: Reuters Information Services.

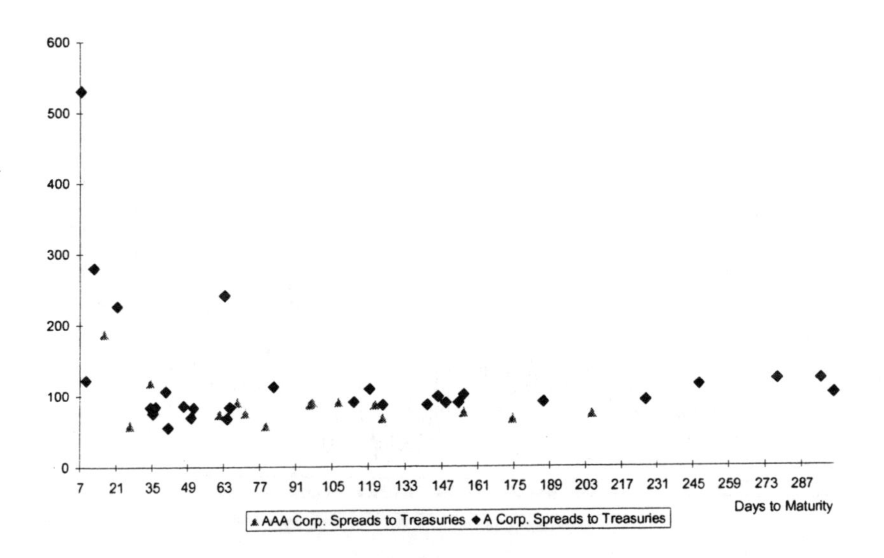

Fig. 6.3. Credit spreads of corporate bonds with S&P ratings of AAA and A for different maturities (in basis points). Source: Reuters Information Services.

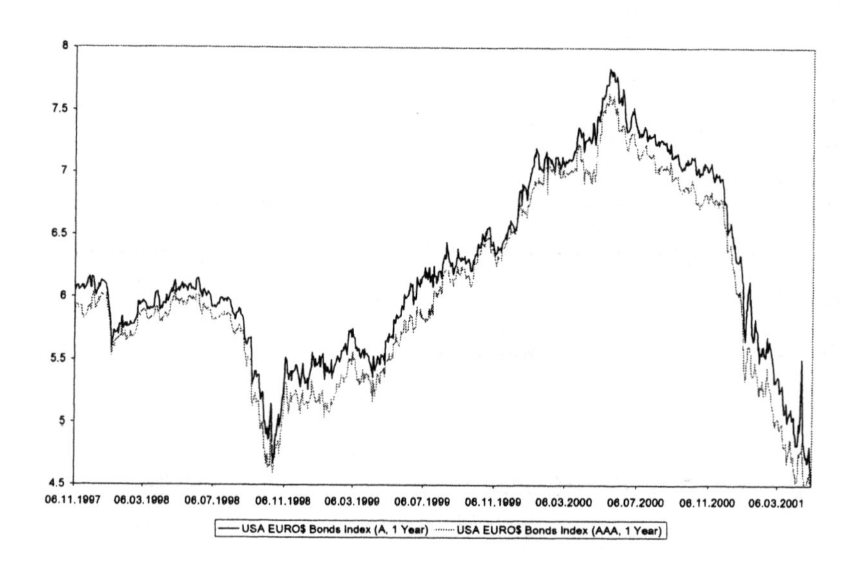

Fig. 6.4. 1 year USA Euro $ bonds indices for the ratings A and AAA (yields in %). Source: Bloomberg.

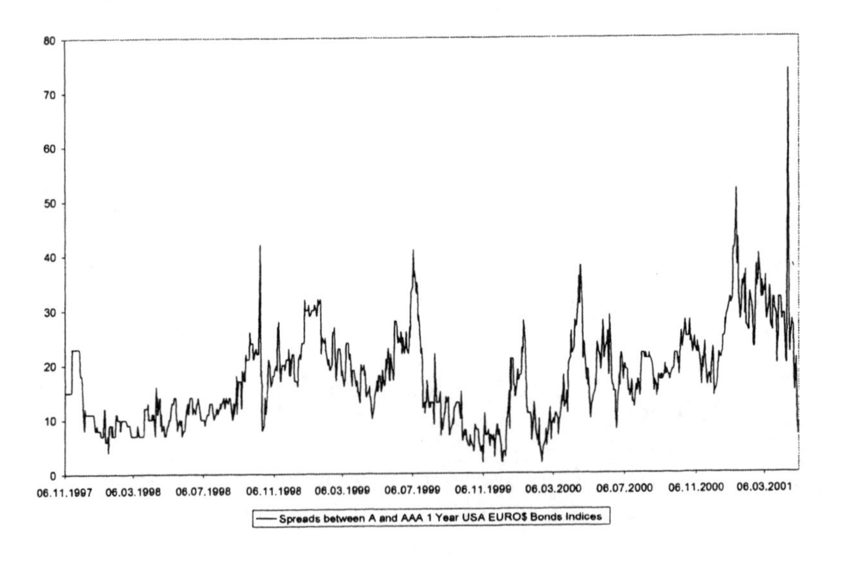

Fig. 6.5. 1 year credit spreads between A and AAA Euro$ bonds indices (in basis points). Source: Bloomberg.

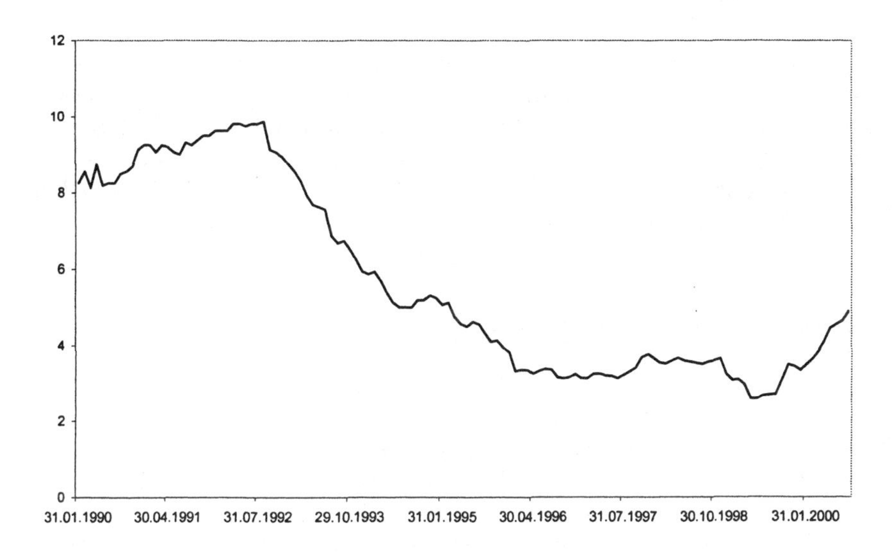

Fig. 6.6. 3-Months LIBOR from 1990 until 2000 (in %). Source: Bloomberg.

determine the credit spread is the so called uncertainty index which can be understood as an aggregation of all information on the quality of the firm currently available: The greater the value of the uncertainty process the lower the quality of the firm. A similar idea to this uncertainty process was first introduced by Cathcart & El-Jahel (1998) in form of a so called signaling process. In their model default is explicitly driven by the signaling process which is assumed to follow a diffusion process. The second underlying process is the non-defaultable short rate which is assumed to follow a mean reverting square root process. Our model differs from Cathcart and El-Jahel in several ways: First, we assume the underlying non-defaultable short rate to follow either a mean reverting Hull-White process or a mean-reverting square root process with time-dependent mean reversion level. The uncertainty or signaling process is assumed to follow a mean reverting square root process. We start our considerations in the "real world" instead of the "risk-adjusted world" and use Girsanov's theorem for the change of measure. Finally, in addition to the non-defaultable short rate and the uncertainty index, we directly model the short rate spread process: we assume that the spread between a defaultable and a non-defaultable bond is considerably driven by the uncertainty index but that there may be additional factors which influence the level of the spreads: at least the contractual provisions, liquidity and the premium demanded in the market for similar instruments have a great impact on credit spreads; we can easily relate credit spreads to business cycles by replacing the uncertainty index by some index of macroeconomic variables without any change of our theoretical framework.

Our approach seems to be reasonable in that credit spreads provide useful observable information on data upon which pricing models can be based. In addition, the model can be fitted directly to match the actual process followed by interest rate credit spreads. The analytical solution obtained for defaultable bonds can be implemented easily in practice, as all the variables and parameters can be implied from market data.

6.2 The Three-Factor Model

6.2.1 The Basic Setup

We want to develop a relative pricing theory based on the no-arbitrage assumption to determine fair values for defaultable securities and credit derivatives. As usual, we assume that markets are frictionless and perfectly competitive, that trading takes place continuously, that there are no taxes, transaction costs, or informational asymmetries, and that investors act as price takers. We fix a terminal time horizon T^*. Uncertainty in the financial market is modelled by a complete probability space $(\Omega, \mathcal{F}, \mathbf{P})$ and a filtration $\mathbf{F} = (\mathcal{F}_t)_{0 \leq t \leq T^*}$ of the three-dimensional standard Brownian motion $W_t^{'} = (W_r(t), W_s(t), W_u(t))$ (v' denotes the transpose of a vector v) satisfying the usual conditions (i.e. $\mathcal{F}_0$ is trivial and contains all the $\mathbf{P}$–null sets of $\mathcal{F}$ and the filtration $\mathbf{F}$ is right continuous). In addition, we assume $\mathcal{F}_{T^*} = \mathcal{F}$. The filtration $\mathbf{F}$ represents the arrival of information over time. We will assume throughout that the price of a security at time t is zero if all dividend payments after time t are zero. All prices are taken to be ex-dividend. All processes are defined on the probability space $(\Omega, \mathcal{F}, \mathbf{P})$. Furthermore, we make the following assumptions:

Assumption 6.2.1
The dynamics of the non-defaultable short rate are given by the following stochastic differential equation (SDE):

$$dr_t = [\theta_r(t) - a_r r_t]\, dt + \sigma_r r_t^{\beta} dW_r(t), \ 0 \leq t \leq T^*, \tag{6.1}$$

where a_r, $\sigma_r > 0$ are positive constants, $\beta = 0$ or $\frac{1}{2}$, and θ_r is a non–negative valued deterministic function.

This specification implies that the current rate r_t is pulled towards $\frac{\theta_r(t)}{a_r}$ with a speed of adjustment a_r, and if $\beta = \frac{1}{2}$ the instantaneous variance of the change in the rate is proportional to its level.

Assumption 6.2.2
The development of the uncertainty index is given by the following stochastic differential equation:

$$du_t = [\theta_u - a_u u_t]\,dt + \sigma_u \sqrt{u_t} dW_u(t)\,,\ 0 \le t \le T^*, \tag{6.2}$$

where a_u, $\sigma_u > 0$ *are positive constants and* θ_u *is a non–negative constant.*

Assumption 6.2.3
The dynamics of the short rate spread (the short rate spread is supposed to be the defaultable short rate minus the non-defaultable short rate) is given by the following stochastic differential equation:

$$ds_t = [b_s u_t - a_s s_t]\,dt + \sigma_s \sqrt{s_t} dW_s(t)\,,\ 0 \le t \le T^*, \tag{6.3}$$

where a_s, $b_s, \sigma_s > 0$ *are positive constants.*

Additionally, we assume that

$$\begin{aligned} Cov\left(dW_r(t), dW_s(t)\right) &= Cov\left(dW_r(t), dW_u(t)\right) \\ &= Cov\left(dW_s(t), dW_u(t)\right) = 0. \end{aligned}$$

Note, that although we assume uncorrelated standard Brownian motions W_r, W_s, and W_u, the short rate spread s_t and the uncertainty index u_t are correlated through the stochastic differential equation for the short rate spread.

First we show that the definition of r, s, and u, makes sense from the mathematical point of view. Afterwards we give some economical motivation and interpretation.

Lemma 6.2.1.
Let r_t, u_t, *and* s_t *be stochastic processes defined by equations (6.1), (6.2), and (6.3), respectively. Assume that* $\theta_r(t)$ *is bounded*[1] *on* $[0,T^*]$, *i.e.* $\theta_r(t) \le \theta_r \forall$ $t \in [0,T^*]$. *In addition, let*

$$dX_t = b(X_t)\,dt + \sigma(X_t)\,dW_t,\ 0 \le t \le T^*, \tag{6.4}$$

define a three-dimensional stochastic differential equation[2], *where*

[1] The assumption that $\theta_r(t)$ is bounded on $[0,T^*]$ is not too restrictive. As we show later, we infer $\theta_r(t)$ from market data by applying methods that ensure that $\theta_r(t)$ is a continuous function on $[0,T^*]$. Hence, in all our cases $\theta_r(t)$ is bounded on $[0,T^*]$.

[2] Let

- $W = (W_1, ..., W_p)$ an $\mathbf{R}^p-$ valued $\mathcal{F}_t-$Brownian motion.
- $b : \mathbf{R}^+ \times \mathbf{R}^n \to \mathbf{R}^n$, $b(s,x) = (b_1(s,x), ..., b_n(s,x))$.
- $\sigma : \mathbf{R}^+ \times \mathbf{R}^n \to \mathbf{R}^{n\times p}$, $\sigma(s,x) = (\sigma_{i,j}(s,x))_{1\le i\le n, 1\le j\le p}$.
- $Z = (Z_1, ..., Z_n)$ an $\mathcal{F}_0-$measurable random variable in $\mathbf{R}^n$.

$$b(X_t) = b(r_t, u_t, s_t)$$
$$= (b_i(r_t, u_t, s_t))_{1 \leq i \leq 3} = \begin{pmatrix} \theta_r(t) - a_r r_t \\ \theta_u - a_u u_t \\ b_s u_t - a_s s_t \end{pmatrix},$$

$$\sigma(X_t) = \sigma(r_t, u_t, s_t)$$
$$= (\sigma_{i,j}(r_t, u_t, s_t))_{1 \leq i,j \leq 3} = \begin{pmatrix} \sigma_r r_t^\beta & 0 & 0 \\ 0 & \sigma_u \sqrt{u_t} & 0 \\ 0 & 0 & \sigma_s \sqrt{s_t} \end{pmatrix},$$

and

$$W_t = (W_r(t), W_u(t), W_s(t))'.$$

Then, given the $\mathcal{F}_0$−measurable random variable (r_0, u_0, s_0), equation (6.4) has a unique weak solution[3] $X = (X_t)_{0 \leq t \leq T^*} = (r_t, u_t, s_t)'_{0 \leq t \leq T^*}$.

Proof.
See appendix B.

Lemma 6.2.2.
r, s and u are $\mathcal{F}_t$−progressively measurable processes on $[0, T^*]$.

Then

$$X_t = Z + \int_0^t b(s, X_s)\,ds + \int_0^t \sigma(s, X_s)\,dW_s$$

is called a multi-dimensional stochastic differential equation. In other words, a multi-dimensional stochastic process is a process $(X_t)_{0 \leq t \leq T^*}$ with values in $\mathbf{R}^n$, adapted to the filtration $(\mathcal{F}_t)_{t \geq 0}$ and such that $\mathbf{P}$ a.s., for any t and for any $i \leq n$

$$X_i(t) = Z_i + \int_0^t b_i(s, X_s)\,ds + \sum_{j=1}^{p} \int_0^t \sigma_{i,j}(s, X_s)\,dW_j(s).$$

[3] A weak solution of the stochastic differential equation (6.4) is a triple (X, W), $(\Omega, \mathcal{F}, \mathbf{P})$, $\mathbf{F} = (\mathcal{F}_t)$, where

1. $(\Omega, \mathcal{F}, \mathbf{P})$ is a probability space, and $\mathbf{F} = (\mathcal{F}_t)$ is a filtration of sub-σ-fields of $\mathcal{F}$ satisfying the usual conditions,
2. X is a continuous, $\mathcal{F}_t$−adapted $\mathbf{R}^3$-valued process, W is a three-dimensional Brownian motion,
3. $\mathbf{P}\left[\int_0^t \{|b_i(X_s)| + \sigma_{ij}^2(X_s)\}\,ds < \infty\right] = 1$ holds for every $1 \leq i, j \leq 3$ and $0 \leq t \leq T^*$,
4. the integral version of (6.4) $X_t = X_0 + \int_0^t b(X_s)\,ds + \int_0^t \sigma(X_s)\,dW_s$, $0 \leq t \leq T^*$, holds almost surely.

Proof.
All three processes are continuous and as such progressively measurable[4] on $[0, T^*]$.

The assumption of the dynamics of the non-defaultable short rate can be replaced by any other short rate specification without major changes. To determine the parameters of the non-defaultable short rate one can use either the swap curve or the Treasury curve. But for the pricing of credit derivatives, the market uses as its base non-defaultable curve the LIBOR-based swap curve rather than the Treasury curve. The uncertainty index can be interpreted as a measure for the quality of the obligor: the larger the value of the uncertainty index the worse the quality. In case of corporates one could use data from rating agencies or firm values to fit the uncertainty index and estimate the parameters. In case of sovereigns one could use data from rating agencies or some kind of macroeconomic data. The model admits various interpretations and is very flexible to capture the characteristics of many different types of obligors. The assumption of mean reversion of the uncertainty index is motivated by several empirical observations. For example, Elsas, Ewert, Krahnen, Rudolph & Weber (1999) mention that a downgrading usually makes the management of a firm start taking actions to improve the firm's quality to reach the previous level again. In addition, the assumption of mean reversion of the short rate spread agrees with the prior beliefs of many academics and practitioners. For example, Taurén (1999) points out, that his empirical studies support the opinion that models which do not have mean reversion should be strongly rejected. Our model generalizes previous models in that the mean reversion level is time-dependent and may depend on the uncertainty index. Thus, we take into consideration that the rating has a great impact on credit spreads. There are some credit spread issues which shouldn't be ignored: Often, the yield spread between defaultable and non-defaultable assets is not a pure credit spread. Loans and bonds can contain explicit or implicit options. In that case, the observed spread must be adjusted for the value of these options. In addition, the observed spread can reflect liquidity or other forms of risk as well. Ideally, these risk components should be taken out of observed credit spreads.

As usual our first task is to find a convenient characterization of the no-arbitrage assumption. Absence of arbitrage is guaranteed by the existence of an equivalent martingale measure $\mathbf{Q}$. By definition, an equivalent martingale measure has to satisfy $\mathbf{Q} \sim \mathbf{P}$ and the discounted security price processes have to be local $\mathbf{Q}$−martingales with respect to a suitable numéraire.

[4] Korn & Korn (1999), p.36: If all paths of a stochastic process are right continuous the process is progressively measurable.

Definition 6.2.1.
The non-defaultable money market account is defined by

$$B(t) = e^{\int_0^t r_l dl}.$$

Taking the non-defaultable money market account as numéraire we have:

Definition 6.2.2.
There are no locally bounded arbitrage opportunities if and only if there is a probability measure $\mathbf{Q}$ *defined on* $(\Omega, \mathcal{F}, \mathbf{P})$ *equivalent to* $\mathbf{P}$ *under which the discounted security price processes become local* $\mathbf{Q}-$*martingales. This measure* $\mathbf{Q}$ *is called an equivalent martingale measure, and for any security price process* $X(t)$ *the discounted price process is defined as* $\frac{X_t}{B_t}$.

In our Brownian setting (basically $\mathcal{F}_t = \sigma(W_\tau, 0 \leq \tau \leq t)$ slightly enlarged to satisfy the usual conditions) we know that each equivalent martingale measure $\mathbf{Q}$ is given in terms of a Radon-Nikodym-Derivative[5]

$$\frac{d\mathbf{Q}}{d\mathbf{P}} \text{ and } \frac{d\mathbf{Q}_t}{d\mathbf{P}_t}$$

where $\mathbf{Q}_t = \mathbf{Q}|\mathcal{F}_t$ and $\mathbf{P}_t = \mathbf{P}|\mathcal{F}_t$ are the restrictions of $\mathbf{Q}$ and $\mathbf{P}$ on $\mathcal{F}_t$, with

$$\frac{d\mathbf{Q}_t}{d\mathbf{P}_t} = \exp\left(-\int_0^t \gamma_\tau' dW(\tau) - \frac{1}{2}\int_0^t \|\gamma_\tau\|^2 d\tau\right),$$

where $\|\cdot\|$ denotes the Euclidean norm and

$$\left\{\gamma_t' = (\gamma_r(t), \gamma_s(t), \gamma_u(t)) : 0 \leq t \leq T^*\right\}$$

is an adapted measurable three-dimensional process satisfying

$$\int_0^{T^*} \gamma_i^2(t)\, dt < \infty \text{ for } i = r, s, u \ \ \mathbf{P} - a.s.$$

More precisely, we assume (e.g., compare to Chen (1996), p.5)

$$\begin{aligned} \gamma_r(t) &= \lambda_r \sigma_r r_t^{1-\beta}, \\ \gamma_s(t) &= \lambda_s \sigma_s \sqrt{s_t}\,, \text{ and} \\ \gamma_u(t) &= \lambda_u \sigma_u \sqrt{u_t},\ 0 \leq t \leq T^*, \end{aligned}$$

where λ_r, λ_s and λ_u are real constants and assume that Novikov's condition

$$E\left(\exp\left(\frac{1}{2}\int_0^{T^*} \|\gamma_\tau\|^2 d\tau\right)\right) < \infty$$

[5] For details see, e.g., Bingham & Kiesel (1998), §5.8.

is satisfied. Then by application of Girsanov's theorem[6] the process

$$\left\{\hat{W}_t^{'} = \left(\hat{W}_r(t), \hat{W}_s(t), \hat{W}_u(t)\right) : 0 \le t \le T^*\right\}$$

defined by

$$\hat{W}_i(t) = W_i(t) + \int_0^t \gamma_i(\tau)\, d\tau, \; i = r, s, u, \; 0 \le t \le T^*,$$

is a three-dimensional standardized Brownian Motion for the filtered probability space $(\Omega, \mathcal{F}, \mathbf{F}, \mathbf{Q})$ restricted to the time interval $[0, T^*]$. Under the measure $\mathbf{Q}$ the dynamics of r_t, s_t and u_t are given by

$$dr_t = [\theta_r(t) - \hat{a}_r r_t]\, dt + \sigma_r r_t^{\beta}\, d\hat{W}_r(t), \tag{6.5}$$

$$ds_t = [b_s u_t - \hat{a}_s s_t]\, dt + \sigma_s \sqrt{s_t} d\hat{W}_s(t), \tag{6.6}$$

$$du_t = [\theta_u - \hat{a}_u u_t]\, dt + \sigma_u \sqrt{u_t} d\hat{W}_u(t), \; 0 \le t \le T^*, \tag{6.7}$$

where $\hat{a}_i = a_i + \lambda_i \sigma_i^2$.

Remark 6.2.1.
Obviously, existence and uniqueness of a solution of equations (6.5), (6.6), and (6.7) in the sense of lemma 6.2.1 is satisfied for the following reasons:

- $Cov\left(d\hat{W}_r(t), d\hat{W}_s(t)\right) = Cov\left(d\hat{W}_r(t), d\hat{W}_u(t)\right)$
  $= Cov\left(d\hat{W}_s(t), d\hat{W}_u(t)\right) = 0.$
- $\theta_r(t)$ doesn't change under the measure transformation.
- The basic structures of $b(r_t, u_t, s_t)$ and $\sigma(r_t, u_t, s_t)$ doesn't change under the measure transformation. We only have to replace the constants a_i, $i = r, s, u$, by the constants $\hat{a}_i$, $i = r, s, u$.

6.2.2 Valuation Formulas For Contingent Claims

In general, a contingent claim (or just a claim) or derivative asset is a financial instrument whose payoffs are exactly determined by the payoffs on one or more underlying assets (for a more detailed definition see, e.g., Ingersoll (1987), pages 50-51). Technically speaking, a simple (European) non-defaultable contingent $T-$claim is a pair (Y, T) consisting of a non-negative $\mathcal{F}_T-$measurable random variable Y and the time $T \le T^*$, at which Y is paid. This concept can be generalized to defaultable claims:

Definition 6.2.3.
A simple (European) defaultable claim is a triple $\left[(Y,T), Z, T^d\right]$ *consisting of*

[6] See, e.g., Karatzas & Shreve (1988), pp. 190-201.

- *a simple (European) non-defaultable contingent claim* (Y, T) *yielding a payoff* Y *at time* T*, provided there has been no default, i.e.* (Y, T) *is describing the obligation of the issuer, and satisfying* $E^{\mathbf{Q}}\left[|Y|^q\right] < \infty$ *for some* $q > 1$;
- *a* $\mathcal{F}_t$*−predictable process* Z *describing the payoff upon default, and satisfying* $E^{\mathbf{Q}}\left[\sup_t |Z_t|^q\right] < \infty$;
- *a* $\mathbf{F}$*−stopping time* T^d *valued in* $[0, \infty)$, *describing the stochastic structure of the default time.*

The default time may be a surprise, i.e. totally inaccessible. With positive probability it can't be foreseen immediately before it occurs. As in section 2.3.4 the default indicator function $H(t) = H_t$ is defined by

$$H_t = 1_{\{T^d \le t\}},\ t \ge 0.$$

In addition, we define

$$L_t = 1 - H_t,\ t \ge 0, \tag{6.8}$$

and let $\tau = \min\left(T, T^d\right)$. In terms of the default indicator function, the cumulative dividend process of the defaultable security is given by

$$\int_0^{t \wedge T} Z_u dH_u + Y 1_{\{T^d > T, t \ge T\}},\ t \ge 0.$$

The arbitrage free price at time t, F_t, of the simple defaultable contingent claim $\left[(Y, T), Z, T^d\right]$ can be obtained as the discounted expected value of the future cash flows[7]. This expectation has to be taken with respect to $\mathbf{Q}$, that is

$$F_t = E^{\mathbf{Q}}\left[\left.\int_t^T e^{-\int_t^u r(l)dl} Z_u dH_u + e^{-\int_t^T r(l)dl}\, Y L_T \right| \mathcal{F}_t\right],\ 0 \le t < \tau, \tag{6.9}$$

and $F_t = 0$ for $t \ge \tau$ since there are no dividends after time τ.

In the following we derive a representation for F which doesn't explicitly involve the stopping time T^d. Therefore we have to state some assumptions, definitions and lemmas first:

Assumption 6.2.4
We assume that the payoff of the claim upon default is given by

$$Z_t(\omega) = Z(\omega, t, F) = w(\omega, t)\, F(\omega, t-),\ 0 \le w(\omega, t) \le 1,$$

where $w(t)$ *is a* $\mathcal{F}_t$*−predictable process (such that* $E^{\mathbf{Q}}\left[\sup_t |Z_t|^q\right] < \infty$*) describing the recovery rate process, i.e. we assume partial recovery of market value*[8].

[7] See, e.g., Duffie et al. (1996).

[8] See section 2.4.

Now we can define a defaultable money market account to be the value of 1 invested at $t = 0$ in a defaultable zero coupon bond of very short maturity, and if there is no default up to t rolled over until that time. Hence, the defaultable money market account can be compared to the non-defaultable money market account B_t in non-defaultable interest rate modelling.

Definition 6.2.4.
The defaultable money market account B^d is defined as

$$B_t^d = \left(1 + \int_0^t (w_l - 1)\, dH_l\right) e^{\int_0^t (r_l + s_l) dl}.$$

Lemma 6.2.3.
$M_t^d = \int_0^t s_l dl + \int_0^t (w_l - 1)\, dH_l$ is a $\mathcal{F}_t$–(local) martingale.

Proof.
According to definition 6.2.2 and our no-arbitrage assumption, under the measure $\mathbf{Q}$, the discounted process

$$\frac{B_t^d}{B_t} = \left(1 + \int_0^t (w_l - 1)\, dH_l\right) e^{\int_0^t s_l dl}$$

must be a (local) martingale. But this is the Doléans Dade exponential[9] of

$$M_t^d = \int_0^t s_l dl + \int_0^t (w_l - 1)\, dH_l,$$

which in turn must also be a (local) martingale[10]: Observe that M_t^d is of finite variation[11]. Hence, the Doléans Dade exponential of M_t^d is given by

[9] See Musiela & Rutkowski (1997), p.245: Let U be a real-valued semimartingale, right continuous with existing left limmits (RCLL, sometimes also called càdlàg), defined on $(\Omega, \mathcal{F}, \mathbf{Q})$ with $U_0 = U_{0-} = 0$. We write $\mathcal{E}(U)$ to denote the Doléans Dade exponential of U; that is, the unique solution of the stochastic differential equation

$$d\mathcal{E}_t(U) = \mathcal{E}_{t-}(U)\, dU_t, \tag{6.10}$$

with $\mathcal{E}_0(U) = 1$.The solution to equation (6.10) is explicitely known, namely

$$\mathcal{E}_t(U) = e^{U_t - \frac{1}{2}[U]_t^c} \prod_{u \le t} (1 + \Delta_u U)\, e^{-\Delta_u U},$$

where $\Delta_u U = U_u - U_{u-}$, and $[U]^c$ is the path-by-path continuous part of $[U]$.

[10] See Lando (1995), p.4: Given a semimartingale U with $U_0 = U_{0-} = 0$. U is a local martingale iff $\mathcal{E}(U)$ is a local martingale.

[11] Jacod & Shiryaev (1987), page 59: If U is of finite variation,

$$\mathcal{E}_t(U) = e^{U_t} \prod_{u \le t} (1 + \Delta_u U)\, e^{-\Delta_u U}.$$

$$\begin{aligned}
\mathcal{E}_t\left(M^d\right) &= e^{M_t^d} \prod_{u \le t} \left(1 + \Delta_u M^d\right) e^{-\Delta_u M^d} \\
&= e^{\int_0^t s_l dl + \int_0^t (w_l - 1) dH_l} \prod_{u \le t} \left(1 + \int_{u-}^{u} (w_l - 1)\, dH_l\right) e^{-\int_{u-}^{u} (w_l - 1) dH_l} \\
&= e^{\int_0^t s_l dl} \prod_{u \le t} \left(1 + \int_{u-}^{u} (w_l - 1)\, dH_l\right) \\
&= \left(1 + \int_0^t (w_l - 1)\, dH_l\right) e^{\int_0^t s_l dl}.
\end{aligned}$$

Lemma 6.2.4.
For a given continuous function $f : \mathbb{R} \times \Omega \times [0, \infty) \to \mathbb{R}$, an $\mathcal{F}_T$-measurable random variable Y, and a process of finite variation $\{D_t : 0 \le t \le T\}$, suppose there is some $q \in [1, \infty)$ such that $E^{\mathbf{Q}}\left[\left(\int_0^T |f(0, \omega, t)|\, dt\right)^q\right] < \infty$, $E^{\mathbf{Q}}\left[|Y|^q\right] < \infty$, and suppose there is some constant $K > 0$ such that

$$|f(x, \omega, t) - f(y, \omega, t)| \le K\, |x - y|$$

for all (ω, t) and all $x, y \in \mathbb{R}$. Then there exists a unique solution V to the recursive stochastic integral equation

$$V_t = E^{\mathbf{Q}}\left[\int_t^T f(V_l, \omega, l)\, dl + \int_t^T dD_l + Y \,\middle|\, \mathcal{F}_t\right], \ t \le T,$$

in the space of all RCLL adapted processes that satisfy $E^{\mathbf{Q}}\left[\left(\int_0^T |V_t|\, dt\right)^q\right] < \infty$.

Proof.
See Duffie & Huang (1996), page 941.

Lemma 6.2.5.
Let M be a $\mathbf{Q}$−martingale. If Y is predictable and $E^{\mathbf{Q}}\left[\sup_t |Y_t|^q\right] < \infty$ for some $q > 1$, then $\int Y dM$ is a $\mathbf{Q}$−martingale.

Proof.
See Duffie et al. (1996), page 1086.

Lemma 6.2.6.
Let Y be a semimartingale (see definition 2.3.3) satisfying $E^{\mathbf{Q}}\left[\int_0^T |Y_t|\, dt\right] < \infty$, let D be a semimartingale satisfying $E^{\mathbf{Q}}\left[\int_0^T |dD_t|\right] < \infty$, where by definition $\int_0^T |dD_t| = \int_0^T d\,|D|_t$ and $|D|_t = VAR_t(D)$ is the variation process of D (see footnote 17), and let G be a progressively measurable process such that $E^{\mathbf{Q}}\left[\int_0^T |G_t|\, dt\right] < \infty$. There exists a martingale m such that

$$dY_t = -G_t dt - dD_t + dm_t,\ t \leq T,$$

if and only if

$$Y_t = E^{\mathbf{Q}} \left[\int_t^T G_u du + \int_t^T dD_u + Y_T \middle| \mathcal{F}_t \right],\ t \leq T.$$

Proof.
See Duffie et al. (1996), page 1086, and Duffie & Huang (1996), page 942.

Corollary 6.2.1.
Let Y be a semimartingale satisfying $E^{\mathbf{Q}} \left[\int_0^T |Y_t| \, dt \right] < \infty$. Suppose that D is an adapted RCLL process of integrable variation[12]*, and ξ is a progressively measurable process. In addition, assume that the properties of ξ guarantee that all processes are well defined*[13]*. Then*

$$dY_t = -dD_t - Y_t \xi_t dt + dm_t,\ t \leq T,$$

for some martingale m, if and only if

$$Y_t = E^{\mathbf{Q}} \left[\int_t^T e^{\int_t^u \xi_v dv} dD_u + e^{\int_t^T \xi_v dv} Y_T \middle| \mathcal{F}_t \right].$$

Proof. See Duffie et al. (1996), page 1086.

Remark 6.2.2.

1. If D is an adapted $RCLL$ process of integrable variation it is a semimartingale satisfying $E^{\mathbf{Q}} \left[\int_0^T |dD_t| \right] < \infty$ (see, e.g., Protter (1992), page 47).
2. If D is of integrable variation or if D satisfies $E^{\mathbf{Q}} \left[\int_0^T |dD_t| \right] < \infty$ it is of finite variation (see, e.g., Jacod & Shiryaev (1987), page 29).

Using lemma 6.2.3 and corollary 6.2.1 we can prove the following important proposition:

Proposition 6.2.1.
Let $E^{\mathbf{Q}} \left[|Y|^q \right] < \infty$ for some $q > 1$. Under assumption 6.2.4 and the dynamics specified by equations (6.5), (6.6) and (6.7) the price process, F_t, given by equation (6.9), satisfies

[12] The stochastic process D is of integrable variation if

$$E^{\mathbf{Q}} \left[VAR(D)_\infty \right] < \infty,$$

where $VAR(D)_\infty = \lim_{t \to \infty} VAR(D)_t$. $VAR(D)_t$ is defined in footnote 17.

[13] See lemma 6.2.6.

$$F_t = V_t 1_{\{t<\tau\}}, \tag{6.11}$$

where the adapted continuous process V is given by

$$V_t = E^{\mathbf{Q}}\left[e^{-\int_t^T (r_l+s_l)dl} Y \mid \mathcal{F}_t\right],\ 0 \le t < T, \tag{6.12}$$

and $V_t = 0$ for $t \ge T$, i.e. if there has been no default until time t, F_t must equal the expected value of riskless cash flows discounted at risky discount rates. Equation (6.12) has a unique solution in the space consisting of every semimartingale, J, such that $E^{\mathbf{Q}}\left[\sup_t |J_t|^q\right] < \infty$ for some $q > 1$.

Proof.
Existence and uniqueness of equation (6.12) follow immediately from lemma 6.2.4. It remains to show that equation (6.11) is valid.

- Because V satisfies the integral equation (6.12) only on $[0, T)$, we define a modified process $\widehat{V}$ by

$$\widehat{V}_t = V_t 1_{\{t<T\}} + Y 1_{\{t \ge T\}},$$

  to extend the definition on the compact interval $[0, T]$.
- Apply corollary 6.2.1 to $\widehat{V}$ with $\xi = -(r+s)$ and $dX_t = 0$ (by lemma 6.2.2 ξ is progressively measurable):

$$\begin{aligned} d\widehat{V}_t &= \widehat{V}_t (r_t + s_t)\, dt + dm_t \\ &= \widehat{V}_t r_t dt + \widehat{V}_t (1 - w_t)\, dH_t + \widehat{V}_t dM_t^d + dm_t \end{aligned}$$

  for some martingale m and M_t^d as defined in lemma 6.2.3.
- Define $U_t = \widehat{V}_t L_t$ with L_t as defined by equation (6.8). Using integration by parts[14],

$$\begin{aligned} dU_t &= \widehat{V}_{t-} dL_t + L_{t-} d\widehat{V}_t + d\left[L, \widehat{V}\right]_t \\ &= \widehat{V}_t dL_t + L_{t-} d\widehat{V}_t \\ &= \widehat{V}_t dL_t + U_t r_t dt + \widehat{V}_t L_{t-} (1 - w_t)\, dH_t + \widehat{V}_t L_{t-} dM_t^d + L_{t-} dm_t \\ &= \widehat{V}_t dL_t + U_t r_t dt + \widehat{V}_t L_{t-} (1 - w_t)\, dH_t + d\widetilde{m}_t. \end{aligned}$$

[14] See, e.g., Protter (1992), p. 60: Let X, Y be two semimartingales. Then XY is a semimartingale and

$$XY = \int X_{\_} dY + \int Y_{\_} dX + [X, Y].$$

The formula follows trivially from the definition of the quadratic covariation $[X, Y]$ of two semimartingales X and Y :

$$[X, Y] = XY - \int X_{\_} dY - \int Y_{\_} dX.$$

as $\widehat{V}$ is continuous and L is a quadratic pure jump semimartingale[15]. Note, that $\widetilde{m}_t$ with $d\widetilde{m}_t = \widehat{V}_t L_{t-} dM_t^d + L_{t-} dm_t$ is a martingale[16].

- Apply corollary 6.2.1 with $\xi = -r$ and $dX_t = -\widehat{V}_t L_{t-} (1 - w_t) dH_t - \widehat{V}_t dL_t$, i.e.

$$dX_t = \begin{cases} 0 & , \text{ if } t \neq T^d, \\ \widehat{V}_{T^d} w_{T^d} & , \text{ if } t = T^d, \end{cases}$$

(by lemma 6.2.2 ξ is progressively measurable; by the definition of H and L, X is an adapted $RCLL$ process; it is of integrable variation as it is adapted $RCLL$, increasing and $E\left(VAR\left(X\right)_\infty\right) \leq \widehat{V}_{T^d} < \infty$):

$$\begin{aligned} & U_t \\ & = E^{\mathbf{Q}}\left[- \int_t^T e^{-\int_t^u r_l dl} \widehat{V}_u \left(L_{u-}\left(1 - w_u\right) - 1\right) dH_u + e^{-\int_t^T r_l dl} Y L_T \middle| \mathcal{F}_t \right] \\ & = E^{\mathbf{Q}}\left[- \int_t^T e^{-\int_t^u r_l dl} \widehat{V}_{u-} \left(L_{u-}\left(1 - w_u\right) - 1\right) dH_u \right. \\ & \quad \left. + e^{-\int_t^T r_l dl} Y L_T \middle| \mathcal{F}_t \right] \\ & = E^{\mathbf{Q}}\left[\int_t^T e^{-\int_t^u r_l dl} \widehat{V}_{u-} w_u dH_u + e^{-\int_t^T r_l dl} Y L_T \middle| \mathcal{F}_t \right]. \end{aligned}$$

- Because $U_t = V_t$ for $t < \tau$,

$$F_t = V_t 1_{\{t<\tau\}}.$$

We calculate equation (6.12) by relating V_t to the solution of a partial differential equation. Under mild regularity conditions (see, e.g., Duffie (1992)) and by using the Feynman-Kac (see, e.g., Duffie (1992)) formula, we get the following lemma:

---

[15] See, e.g., Protter (1992), p.68: Let L be a quadratic pure jump semimartingale. Then for any semimartingale $\widehat{V}$ we have

$$\left[L, \widehat{V}\right]_t = L_0 \widehat{V}_0 + \sum_{0<s\leq t} \Delta L_s \Delta \widehat{V}_s.$$

A semimartingale L will be called quadratic pure jump if $[L, L]^c = 0$.

[16] $\int L_{t-} dm_t$ is a martingale since L is bounded and m is a martingale (see, e.g., Jacod & Shiryaev (1987), p. 47, 4.34 (b)). $\int \widehat{V}_t L_{t-} dM_t^d$ is a martingale by lemma 6.2.5: because V is continuous on $(0, T)$ it must be predictable. In addition, V is in the space of semimartingales, Y, such that $E^{\mathbf{Q}}\left[\sup_t |Y_t|^q\right] < \infty$. By proposition 2.6 in Jacod & Shiryaev (1987), p.17, L_{t-} is predictable.

Lemma 6.2.7.
Assuming the dynamics specified by equations (6.5), (6.6), and (6.7), the pre-default value at time t, $V(r,s,u,t)$, of a defaultable claim $\left[(Y,T),Z,T^d\right]$ with Z like in assumption 6.2.4, is the solution to the partial differential equation (PDE)[17]

$$\begin{aligned} 0 = &\frac{1}{2}\left[\sigma_r^2 r^{2\beta} V_{rr} + \sigma_s^2 s V_{ss} + \sigma_u^2 V_{uu}\right] \\ &+ \left[\theta_r(t) - \hat{a}_r r\right] V_r + \left[b_s u - \hat{a}_s s\right] V_s \\ &+ \left[\theta_u - \hat{a}_u u\right] V_u + V_t - (r+s) V, \quad (r,s,u,t) \in \mathbb{R}^3 \times [0,\mathrm{T}), \end{aligned} \tag{6.13}$$

with boundary condition

$$V(r,s,u,T) = Y(r,s,u), \quad (r,s,u) \in \mathbb{R}^3. \tag{6.14}$$

Similarly, assuming the dynamics specified by equation (6.5) the time t value, $F(r,t)$, of a non-defaultable claim (Y,T) is the solution to the PDE

$$0 = \frac{1}{2}\sigma_r^2 r^{2\beta} F_{rr} + \left[\theta_r(t) - \hat{a}_r r\right] F_r + F_t - rF, \quad (r,t) \in \mathbb{R} \times [0,\mathrm{T}),$$

with boundary condition

$$F(r,T) = Y(r), \quad r \in \mathbb{R}.$$

Hence, the time t value of a non-defaultable discount bond with maturity T, $P(t,T) = P(r,t,T)$, is given by (compare to Vasicek (1977), Hull & White (1990), and Hull (1997))

$$P(r,t,T) = A(t,T)\, e^{-B(t,T)r}, \tag{6.15}$$

where $A(t,T)$ and $B(t,T)$ are defined by

[17] If we assumed

$$\begin{aligned} Cov\left(dW_r(t), dW_s(t)\right) &= \rho_{rs} dt, \\ Cov\left(dW_r(t), dW_u(t)\right) &= \rho_{ru} dt, \\ Cov\left(dW_s(t), dW_u(t)\right) &= \rho_{su} dt, \end{aligned}$$

for some constants $\rho_{rs}, \rho_{ru}, \rho_{su}$, instead of PDE (6.13) we would have to consider the following PDE (see section 6.5.3):

$$\begin{aligned} 0 = &\frac{1}{2}\left[\sigma_r^2 r^{2\beta} V_{rr} + \sigma_s^2 s V_{ss} + \sigma_u^2 V_{uu}\right] \\ &+ \rho_{rs}\sigma_r\sigma_s r^{\beta}\sqrt{s} V_{rs} + \rho_{ru}\sigma_r\sigma_u r^{\beta}\sqrt{u} V_{ru} + \rho_{us}\sigma_u\sigma_s\sqrt{su} V_{us} \\ &+ \left[\theta_r(t) - \hat{a}_r r\right] V_r + \left[b_s u - \hat{a}_s s\right] V_s \\ &+ \left[\theta_u - \hat{a}_u u\right] V_u + V_t - (r+s) V, \quad (r,s,u,t) \in \mathbb{R}^3 \times [0,\mathrm{T}). \end{aligned}$$

$$B(t,T) = \begin{cases} \frac{1}{\hat{a}_r}\left[1 - e^{-\hat{a}_r(T-t)}\right], & \text{if } \beta = 0, \\ \frac{1-e^{-\delta_r(T-t)}}{\kappa_1^{(r)} - \kappa_2^{(r)} e^{-\delta_r(T-t)}}, & \text{if } \beta = \frac{1}{2}, \end{cases} \text{, and,} \tag{6.16}$$

$$\ln A(t,T) = \begin{cases} \int_t^T \left(\frac{1}{2}\sigma_r^2 B(\tau,T)^2 - \theta_r(\tau) B(\tau,T)\right) d\tau \\ = \ln \frac{P(0,T)}{P(0,t)} - B(t,T) \frac{\partial \ln P(0,t)}{\partial t} & \text{if } \beta = 0, \\ -\frac{\sigma_r^2}{4\hat{a}_r^3}\left(e^{-\hat{a}_r T} - e^{-\hat{a}_r t}\right)^2 \left(e^{2\hat{a}_r t} - 1\right), \\ -\int_t^T \theta_r(\tau) B(\tau,T) d\tau, & \text{if } \beta = \frac{1}{2}, \end{cases} \tag{6.17}$$

with

$$\delta_x = \sqrt{\hat{a}_x^2 + 2\sigma_x^2} \text{ and } \kappa_{1/2}^{(x)} = \frac{\hat{a}_x}{2} \pm \frac{1}{2}\delta_x. \tag{6.18}$$

Equations (6.15), (6.16) and (6.17) define the price of a discount bond at a future time t in terms of the short rate at time t and the prices of non-defaultable discount bond prices today. The latter can be calculated from today's term structure. The partial derivative $\frac{\partial \ln P(0,t)}{\partial t}$ in equation (6.17) can be approximated by $\frac{\ln P(0,t+\varepsilon) - \ln P(0,t-\varepsilon)}{2\varepsilon}$ where ε is a small length of time. In addition, if $\beta = 0$, the deterministic function $\theta_r(t)$ is given by

$$\theta_r(t) = f_t(0,t) + \hat{a}_r f(0,t) + \frac{\sigma_r^2}{2\hat{a}_r}\left(1 - e^{-2\hat{a}_r t}\right),$$

where $f(0,t)$ denotes the instantaneous forward rate as seen at time 0 for a contract maturing at time t. If $\beta = \frac{1}{2}$, $\theta_r(t)$ can be obtained iteratively from

$$-\int_0^t \theta_r(\tau) B(\tau,t) d\tau = \ln A(0,t)$$

by time discretization.

6.3 The Pricing of Defaultable Fixed and Floating Rate Debt

6.3.1 Introduction

In the following we determine pricing formulas for defaultable zero coupon bonds, coupon bonds, floating rate notes, and interest rate swaps. If not otherwise stated we always assume face values of 1.

6.3.2 Defaultable Discount Bonds

A defaultable discount bond is a financial instrument promising to pay 1 at some pre-defined maturity date T. Trivially, such a financial instrument satisfies definition 6.2.3 of a simple defaultable claim. So we can use the results of the previous section to determine the time t ($t < T$) price of the claim. Applying lemma 6.2.7 to a defaultable discount bond yields the following theorem:

Theorem 6.3.1.
Assuming the dynamics specified by equations (6.5), (6.6) and (6.7), the value at time $t < \tau$, $P^d(t,T) = P^d(r,s,u,t,T)$ of a defaultable discount bond is given by

$$P^d(t,T) = A^d(t,T)\, e^{-B(t,T)r_t - C^d(t,T)s_t - D^d(t,T)u_t} \tag{6.19}$$

where

$$\begin{aligned} C^d(t,T) &= \frac{1-e^{-\delta_s(T-t)}}{\kappa_1^{(s)} - \kappa_2^{(s)} e^{-\delta_s(T-t)}}, \\ D^d(t,T) &= \frac{-2v'(t,T)}{\sigma_u^2 v(t,T)}, \\ \ln A^d(t,T) &= \ln A(t,T) + \frac{2\theta_u}{\sigma_u^2} \ln \left| \frac{v(T,T)}{v(t,T)} \right|, \end{aligned}$$

where $v(t,T)$ is defined as in equations (6.31) and (6.33) below.

Proof.
According to lemma 6.2.7, the value at time t, $P^d(r,s,u,t,T)$, of a defaultable discount bond is the solution to the PDE

$$\begin{aligned} 0 = &\frac{1}{2}\sigma_r^2 r^{2\beta} P_{rr}^d + \frac{1}{2}\sigma_s^2 s P_{ss}^d + \frac{1}{2}\sigma_u^2 u P_{uu}^d \\ &+ [\theta_r - \hat{a}_r r] P_r^d + [b_s u - \hat{a}_s s] P_s^d \\ &+ [\theta_u - \hat{a}_u u] P_u^d + P_t^d - (r+s) P^d, \end{aligned}$$

with boundary condition $P^d(r,s,u,T,T) = 1$.
If $\beta = 0$, plugging in the partial derivatives of P^d yields the following system of ODEs

$$\hat{a}_r B^d - B_t^d - 1 = 0 \tag{6.20}$$

$$\frac{1}{2}\sigma_s^2 \left(C^d\right)^2 + \hat{a}_s C^d - C_t^d - 1 = 0 \tag{6.21}$$

$$\frac{1}{2}\sigma_u^2 \left(D^d\right)^2 + \hat{a}_u D^d - D_t^d - b_s C^d = 0 \tag{6.22}$$

$$A^d \left(\theta_r B^d + \theta_u D^d - \frac{1}{2}\sigma_r^2 \left(B^d\right)^2 \right) - A_t^d = 0. \tag{6.23}$$

Equation (6.21) has Ricatti form $C_t^d = \frac{1}{2}\sigma_s^2 \left(C^d\right)^2 + \hat{a}_s C^d - 1$. The solution is

$$C^d(t,T) = \frac{-2w'(t,T)}{\sigma_s^2 w(t,T)}$$

where $w(t,T)$ satisfies

$$w'' - \hat{a}_s w' - \frac{1}{2}\sigma_s^2 w = 0. \tag{6.24}$$

From $\hat{a}_s^2 + 2\sigma_s^2 > 0$ it follows that all solutions of the ODE (6.24) are given by $w = \alpha_1 e^{\kappa_1^{(s)} t} + \alpha_2 e^{\kappa_2^{(s)} t}$, where α_1 and α_2 are some constants and $\kappa_1^{(s)}$ and $\kappa_2^{(s)}$ are given by

$$\kappa_{1/2}^{(x)} = \frac{\hat{a}_x}{2} \pm \frac{1}{2}\sqrt{\hat{a}_x^2 + 2\sigma_x^2}. \tag{6.25}$$

Then all solutions to equation (6.21) are given by

$$C^d(t,T) = -\frac{2}{\sigma_s^2} \frac{\alpha_1 \kappa_1^{(s)} e^{\kappa_1^{(s)} t} + \alpha_2 \kappa_2^{(s)} e^{\kappa_2^{(s)} t}}{\alpha_1 e^{\kappa_1^{(s)} t} + \alpha_2 e^{\kappa_2^{(s)} t}}. \tag{6.26}$$

In order to satisfy the boundary condition

$$C^d(T,T) = 0,$$

the constant α_1 equals

$$\alpha_1 = -\alpha_2 \frac{\kappa_2^{(s)}}{\kappa_1^{(s)}} e^{-\delta_s T},$$

and therefore

$$C^d(t,T) = \frac{1 - e^{-\delta_s (T-t)}}{\kappa_1^{(s)} - \kappa_2^{(s)} e^{-\delta_s (T-t)}}, \tag{6.27}$$

where

$$\delta_x = \sqrt{\hat{a}_x^2 + 2\sigma_x^2}. \tag{6.28}$$

Equation (6.22) has Ricatti form $D_t^d = \frac{1}{2}\sigma_u^2 \left(D^d\right)^2 + \hat{a}_u D^d - b_s C^d$. The solution is

$$D^d(t,T) = \frac{-2v'(t,T)}{\sigma_u^2 v(t,T)}, \tag{6.29}$$

where $v(t,T)$ satisfies

$$v'' - \hat{a}_u v' - \frac{1}{2} b_s \sigma_u^2 C^d v = 0. \tag{6.30}$$

The solutions to equation (6.30) are given by

$$\begin{aligned} v(t,T) = {} & \vartheta_1 \left(\sigma_u^2 e^{-\delta_s (T-t)}\right)^{\frac{\hat{a}_u}{2\delta_s} - \phi\left(\kappa_1^{(s)}\right)} F_1(t,T) \\ & + \vartheta_2 \left(\sigma_u^2 e^{-\delta_s (T-t)}\right)^{\frac{\hat{a}_u}{2\delta_s} + \phi\left(\kappa_1^{(s)}\right)} F_3(t,T), \end{aligned} \tag{6.31}$$

where ϑ_1 and ϑ_2 are some constants and

$$\begin{aligned} F_1(t,T) = {} & F\left(-\phi\left(\kappa_1^{(s)}\right) - \phi\left(\kappa_2^{(s)}\right), -\phi\left(\kappa_1^{(s)}\right) + \phi\left(\kappa_2^{(s)}\right),\right. \\ & \left. 1 - 2\phi\left(\kappa_1^{(s)}\right), \kappa_2^{(s)}/\kappa_1^{(s)} e^{-\delta_s (T-t)}\right), \\ F_3(t,T) = {} & F\left(\phi\left(\kappa_1^{(s)}\right) - \phi\left(\kappa_2^{(s)}\right), \phi\left(\kappa_1^{(s)}\right) + \phi\left(\kappa_2^{(s)}\right),\right. \\ & \left. 1 + 2\phi\left(\kappa_1^{(s)}\right), \kappa_2^{(s)}/\kappa_1^{(s)} e^{-\delta_s (T-t)}\right), \end{aligned}$$

with

$$\phi(g) = \sqrt{\frac{\hat{a}_u^2 g + 2b_s\sigma_u^2}{4\delta_s^2 g}}$$

and $F(a,b,c,z)$ is the hypergeometric function[18].
Differentiation of $v(t,T)$ yields

$$v'(t,T) = y_1(t,T) + y_2(t,T), \tag{6.32}$$

where

$$y_1(t,T) = \vartheta_1\left(\sigma_u^2 e^{-\delta_s(T-t)}\right)^{\frac{\hat{a}_u}{2\delta_s}-\phi\left(\kappa_1^{(s)}\right)} \varphi_2(t,T),$$

$$y_2(t,T) = -\vartheta_2\left(\sigma_u^2 e^{-\delta_s(T-t)}\right)^{\frac{\hat{a}_u}{2\delta_s}+\phi\left(\kappa_1^{(s)}\right)} \varphi_1(t,T),$$

with

$$\varphi_1(t,T) = \zeta_2 e^{-\delta_s(T-t)} F_4(t,T) - \xi_1 F_3(t,T),$$
$$\varphi_2(t,T) = \xi_2 F_1(t,T) - \zeta_1 e^{-\delta_s(T-t)} F_2(t,T),$$

where

$$F_2(t,T) = F\left(1-\phi\left(\kappa_1^{(s)}\right)-\phi\left(\kappa_2^{(s)}\right), 1-\phi\left(\kappa_1^{(s)}\right)+\phi\left(\kappa_2^{(s)}\right),\right.$$
$$\left.2-2\phi\left(\kappa_1^{(s)}\right), \kappa_2^{(s)}/\kappa_1^{(s)} e^{-\delta_s(T-t)}\right),$$

$$F_4(t,T) = F\left(1+\phi\left(\kappa_1^{(s)}\right)-\phi\left(\kappa_2^{(s)}\right), 1+\phi\left(\kappa_1^{(s)}\right)+\phi\left(\kappa_2^{(s)}\right),\right.$$
$$\left.2+2\phi\left(\kappa_1^{(s)}\right), \kappa_2^{(s)}/\kappa_1^{(s)} e^{-\delta_s(T-t)}\right),$$

and

[18] The hypergeometric function, usually denoted by F, has series expansion

$$F(a,b,c,z) = \sum_{k=0}^{\infty} \frac{(a)_k (b)_k}{(c)_k} \frac{z^k}{k!},$$

where$(a)_0 = 1$, $(a)_n = a(a+1)(a+2)\ldots(a+n-1)$, $n \in \mathbf{N}$, and is the solution of the hypergeometric differential equation

$$z(1-z)y'' + [c-(a+b+1)z]y' - aby = 0.$$

The hypergeometric function can be written as an integral

$$F(a,b,c,z) = \frac{\Gamma(c)}{\Gamma(b)\Gamma(c-b)} \int_0^1 t^{b-1}(1-t)^{c-b-1}(1-tz)^{-a}\,dt, \quad (c>b>0),$$

and is also known as the Gauss series or the Kummer series.

$$\xi_{1/2} = \left(\frac{\hat{a}_u}{2} \pm \delta_s \phi \left(\kappa_1^{(s)} \right) \right), \quad \zeta_{1/2} = \delta_s \frac{\kappa_2^{(s)}}{\kappa_1^{(s)}} \frac{\phi^2 \left(\kappa_2^{(s)} \right) - \phi^2 \left(\kappa_1^{(s)} \right)}{1 \mp 2\phi \left(\kappa_1^{(s)} \right)}.$$

In order to satisfy the boundary condition

$$D^d (T, T) = 0,$$

ϑ_1 equals

$$\vartheta_1 = \vartheta_2 \left(\sigma_u^2 \right)^{2\phi(\kappa_1(s))} \frac{\varphi_1}{\varphi_2}, \tag{6.33}$$

where $\varphi_1 = \varphi_1 (T, T)$ and $\varphi_2 = \varphi_2 (T, T)$. From equations (6.29), (6.31), (6.32) and (6.33), $D^d (t, T)$ can be determined as

$$D^d (t, T) = \frac{2}{\sigma_u^2} \cdot \frac{\varphi_2 e^{-2\delta_s \phi \left(\kappa_1^{(s)} \right)(T-t)} \varphi_1 (t, T) - \varphi_1 \varphi_2 (t, T)}{\varphi_1 F_1 (t, T) + \varphi_2 e^{-2\delta_s \phi \left(\kappa_1^{(s)} \right)(T-t)} F_3 (t, T)}.$$

Consider the case $\beta = 0$:
The solution of equation (6.20) that satisfies the boundary condition

$$B^d (T, T) = 0$$

is

$$B^d (t, T) = \frac{1}{\hat{a}_r} \left[1 - e^{-\hat{a}_r (T-t)} \right].$$

By direct substitution, the solution of equation (6.23) for A^d that satisfies the boundary condition

$$A^d (T, T) = 1$$

is

$$A^d (t, T) = \exp \left[- \int_t^T \left(\theta_r (\tau) B^d (\tau, T) + \theta_u D^d (\tau, T) - \frac{1}{2} \sigma_r^2 \left(B^d (\tau, T) \right)^2 \right) d\tau \right].$$

According to equation (6.17),

$$- \int_t^T \left(\theta_r (\tau) B^d (\tau, T) - \frac{1}{2} \sigma_r^2 \left(B^d (\tau, T) \right)^2 \right) d\tau = \ln A (t, T),$$

and in addition, according to equation (6.29),

$$\theta_u \int_t^T D^d (\tau, T) d\tau = \frac{-2\theta_u}{\sigma_u^2} \left[\ln \left| \frac{v (T, T)}{v (t, T)} \right| \right],$$

where $\vartheta_1 = \vartheta_2 \left(\sigma_u^2 \right)^{2\phi(\kappa_1(s))} \frac{\varphi_1}{\varphi_2}$.

The case $\beta = \frac{1}{2}$ is considered in appendix B.2.

Remark 6.3.1.
If we extend the definition of the uncertainty index from a constant parameter θ_u to a time-dependent parameter $\theta_u(t)$, the values of $\theta_u(t)$ can be determined iteratively from

$$\ln \frac{A^d(0,t)}{A(0,t)} = \int_0^t \theta_u(\tau) D^d(\tau,t)\, d\tau.$$

Although the bond pricing and credit spread formulas are rather complex and there are many parameters involved it is easy to prove that they show realistic features: $\lim_{T\to\infty} P^d(r,s,u,t,T) = 0$; for $T > t$: $lim_{r\to\infty} P^d(r,s,u,t,T) = \lim_{s\to\infty} P^d(r,s,u,t,T) = 0$, since $B(t,T) > 0$ and $C^d(t,T) > 0$; the price of a defaultable bond is a decreasing function of r and a decreasing function of s; further more, it is a convex function of r and a convex function of s.

Factor Loadings. The yield to maturity of a defaultable discount bond is given by

$$\begin{aligned} &R^d(r,s,u,t,T) \\ &= -\frac{1}{T-t}\left[\ln A^d(t,T) - B(t,T)r - C^d(t,T)s - D^d(t,T)u\right]. \end{aligned}$$

The functions B, C^d and D^d determine the sensitivity of a defaultable zero coupon bond's yield to the factors r, s, and u. We call them factor loadings for r, s, and u, respectively.
Figure 6.7 shows that the sensitivity of a bond's yield to the factors r and s is greater for short and intermediate maturities. The sensitivity to u is greatest for intermediate maturities. All sensitivities die out as maturities approach infinity. This is consistent with the well-known effect of flattening yield curves for long maturities.

The Term Structure of Credit Spreads. The credit spread is defined as the difference between the yields of a defaultable and its corresponding non-defaultable bond. Since $A^d(t,T) = A(t,T)\, e^{\int_t^T \theta_u D^d(\tau,T)d\tau}$, the defaultable bond pricing formula can be written as

$$P^d(r,s,u,t,T) = P(r,t,T)\, e^{\int_t^T \theta_u D^d(\tau,T)d\tau - C^d(t,T)s - D^d(t,T)u}. \tag{6.34}$$

Hence, the default premium of the defaultable bond is given by

$$P(r,t,T)\left(1 - e^{\int_t^T \theta_u D^d(\tau,T)d\tau - C^d(t,T)s - D^d(t,T)u}\right).$$

If we denote the time t yield to maturity of a non-defaultable discount bond with maturity T by $R(r,t,T)$ and the time t yield to maturity of a defaultable discount bond with maturity T by $R^d(r,s,u,t,T)$, the credit spread, $S(r,s,u,t,T)$, is given by

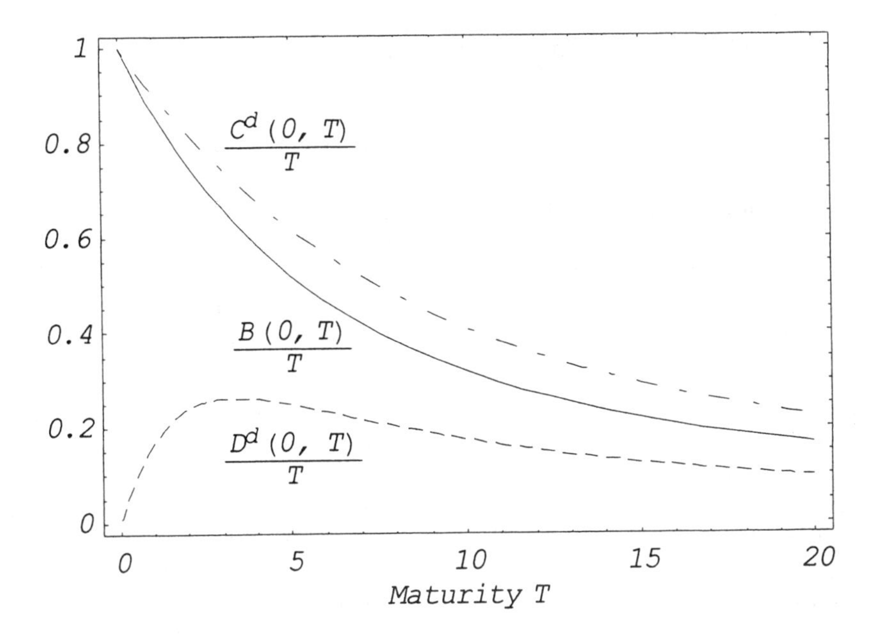

Fig. 6.7. Term structure of $\frac{B(0,T)}{T}$, $\frac{C^d(0,T)}{T}$, and $\frac{D^d(0,T)}{T}$. The values of the parameters are $\hat{a}_r = 0.3$, $\hat{a}_s = 0.2$, $\sigma_s = 0.1$, $b_s = 0.5$, $\hat{a}_u = 1$, $\sigma_u = 0.4$, $\theta_u = 1.5$.

$$\begin{aligned} & S(r, s, u, t, T) \\ &= R^d(r, s, u, t, T) - R(r, t, T) \\ &= \frac{1}{T-t}\left[-\int_t^T \theta_u D^d(\tau, T)\, d\tau + C^d(t, T)\, s + D^d(t, T)\, u\right]. \end{aligned} \tag{6.35}$$

It is obvious from equation (6.35) that the credit spreads implied by our model are independent of the level of the non-defaultable short rate, although the influence of the non-defaultable short rate to the prices of defaultable and non-defaultable bonds is crucial. In addition, the credit spread tends to $\frac{-\int_t^T \theta_u D^d(\tau,T)d\tau}{T-t}$ as s and u tend to 0. $\frac{-\int_t^T \theta_u D^d(\tau,T)d\tau}{T-t}$ can be either positive or negative depending on the values of the parameters. It reflects the fact that there is a certain probability that the quality of the firm under consideration will decrease (i.e. the uncertainty index will increase) between time t and maturity although its current quality is excellent and therefore the current short rate spread is 0. There will always remain a certain amount of default risk and the buyer of the defaultable bond should get a compensation for it. Nevertheless, if $\theta_u \equiv 0$, zero becomes an absorbing state for u. Then, if s tends to 0 the yield spread will also tend to 0. Finally, the model is consistent in that the short rate spread implied by the credit spreads equals[19]

[19] $\lim_{T \to t} \frac{-\int_t^T \theta_u D^d(\tau,T)d\tau}{T-t} = -\theta_u D^d(T, T) = 0,$

$$\lim_{T \to t} [S(r, s, u, t, T)] = s_t. \tag{6.36}$$

Two immediate implications can be drawn from the wider set of parameters influencing the term structure of credit spreads compared to previous models. First, we can expect larger spreads which are closer to those observed in practice. Second, our model captures more complex properties of credit spreads. In order to better understand these properties we consider the following specific parameter values and study the effects of varying them: $s(0) = 0.0008$, $u(0) = 3$, $\theta_u = 1$, $\hat{a}_u = 1$, $\sigma_u^2 = 0.16$, $b_s = 0.0001$, $\hat{a}_s = 0.1$, $\sigma_s^2 = 0.01$. In the following we will refer to these parameters as the base case parameters.

The Effect of the Mean Reversion Parameter θ_u. Figure 6.8 shows that the parameter θ_u and therefore the mean-reversion level of the uncertainty index behave as expected: an increase in θ_u widens the credit spread. The parameter represents the long term expectation of the quality of the firm. The lower this expectation is the smaller the spreads are. As the actual quality of the firm ($u(0) = 3$) is much worse than the long-run levels of the uncertainty index: $\frac{\theta_u}{\hat{a}_u} = 1.5, 1.0, 0.5, 0.1$, the credit spreads narrow for long maturities. For small values of θ_u the credit spread term structures are humped shaped. For larger values of θ_u, i.e. for values of θ_u such that $\frac{\theta_u}{\hat{a}_u}$ is closer to $u(0)$, the term structure changes to an upward-sloping shape.

The Effect of the Speed of Mean Reversion Parameter $\hat{a}_u$. Figure 6.9 shows that an increase of the mean reversion speed parameter $\hat{a}_u$ narrows the credit spreads. In the cases $\hat{a}_u = 1.5$, 1.0, and 0.5, the actual quality of the firm ($u_0 = 3$) is much worse than the long-run levels of the uncertainty index: $\frac{\theta_u}{\hat{a}_u} = \frac{2}{3}$, 1, and 2. For $\hat{a}_u = 1.5$ and 1.0 the credit spreads widen for short maturities and narrow for long maturities. The credit spread term structure is humped shaped. In case of $\hat{a}_u = 0.1$, the long-run level of the uncertainty index is worse than the actual quality of the firm. Hence, the long-run expectation on the quality of the firm is to deteriorate further. We get an upward-sloping credit spread term structure. The larger $\hat{a}_u$ the faster is the returning to the mean reversion level. In the case of $\hat{a}_u = 0.1$, this increases the credit spread widening effect even more. For $\hat{a}_u = 1.5$, 1.0, and 0.5, the acceleration effect increases the credit spread narrowing.

The Effect of the Uncertainty Index u_0. Figure 6.10 shows that credit spreads of short term debt are sensitive to the current quality of the firm. For long term debt the long-run level of the quality of the firm is more important than its current quality. The bigger the actual uncertainty index values the bigger the credit spreads. The maximal credit spread is observed for a defaultable bond with a maturity of approximately four years.

$\lim_{T\to t} \frac{C^d(t,T)}{T-t} = \lim_{T\to t} \frac{\left(1-e^{-\delta_s(T-t)}\right)s(t)}{\left(\kappa_1^{(s)}-\kappa_2^{(s)}e^{-\delta_s(T-t)}\right)(T-t)} \overset{L'Hospital}{=} s(t)$,

$\lim_{T\to t} \frac{D^d(t,T)u(t)}{T-t} = \lim_{T\to t} \frac{-2v'(t,T)u(t)}{\sigma_u^2 v(t,T)(T-t)} \overset{L'Hospital}{=} 0$, as

$v''(t,t) = -\frac{1}{2}\sigma_u^2 \left(\hat{a}_u D^d(t,t) - b_s C^d(t,t)\right) v(t,t) = 0$ and $v(t,t) \neq 0$.

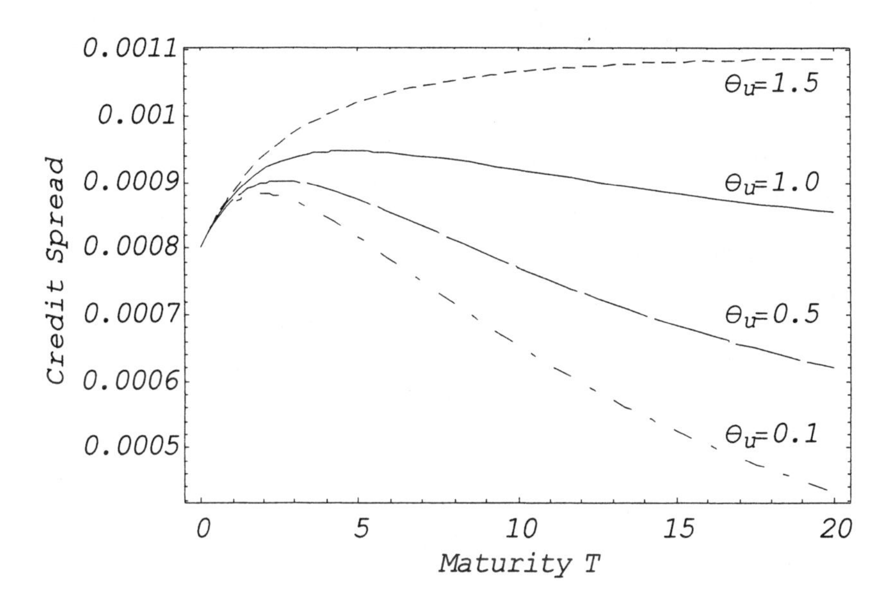

Fig. 6.8. Credit spread term structure vs. θ_u.

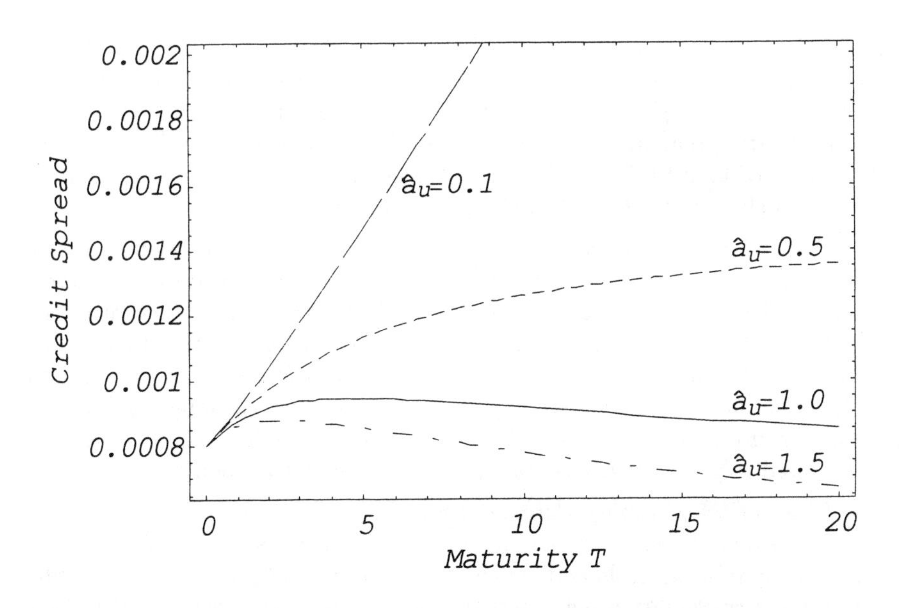

Fig. 6.9. Credit spread term structure vs. $\hat{a}_u$.

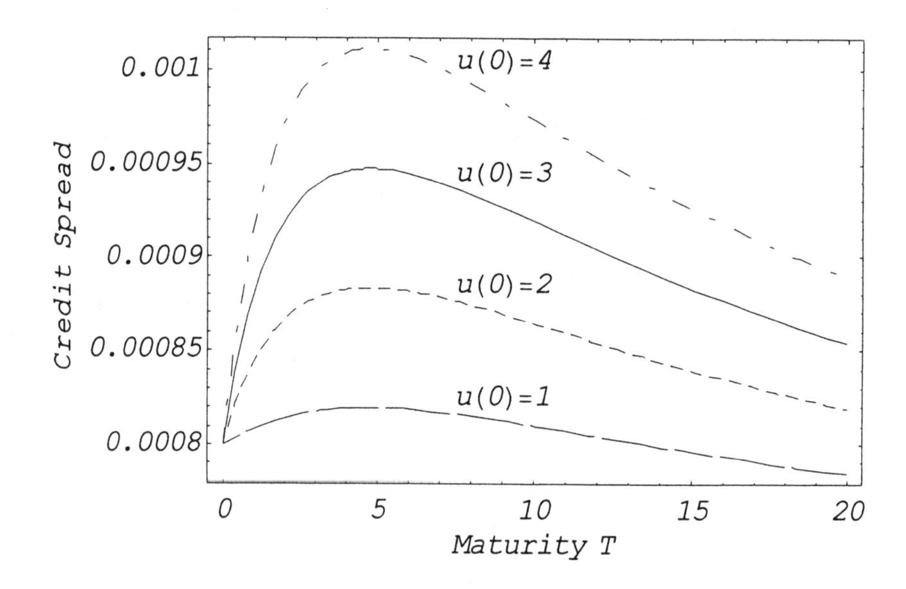

Fig. 6.10. Credit spread term structure vs. u_0.

The Effect of the Parameter b_s. Figure 6.11 shows that an increase in b_s has a great impact on the term structure of credit spreads. In general, the larger b_s the greater the influence of the uncertainty index on the term structure of the short rate spread and the credit spread. For $b_s = 0.0001$ and $b_s = 0.00015$ the curves of figure 6.11 are humped-shaped: for $b_s = 0.00001$ the curve is downward-sloping, for $b_s = 0.0002$ upward-sloping.

The Effect of the Parameter $\hat{a}_s$. The larger $\hat{a}_s$ the narrower the credit spreads. In the cases $\hat{a}_s = 0.2$ and 0.5, the actual short rate credit spread $s(0) = 0.0008$ is wider than the long-run levels of $\frac{b_s}{\hat{a}_s}\frac{\theta_u}{\hat{a}_u} = 0.0005$ and 0.0002. Hence, the credit spreads widen for short maturities and narrow for long maturities. The credit spread term structure is humped shaped. In the cases of $\hat{a}_s = 0.1$, and 0.01, the actual short rate credit spread is narrower than the long-run levels of the long-run level of $\frac{b_s}{\hat{a}_s}\frac{\theta_u}{\hat{a}_u} = 0.001$ and 0.01. Hence, the long-run expectation on the credit spreads is not to narrow much or even to widen further.

The Effect of the Volatility Parameter σ_s *of the Short Rate Spread.* For our set of base case parameters an increase of σ_s narrows the credit spreads. This can mainly be explained by the fact that the actual short rate credit spread and firm quality are worse than the long-run levels. A larger volatility parameter increases the probability of a fast return to this level.

The Effect of the Short Rate Spread s_0. Figure 6.14 shows that credit spreads are an increasing function in s. This fits to the fact that the price of a defaultable bond is a decreasing function in s. In contrast with Merton's and

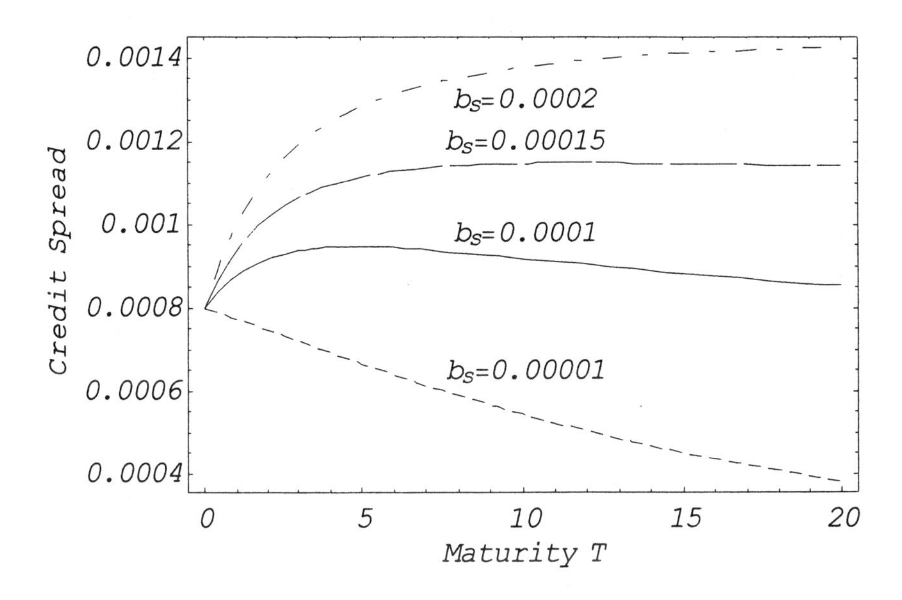

Fig. 6.11. Credit spread term structure vs. b_s.

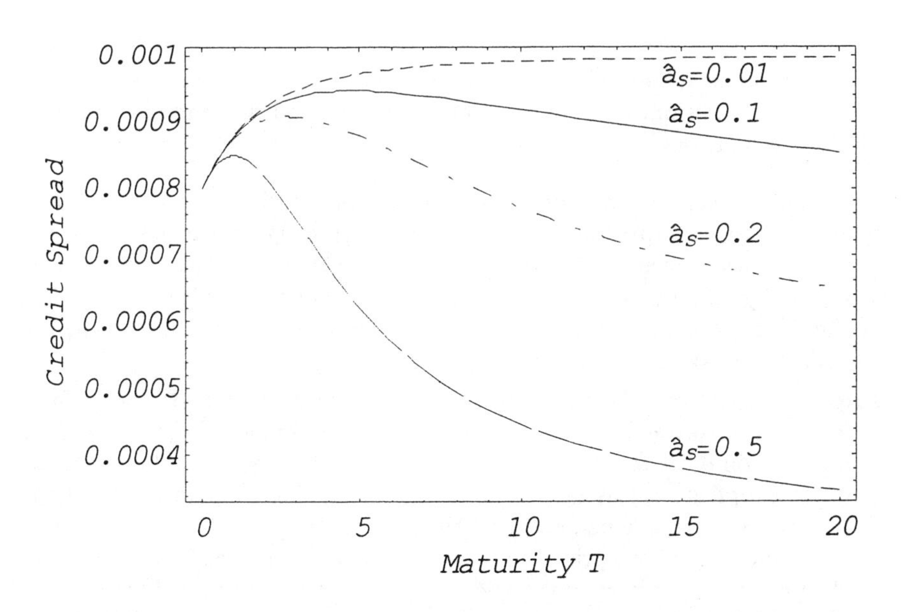

Fig. 6.12. Credit spread term structure vs. $\hat{a}_s$.

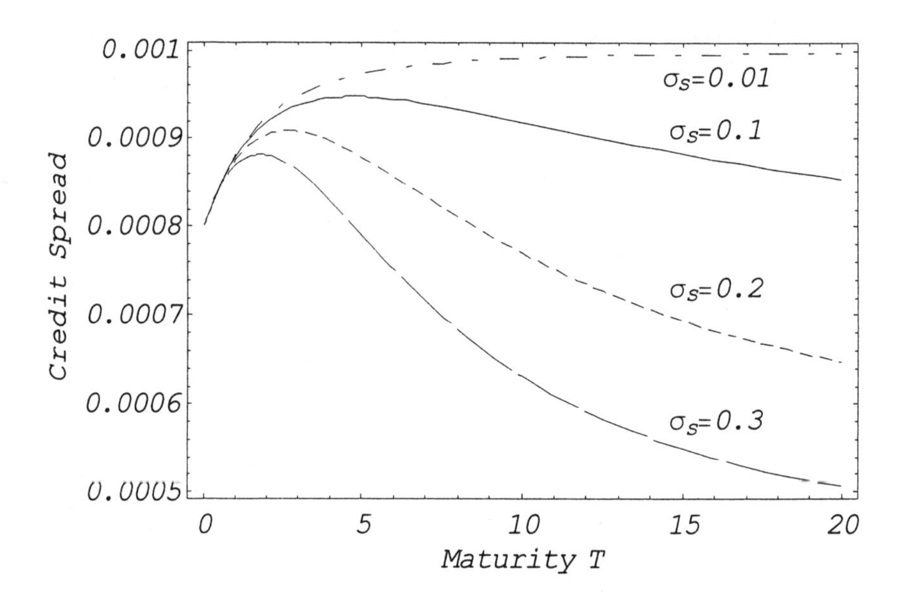

Fig. 6.13. Credit spread term structure vs. σ_s.

many other models we get positive credit spreads even for very short maturities. Thus, our model takes into consideration that instantaneous default is possible. In addition, expected as well as unexpected defaults are captured by the methodology.

Our model is capable of producing upward-sloping, downward-sloping and humped term structures of credit spreads meeting the properties suggested by the empirical results of Sarig & Warga (1989). We are able to produce a wide range of levels of credit spreads and are not faced with the problem of the traditional Merton model which can't reproduce the entire range of spreads observed in the market.

Implied Recovery Rates. One of the key components of the default process is the recovery by bond holders in the event of default. According to assumption 6.2.4 we assume fractional recovery of market value. In addition, for simplicity we assume that the stochastic recovery rate w and the default process are independent and that the recovery rate is paid at maturity (if there is a default prior to maturity). Then we can easily determine an implicit expected recovery rate. These assumptions are not too restrictive as most models assume constant recovery rates. By our assumptions, the price of a risky bond is given by

$$\begin{aligned} P^d(t,T) &= P(t,T)\, E^{\mathbf{Q}}\left[w 1_{\{\tau \le T\}} + 1_{\{\tau > T\}} | \mathcal{F}_t\right] \\ &= P(t,T)\left[E^{\mathbf{Q}}\left[w|\mathcal{F}_t\right](1-\mathbf{Q}(\tau > T|\mathcal{F}_t)) + \mathbf{Q}(\tau > T|\mathcal{F}_t)\right], \end{aligned} \tag{6.37}$$

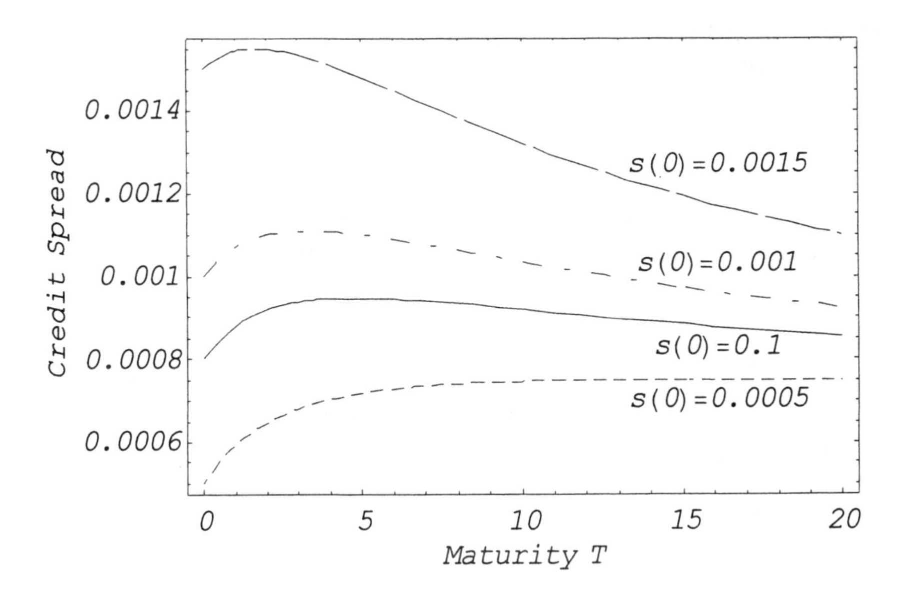

Fig. 6.14. Credit spread term structure vs. s_0.

where $P(t,T)$ denotes the value at time t of a non-defaultable discount bond assuming the dynamics of the riskless short rate specified by equation (6.1). Using equations (6.34) and (6.37), we get

$$E^{\mathbf{Q}}[w_t|\mathcal{F}_t] = \frac{e^{\int_t^T \theta_u D^d(\tau,T)d\tau - C^d(t,T)s - D^d(t,T)u} - \mathbf{Q}(\tau > T|\mathcal{F}_t)}{1 - \mathbf{Q}(\tau > T|\mathcal{F}_t)}. \tag{6.38}$$

If we assumed a constant recovery rate (as we do, e.g., in section 6.4), the right hand side of equation 6.38 is exactly the recovery rate. Thus, our model yields an alternative way for finding recovery rates if there is only scarce empirical data available (in chapter 6.6 we show how the risk-neutral survival probabilities can be calculated).

6.3.3 Defaultable (Non-Callable) Fixed Rate Debt

The theory for the pricing of simple (European) defaultable contingent claims can't be used to price defaultable fixed rate bonds. Therefore, we extend our theory to the pricing of general (European) defaultable contingent claims. A general (European) non-defaultable contingent claim is a triple (Y, D, T) consisting of a cumulative dividend process D (any $\mathcal{F}_t$−adapted process of integrable variation), the $\mathcal{F}_T$−measurable random variable Y and the time $T \leq T^*$, at which Y is paid. This concept can be generalized to defaultable claims:

Definition 6.3.1.
A general (European) defaultable claim is a triple $\left[(Y,D,T),Z,T^d\right]$ *consisting of*

- *a general (European) non-defaultable contingent claim* (Y,D,T) *yielding payoffs* $\int_t^{T\wedge T^d} dD_u$ *over the time interval* $\left[t,T\wedge T^d\right]$, *and a final payoff* Y *at time* T, *provided there has been no default until time* T, *i.e.* (Y,D,T) *is describing the obligation of the issuer, and satisfying*

$$E^{\mathbf{Q}}\left[\int_0^T |dD_t|^q\right] < \infty \text{ as well as } E^{\mathbf{Q}}\left[|Y|^q\right] < \infty \text{ for some } q > 1;$$

- *a* $\mathcal{F}_t$*−predictable process* Z *describing the payoff upon default, and satisfying* $E^{\mathbf{Q}}\left[\sup_t |Z_t|^q\right] < \infty$;
- *a* $\mathbf{F}$*−stopping time* T^d *valued in* $[0,\infty)$, *describing the stochastic structure of the default time.*

Hence, a simple (European) defaultable contingent claim $\left[(Y,T),Z,T^d\right]$ equals a general (European) defaultable contingent claim $\left[(Y,D,T),Z,T^d\right]$, where $dD_t = 0$, $0 \le t \le T$. For simplicity, we will refer to a general defaultable contingent claim as a defaultable contingent claim. In terms of the default indicator function, the cumulative dividend process of the defaultable security is given by

$$\int_0^{t\wedge T} Z_u dH_u + \int_0^{t\wedge T} L_u dD_u + Y 1_{\{T^d>T,t\ge T\}},\ t \ge 0.$$

The arbitrage free price at time t, $F(t)$, of the defaultable contingent claim $\left[(Y,D,T),Z,T^d\right]$ can be obtained as the discounted expected value of the future cash flows. This expectation has to be taken with respect to $\mathbf{Q}$, that is for $0 \le t < \tau$,

$$\begin{aligned} &F_t \\ &= E^{\mathbf{Q}}\left[\int_t^T e^{-\int_t^u r_l dl} Z_u dH_u + \int_t^T e^{-\int_t^u r_l dl} L_u dD_u \right. \\ &\quad \left. + e^{-\int_t^T r_l dl}\, Y L_T \right| \mathcal{F}_t\Big], \end{aligned} \tag{6.39}$$

and $F_t = 0$ for $t \ge \tau$ since there are no dividends after time τ.

Proposition 6.3.1.
Under assumption 6.2.4 and the dynamics specified by equations (6.5), (6.6) and (6.7) the price process, F_t, *given by equation (6.39), satisfies*

$$F_t = V_t 1_{\{t<\tau\}}, \tag{6.40}$$

if V is predictable or if $\Delta V_{T^d} = 0$ a.s., and where the adapted process V is given by

$$V_t = E^{\mathbf{Q}}\left[\left.\int_t^T e^{-\int_t^u (r_l+s_l)dl}dD_u + e^{-\int_t^T (r_l+s_l)dl}Y\right| \mathcal{F}_t\right], \quad 0 \leq t < T, \quad (6.41)$$

and $V_t = 0$ for $t \geq T$, i.e. if there has been no default until time t, F_t must equal the expected value of riskless cash flows discounted at risky discount rates. Equation (6.41) has a unique solution in the space consisting of every semimartingale, J, such that $E^{\mathbf{Q}}\left[\sup_t |J_t|^q\right]$ for some $q > 1$.

Proof.
All prerequisites of lemma 6.2.4 are satisfied by the definition of a defaultable claim and the characteristics of r and s. Hence, existence and uniqueness of equation (6.41) follows immediately from lemma 6.2.4. It remains to show that equation (6.40) is valid.

- Because V satisfies the integral equation (6.12) only on $[0, T)$, we define a modified process $\widehat{V}$ by

$$\widehat{V}_t = V_t 1_{\{t<T\}} + 1_{\{t \geq T\}} dD_T + Y 1_{\{t \geq T\}},$$

  to extend the definition on the compact interval $[0, T]$.
- Apply corollary 6.2.1 to $\widehat{V}$ with $\xi = -(r+s)$ and $dX_t = dD_t$ (by lemma 6.2.2 ξ is progressively measurable):

$$\begin{aligned} d\widehat{V}_t &= -dD_t + \widehat{V}_t (r_t + s_t)\, dt + dm_t \\ &= -dD_t + \widehat{V}_t r_t dt + \widehat{V}_t (1 - w_t)\, dH_t + \widehat{V}_t dM_t^d + dm_t \end{aligned}$$

  for some martingale m and M_t^d as defined in lemma 6.2.3.
- Define $U_t = \widehat{V}_t L_t$ with L_t as defined above. Using integration by parts,

$$\begin{aligned} dUt &= \widehat{V}_{t-} dL_t + L_{t-} d\widehat{V}_t + d\left[L, \widehat{V}\right]_t \\ &= \widehat{V}_t dL_t + L_{t-} d\widehat{V}_t \\ &= \widehat{V}_t dL_t - L_{t-} dD_t + U_t r_t dt + \\ &\quad \widehat{V}_t L_{t-} (1 - w_t)\, dH_t + \widehat{V}_t L_{t-} dM_t^d + L_{t-} dm_t \\ &= \widehat{V}_t dL_t - L_{t-} dD_t + U_t r_t dt + \\ &\quad \widehat{V}_t L_{t-} (1 - w_t)\, dH_t + d\widetilde{m}_t. \end{aligned}$$

  As in the proof of proposition 6.2.1 $\widetilde{m}_t$ with $d\widetilde{m}_t = \widehat{V}_t L_{t-} dM_t^d + L_{t-} dm_t$ is a martingale.
- Apply corollary 6.2.1 with $\xi = -r$ and

$$dX_t = L_{t-} dD_t - \widehat{V}_t L_{t-} (1 - w_t)\, dH_t - \widehat{V}_t dL_t,$$

(by lemma 6.2.2 ξ is progressively measurable; by the definition of H and L, X is an adapted $RCLL$ process; it is of integrable variation as it is adapted $RCLL$, increasing and $E\left(Var\left(X\right)_{\infty}\right) \leq \widehat{V}_{T^d} < \infty$):

$$
\begin{aligned}
& U(t) \\
&= E^{\mathbf{Q}}\left[-\int_t^T e^{-\int_t^u r_l dl}\widehat{V}_u\left(L_{u-}\left(1-w_u\right)-1\right)dH_u\right. \\
&\quad \left.+\int_t^T e^{-\int_t^u r_l dl}L_{u-}dD_u + e^{-\int_t^T r_l dl}YL_T\right|\mathcal{F}_t\Bigg] \\
&= E^{\mathbf{Q}}\left[\int_t^T e^{-\int_t^u r_l dl}\widehat{V}_{u-}w_u dH_u\right. \\
&\quad \left.+\int_t^T e^{-\int_t^u r_l dl}L_{u-}dD_u + e^{-\int_t^T r_l dl}YL_T\right|\mathcal{F}_t\Bigg],
\end{aligned}
$$

- Because $U_t = V_t$ for $t < \tau$,

$$F_t = V_t 1_{\{t<\tau\}}.$$

Remark 6.3.2.
The assumption $\Delta V_{T^d} = 0$ *a.s.* is not very restrictive. E.g., in case of defaultable coupon bonds default is usually not announced exactly at a coupon payment date but some time earlier.

Suppose we want to price a defaultable coupon bond promising to make discrete coupon payments c_i occurring at dates $t < t_i \leq T$, $i = 1, ..., n$, and a final payment of 1 at maturity T. Trivially, a defaultable coupon bond satisfies the characteristics of a defaultable claim. Hence, by using proposition 6.3.1 we find easily that the time t price of a coupon bond that hasn't defaulted yet, $P_c^d(t,T) = P_c^d(r,s,u,t,T)$, is given by

$$P_c^d(t,T) = \sum_{i=1}^{n} c_i P^d(t,t_i) + P^d(t,T). \tag{6.42}$$

6.3.4 Defaultable Callable Fixed Rate Debt

Many corporate bonds are callable. Thus, it is important to extend our pricing theory to capture fixed rate bonds with embedded call options. We assume that it is optimal for the bond issuer to call the bond when its market value is minimal. If we denote the feasible call policies in the time interval $[t,T]$ by $\mathcal{CP}(t,T)$ and assume that the bond hasn't been called and hasn't defaulted until time t, the time t price of a defaultable callable fixed rate bond with discrete coupon payments c_i occurring at dates $t < t_i \leq T$, $i = 1, ..., n$, and maturity T, $P^d_{cc}(t,T) = P^d_{cc}(r,s,u,t,T)$, is given by

$$P^d_{cc}(t,T) \tag{6.43}$$
$$= \min_{T^c \in \mathcal{CP}(t,T)} E^{\mathbf{Q}} \left(\sum_{t < t_i \leq T^c} c_i e^{-\int_t^{t_i}(r_l+s_l)dl} + e^{-\int_t^{T^c}(r_l+s_l)dl} \middle| \mathcal{F}_t \right),$$

assuming that the minimum in the equation above is attained by some stopping time. Technical regularity conditions are given in Karatzas (1988). This stopping time approach to callable bond pricing in a quite different context was first proposed by Merton (1974). Equation (6.43) can be solved, e.g., by using discretization techniques (see, e.g., Duffie & Singleton (1997)).

6.3.5 Building a Theoretical Framework for Pricing One-Party Defaultable Interest Rate Derivatives

We consider a simple (European) defaultable claim $\left[(Y,T), Z, T^d\right]$ as stated in definition 6.2.3 and assume recovery of market value (see assumption 6.2.4). As usual we denote the price of the defaultable claim at time t by F_t, where F_t is given in proposition 6.2.1. V is the solution of equation

$$\begin{aligned} 0 = & \frac{1}{2}\left[\sigma_r^2 r^{2\beta} V_{rr} + \sigma_s^2 s V_{ss} + \sigma_u^2 \mathrm{V}_{uu}\right] \\ & + \left[\theta_r(t) - \hat{a}_r r\right] V_r + \left[b_s u - \hat{a}_s s\right] V_s \\ & + \left[\theta_u - \hat{a}_u u\right] V_u + V_t - (r+s) V, \quad (r,s,u,t) \in \mathbb{R}^3 \times [0, \mathrm{T}), \end{aligned} \tag{6.44}$$

with boundary condition

$$V(r,s,u,T) = Y(r,s,u), \quad (r,s,u) \in \mathbb{R}^3. \tag{6.45}$$

Now assume that $G\left(y, \check{t}, z, t\right)$ is the solution to the problem

$$\begin{aligned} 0 = & \frac{1}{2}\sigma_r^2 r^{2\beta} G_{rr} + \frac{1}{2}\sigma_s^2 s G_{ss} + \frac{1}{2}\sigma_u^2 u G_{uu} \\ & + \left[\theta_r(t) - \hat{a}_r r\right] G_r + \left[b_s u - \hat{a}_s s\right] G_s \\ & + \left[\theta_u - \hat{a}_u u\right] G_u + G_t - (r+s) G, \end{aligned} \tag{6.46}$$

with boundary condition

$$G\left(y, \check{t}, z, \check{t}\right) = \delta(r - y_r)\, \delta(s - y_s)\, \delta(u - y_u), \tag{6.47}$$

where $\delta(.)$ is the Dirac function[20], and $x = (x_1, x_2, x_3)$, $y = (y_r, y_s, y_u)$ and $z = (r, s, u)$. Then it is straightforward to show[21] that the solution to the PDE (6.44) with boundary condition (6.45) can be written as

$$V(z,t) = \iiint_{\mathbf{R}^3} G(y,T,z,t)\, Y(y)\, dy. \tag{6.48}$$

As a first step we show that V - as defined in equation (6.48) - satisfies the PDE (6.44):

$$\begin{aligned}
&\frac{1}{2}\left[\sigma_r^2 r^{2\beta} V_{rr} + \sigma_s^2 s V_{ss} + \sigma_u^2 V_{uu}\right] + \left[\theta_r(t) - \hat{a}_r r\right] V_r + \left[b_s u - \hat{a}_s s\right] V_s \\
&+ \left[\theta_u - \hat{a}_u u\right] V_u + V_t - (r+s) V \\
&= \iiint_{\mathbf{R}^3} \Big\{ \frac{1}{2}\left[\sigma_r^2 r^{2\beta} G_{rr}(y,T,z,t) + \sigma_s^2 s G_{ss}(y,T,z,t) + \sigma_u^2 u G_{uu}(y,T,z,t)\right] \\
&+ \left[\theta_r(t) - \hat{a}_r r\right] G_r(y,T,z,t) + \left[b_s u - \hat{a}_s s\right] G_s(y,T,z,t) \\
&+ \left[\theta_u - \hat{a}_u u\right] G_u(y,T,z,t) + G_t(y,T,z,t) - (r+s) G(y,T,z,t) \Big\}\, Y(y)\, dy \\
&= \iiint_{\mathbf{R}^3} 0 \cdot Y(y)\, dy = 0.
\end{aligned}$$

The first equality follows by interchanging the order of integration and differentiation. Therefore G must satisfy some appropriate smoothness conditions. The second equality follows directly from equation (6.46).

[20] We use the definition of the Dirac function given in Wilmott, Dewynne & Howison (1993), pp.94 ff.: The Dirac function δ is not in fact a function, but is rather a generalized function. It is the limit as $\varepsilon \to 0$ of any one-parameter family of functions $\delta_\epsilon(t)$ with the following properties:

- for each ϵ, $\delta_\epsilon(t)$ is piecewise smooth;
- $\int_{-\infty}^{\infty} \delta_\epsilon(t)\, dt = 1$;
- for each $t \neq 0$, $\lim_{\epsilon \to 0} \delta_\epsilon(t) = 0$.

Such a sequence of functions is called a delta-sequence. For any smooth function $f(x)$,

$$\int_{-\infty}^{\infty} \delta(x - x_0)\, f(x)\, dx = f(x_0),$$

so multiplying by $\delta(x - x_0)$ and integrating picks out the value of f at x_0. Furthermore, if $\mathcal{H}(x)$ is the unit step function,

$$\mathcal{H}(x) = \begin{cases} 0 \text{ for } x < 0, \\ 1 \text{ for } x \geq 0, \end{cases}$$

then $\int_{-\infty}^{x} \delta(\xi)\, d\xi = \mathcal{H}(x)$, and conversely, $\mathcal{H}'(x) = \delta(x)$. Sometimes $\mathcal{H}(x)$ is called Heaviside function.

[21] Compare, e.g., to Chen (1996) or Cathcart & El-Jahel (1998). Chen (1996) uses a similar approach for the pricing of non-defaultable bonds.

As a second step we check that the boundary condition (6.45) is satisfied:

$$\begin{aligned} V(z,T) &= \iiint_{\mathbf{R}^3} G(y,T,z,T)\,Y(y)\,dy \\ &= \iiint_{\mathbf{R}^3} \delta(r-y_r)\,\delta(s-y_s)\,\delta(u-y_u)\,Y(y)\,dy \qquad (6.49) \\ &= Y(z). \qquad (6.50) \end{aligned}$$

Equation (6.49) uses equation (6.47), and equation (6.50) follows immediately from the properties of the Dirac function.

Corollary 6.3.1.

$$F(z,t) = 1_{\{t<\tau\}} \iiint_{\mathbf{R}^3} G(y,T,z,t)\,Y(y)\,dy.$$

Especially, if $t<\tau$,

$$P^d(z,t,T) = \iiint_{\mathbf{R}^3} G(y,T,z,t)\,dy.$$

Hence, if $t<\tau$,

$$\iiint_{\mathbf{R}^3} G(y,T,z,t)\,dy = A^d(t,T)\,e^{-B(t,T)r_t - C^d(t,T)s_t - D^d(t,T)u_t}.$$

If we define

$$\widehat{L}(y,T,z,t) = \frac{G(y,T,z,t)}{P^d(z,t,T)},$$

we get

$$F(z,t) = 1_{\{t<\tau\}} P^d(z,t,T) \iiint_{\mathbf{R}^3} \widehat{L}(y,T,z,t)\,Y(y)\,dy.$$

By construction $\widehat{L}(y,T,z,t)$ is positive and integrates to 1 with respect to y. Therefore we can think of $\widehat{L}$ as of a probability density, the joint density of r_T, s_T and u_T under some measure $\mathbf{Q}^{\widehat{L}}$. Hence,

$$F(z,t) = 1_{\{t<\tau\}} P^d(z,t,T)\,E^{\mathbf{Q}^{\widehat{L}}}\left[Y(y)\,|\mathcal{F}_t\right].$$

Corollary 6.3.2.
The arbitrage free price at time t, F_t, of the simple defaultable contingent claim $\left[(Y,T), Z, T^d\right]$ is given by

$$
\begin{aligned}
F_t &= E^{\mathbf{Q}}\left[\int_t^T e^{-\int_t^u r_l dl} Z_u dH_u + e^{-\int_t^T r_l dl}\, Y L_T \middle| \mathcal{F}_t\right] \\
&= 1_{\{t<\tau\}} E^{\mathbf{Q}}\left[e^{-\int_t^T (r_l+s_l)dl} Y \,\middle|\, \mathcal{F}_t\right] \\
&= 1_{\{t<\tau\}} \iiint_{\mathbf{R}^3} G\left(y,T,z,t\right) Y\left(y\right) dy \\
&= 1_{\{t<\tau\}} P^d\left(z,t,T\right) E^{\mathbf{Q}^L}\left[Y\left(y\right)|\mathcal{F}_t\right].
\end{aligned}
$$

So we can conclude that if we know the function G we can determine the price of the derivative easily.

Lemma 6.3.1.
$G\left(y,\check{t},z,t\right)$, is given by

$$
G\left(y,\check{t},z,t\right) = \frac{1}{\left(2\pi\right)^{3/2}} \iiint_{\mathbf{R}^3} e^{-ixy'} \tilde{G}\left(x,\check{t},z,t\right) dx \tag{6.51}
$$

where

$$
\tilde{G}\left(x,\check{t},z,t\right) = A^{\tilde{G}}\left(x,t,\check{t}\right) e^{-B^{\tilde{G}}\left(x,t,\check{t}\right)r - C^{\tilde{G}}\left(x,t,\check{t}\right)s - D^{\tilde{G}}\left(x,t,\check{t}\right)u}
$$

with $z = \left(r,s,u\right),\ x = \left(x_1,x_2,x_3\right),\ y = \left(y_r,y_s,y_u\right),$ *and*

$$
B^{\tilde{G}}\left(x,t,\check{t}\right) = \begin{cases} -ix_1 e^{-\hat{a}_r\left(\check{t}-t\right)} + \frac{1}{\hat{a}_r}\left(1 - e^{-\hat{a}_r\left(\check{t}-t\right)}\right), & \text{if } \beta = 0, \\ -\frac{2}{\sigma_r^2} \frac{\varpi(x_1)\kappa_1^{(r)} e^{-\kappa_1^{(r)}(\check{t}-t)} + \kappa_2^{(r)} e^{-\kappa_2^{(r)}(\check{t}-t)}}{\varpi(x_1) e^{-\kappa_1^{(r)}(\check{t}-t)} + e^{-\kappa_2^{(r)}(\check{t}-t)}}, & \text{if } \beta = \frac{1}{2}, \end{cases} \tag{6.52}
$$

$$
\begin{aligned}
C^{\tilde{G}}\left(x,t,\check{t}\right) &= \frac{\kappa_3^{(s)}\left(x\right) - e^{-\delta_s\left(\check{t}-t\right)}}{\kappa_4^{(s)}\left(x\right) - \kappa_5^{(s)}\left(x\right) e^{-\delta_s\left(\check{t}-t\right)}}, \\
D^{\tilde{G}}\left(x,t,\check{t}\right) &= \frac{-2}{\sigma_u^2}\left(\pi_1\left(x\right)\varphi_2\left(x,t,\check{t}\right)\right. \\
&\quad \left. - e^{-2\delta_s \phi\left(\kappa_4^{(s)}(x),\kappa_3^{(s)}(x)\right)\left(\check{t}-t\right)} \pi_2\left(x\right)\varphi_1\left(x,t,\check{t}\right)\right) / \\
&\quad \left(\pi_1\left(x\right) F_1^{\tilde{G}}\left(x,t,\check{t}\right)\right. \\
&\quad \left. + e^{-2\delta_s \phi\left(\kappa_4^{(s)}(x),\kappa_3^{(s)}(x)\right)\left(\check{t}-t\right)} \pi_2\left(x\right) F_3^{\tilde{G}}\left(x,t,\check{t}\right)\right),
\end{aligned}
$$

if $\beta = \frac{1}{2}$:

$$
A^{\tilde{G}}\left(x,t,\check{t}\right) = \tilde{A}\left(x,t,\check{t}\right),
$$

with

$$
\ln \tilde{A}\left(x,t,\check{t}\right) = \int_{\check{t}}^t \left(\theta_r\left(\tau\right) B^{\tilde{G}}\left(x,\tau,\check{t}\right) + \theta_u D^{\tilde{G}}\left(x,\tau,\check{t}\right)\right) d\tau,
$$

if $\beta = 0$:

$$A^{\tilde{G}}\left(x,t,\check{t}\right) = \tilde{A}\left(x,t,\check{t}\right) e^{-\int_{\check{t}}^{t} \frac{1}{2}\sigma_\tau^2 \left(B^{\tilde{G}}\left(x,\tau,\check{t}\right)\right)^2 d\tau},$$

and $F(a,b,c,z)$ *is the hypergeometric function.* ϖ, $\kappa_3^{(s)}$, $\kappa_4^{(s)}$, $\kappa_5^{(s)}$, π_1, π_2, φ_1, *and* φ_2 *are defined in the proof below.* δ_s, $\kappa_1^{(r)}$, *and* $\kappa_2^{(r)}$ *are given by equation (6.18).*

Proof.
We apply the proof of appendix B.3 with $\alpha = 1$. Then

$$\begin{array}{ll}
G\left(y,\check{t},z,t\right) = G^{(1)}\left(y,\check{t},z,t\right), & \tilde{G}\left(x,\check{t},z,t\right) = \widetilde{G}^{(1)}\left(x,\check{t},z,t\right), \\
A^{\tilde{G}}\left(x,t,\check{t}\right) = A^{\widetilde{G}^{(1)}}\left(x,t,\check{t}\right), & B^{\tilde{G}}\left(x,t,\check{t}\right) = B^{\widetilde{G}^{(1)}}\left(x,t,\check{t}\right), \\
C^{\tilde{G}}\left(x,t,\check{t}\right) = C^{\widetilde{G}^{(1)}}\left(x,t,\check{t}\right), & D^{\tilde{G}}\left(x,t,\check{t}\right) = D^{\widetilde{G}^{(1)}}\left(x,t,\check{t}\right), \\
\pi_1(x) = \pi_1^{(1)}(x), & \pi_2(x) = \pi_2^{(1)}(x), \\
\varphi_1\left(x,t,\check{t}\right) = \varphi_1^{(1)}\left(x,t,\check{t}\right), & \varphi_2\left(x,t,\check{t}\right) = \varphi_2^{(1)}\left(x,t,\check{t}\right), \\
\kappa_3^{(s)}(x) = \kappa_3^{(s,1)}(x), & \kappa_4^{(s)}(x) = \kappa_4^{(s,1)}(x), \\
\kappa_5^{(s)}(x) = \kappa_5^{(s,1)}(x), &
\end{array}$$

and ϖ is given by equation (B.21).

In the following we summarize the steps necessary to calculate prices of defaultable interest rate derivatives under the assumption that the prices for defaultable bonds are given[22]:

1. Determine the Fourier transform[23] $\tilde{G}$ of G by using lemma 6.3.1.
2. Determine the probability density function $\widehat{L}$ in the following way:
   a) Calculate the Fourier transform $\widetilde{L}$ of $\widehat{L}$ by using $\tilde{G}$:

$$\widetilde{L}\left(x,\check{t},z,t\right) = \frac{1}{(2\pi)^{3/2}} \frac{\tilde{G}\left(x,\check{t},z,t\right)}{\tilde{G}\left(0,\check{t},z,t\right)},$$

   where 0 denotes the three-dimensional zero vector.

[22] If defaultable bond prices are not available we can use theorem 6.3.1 to determine them. Therefore the three factor defaultable term structure model must be calibrated first.

[23] The Fourier transform of a function $f : \mathbb{R} \to \mathbb{R}$, which satisfies Dirichlet's conditions (see, e.g., Bronstein & Semendjajew (1991), page 608) in each finite subinterval of $\mathbb{R}$ and $\int_{-\infty}^{\infty} |f(x)| < \infty$, is defined by

$$\frac{1}{\sqrt{2\pi}} \int_{-\infty}^{\infty} f(x)\, e^{iyx}\, dx.$$

Proof.

$$
\begin{aligned}
\tilde{G}\left(x,\check{t},z,t\right) &= \frac{1}{\left(2\pi\right)^{3/2}} \iiint\limits_{\mathbf{R}^3} e^{ixy'} G\left(y,\check{t},z,t\right) dy \\
&= \frac{1}{\left(2\pi\right)^{3/2}} \iiint\limits_{\mathbf{R}^3} e^{ixy'} \widehat{L}\left(y,\check{t},z,t\right) P^d\left(z,t,\check{t}\right) dy \\
&= \frac{1}{\left(2\pi\right)^{3/2}} P^d\left(z,t,\check{t}\right) \iiint\limits_{\mathbf{R}^3} e^{ixy'} \widehat{L}\left(y,\check{t},z,t\right) dy \\
&= P^d\left(z,t,\check{t}\right) \widetilde{L}\left(x,\check{t},z,t\right).
\end{aligned}
$$

Hence,

$$
\widetilde{L}\left(x,\check{t},z,t\right) = \frac{\tilde{G}\left(x,\check{t},z,t\right)}{P^d\left(z,t,\check{t}\right)}.
$$

Finally,

$$
\begin{aligned}
P^d\left(z,t,\check{t}\right) &= \iiint\limits_{\mathbf{R}^3} G\left(y,\check{t},z,t\right) dy \\
&= \left(2\pi\right)^{3/2} \tilde{G}\left(0,\check{t},z,t\right).
\end{aligned}
$$

b) Evaluate the Fourier inverse $\widehat{L}$ of $\widetilde{L}$:

$$
\widehat{L}\left(y,\check{t},z,t\right) = \frac{1}{\left(2\pi\right)^{3/2}} \iiint\limits_{\mathbf{R}^3} e^{-ixy'} \widetilde{L}\left(x,\check{t},z,t\right) dx.
$$

If an analytical result can't be determined we can use numerical techniques such as the evaluation of $\widehat{L}$ by numerical inversion of the characteristic function.

3. Use defaultable bond prices and L to get prices of defaultable interest rate derivatives:

$$
F\left(z,t\right) = 1_{\{t<\tau\}} P^d\left(z,t,T\right) \iiint\limits_{\mathbf{R}^3} \widehat{L}\left(y,T,z,t\right) Y\left(y\right) dy.
$$

6.3.6 Defaultable Floating Rate Debt

In the following we consider a defaultable floating rate note with a given sequence of reset dates $t_1 \leq t < t_i \leq T$, $i = 2,...,n$, and a promised sequence of payment dates $t < t_p\left(t_i\right) \leq T$, $i = 1,...,n$, such that $t_i \leq t_p\left(t_i\right)$ for all i. Let $\tau_p^1 = t_p\left(t_1\right) - t$ and $\tau_p^i = t_p\left(t_i\right) - t_p\left(t_{i-1}\right)$, $i = 2,...,n$. In practice, floating rate debt spreads are usually relative to some agreed-upon floating index such as the three-month LIBOR, that need not be purely default-free.

There is no problem in the further analysis if the pure default-free floating rate and the reference differ by a constant. We assume that the periodic floating payments float with LIBOR plus a constant spread S. The index yield fixed at time t_i and the time to maturity of the index are denoted by $I(t_i)$ and τ^I, respectively. For the annualized LIBOR, we have

$$I(t_i) = \frac{1}{\tau^I}\left(\frac{1}{P(t_i, t_i + \tau^I)} - 1\right),$$

where the non-defaultable bond price $P(t_i, t_i + \tau^I)$ is given by equation (6.15). At time $t_p(t_i)$ there is a floating rate payment of $\tau_p^i I(t_i) + S$. Without any loss of generality we assume the standard market practice of $\tau_p^i = \tau^I$ for all $i = 1, ..., n$. Trivially, defaultable floating rate notes qualify as defaultable claims. Hence, we can apply the framework of section 6.3.5 to the pricing of this financial instrument. Let us first consider a special case where the reset dates equal the payment dates:

Special case: There are only payments $t_p(t_i) = t_i$ for $i = 2, ..., n$:. Applying proposition 6.3.1, the time t price of $\tau^I I(t_i)$ (denoted by $f_i(t)$) is given by

$$\begin{aligned} f_i(t) &= E^{\mathbf{Q}}\left[e^{-\int_t^{t_i}(r_l+s_l)dl}\tau^I I(t_i) \Big| \mathcal{F}_t\right] \\ &= E^{\mathbf{Q}}\left[e^{-\int_t^{t_i}(r_l+s_l)dl}\left(\frac{1}{P(t_i, t_i+\tau^I)} - 1\right)\Big| \mathcal{F}_t\right]. \end{aligned}$$

We determine the expectation by using the PDE techniques described in section 6.3.5. By the Feynman-Kac Formula, f_i satisfies the PDE

$$\begin{aligned} 0 &= \frac{1}{2}\sigma_r^2 r^{2\beta}(f_i)_{rr} + \frac{1}{2}\sigma_s^2 s(f_i)_{ss} + \frac{1}{2}\sigma_u^2 u(f_i)_{uu} \\ &\quad + [\theta_r(t) - \hat{a}_r r](f_i)_r + [b_s u - \hat{a}_s s](f_i)_s \\ &\quad + [\theta_u - \hat{a}_u u](f_i)_u + (f_i)_t - (r+s) f_i, \quad (r,s,u,t) \in \mathbb{R}^3 \times [0, \mathrm{T}) \end{aligned} \tag{6.53}$$

with boundary condition

$$f_i(z, t_i) = \frac{1}{P(z, t_i, t_i + \tau^I)} - 1. \tag{6.54}$$

The solution of equations (6.53) and (6.54) is given by

$$f_i(t) = f_i(z,t) = \iiint_{\mathbf{R}^3} G(y, t_i, z, t)\left(\frac{1}{P(y, t_i, t_i + \tau^I)} - 1\right) dy. \tag{6.55}$$

With

$$\widehat{L}(y, t_i, z, t) = \frac{G(y, t_i, z, t)}{P^d(t, t_i)}, \tag{6.56}$$

$$f_i(t) = f_i(z,t) = P^d(t,t_i) \iiint_{\mathbf{R}^3} \widehat{L}(y,t_i,z,t) \left(\frac{1}{P(y,t_i,t_i+\tau^I)} - 1 \right) dy.$$

Finally, the time t price of the floating rate note is given by

$$P_f^d(t,T) = 1_{\{t<\tau\}} \sum_{i=2}^{n} \left(f_i(t) + S \cdot P^d(t,t_i) \right) + 1_{\{t<\tau\}} P^d(t,T). \qquad (6.57)$$

General case:. For $i=1$ the payment of $\tau^I I(t_1)$ is already known at time t and the pricing problem reduces to the pricing of a fixed rate coupon. Hence, the time t price of $\tau^I I(t_1)$ (denoted by $f_1(t)$) is given by

$$f_1(t) = P^d(t,t_p(t_1)) \left(\frac{1}{P(t_1,t_1+\tau^I)} - 1 \right).$$

For $i \geq 2$, the situation is more complicated. The time t price of $\tau^I I(t_i)$ (denoted by $f_i(t)$) is given by

$$\begin{aligned} f_i(t) &= E^{\mathbf{Q}} \left[e^{-\int_t^{t_p(t_i)} (r_l+s_l)dl} \tau^I I(t_i) \right| \mathcal{F}_t \Big] \\ &= E^{\mathbf{Q}} \left[e^{-\int_t^{t_i} (r_l+s_l)dl} \left(\frac{P^d(t_i,t_p(t_i))}{P(t_i,t_i+\tau^I)} - P^d(t_i,t_p(t_i)) \right) \right| \mathcal{F}_t \Big]. \end{aligned}$$

Proof.

$$\begin{aligned} & E^{\mathbf{Q}} \left[e^{-\int_t^{t_p(t_i)} (r_l+s_l)dl} \tau^I I(t_i) \right| \mathcal{F}_{t_i} \Big] \\ &= E^{\mathbf{Q}} \left[e^{-\int_t^{t_i} (r_l+s_l)dl} e^{-\int_{t_i}^{t_p(t_i)} (r_l+s_l)dl} \tau^I I(t_i) \right| \mathcal{F}_{t_i} \Big] \\ &= e^{-\int_t^{t_i} (r_l+s_l)dl} \tau^I I(t_i) P^d(t_i,t_p(t_i)). \end{aligned}$$

By using the tower law we get

$$\begin{aligned} & E^{\mathbf{Q}} \left[e^{-\int_t^{t_p(t_i)} (r_l+s_l)dl} \tau^I I(t_i) \right| \mathcal{F}_t \Big] \\ &= E^{\mathbf{Q}} \left[e^{-\int_t^{t_i} (r_l+s_l)dl} \tau^I I(t_i) P^d(t_i,t_p(t_i)) \right| \mathcal{F}_t \Big]. \end{aligned}$$

Solving PDE (6.53) with boundary condition

$$f_i(z,t_i) = \frac{P^d(z,t_i,t_p(t_i))}{P(z,t_i,t_i+\tau^I)} - P^d(z,t_i,t_p(t_i))$$

by applying the results of section 6.3.5 yields

$$
\begin{aligned}
f_i(t) &= f_i(z,t) \\
&= P^d(t,t_i) \iiint_{\mathbf{R}^3} \widehat{L}(y,t_i,z,t)\left(\frac{P^d(z,t_i,t_p(t_i))}{P(z,t_i,t_i+\tau^I)} - P^d(z,t_i,t_p(t_i))\right) dy.
\end{aligned}
$$

Finally, the time t price of the floating rate note is given by

$$
P_f^d(t,T) = 1_{\{t<\tau\}} \sum_{i=1}^{n} \left(f_i(t) + S \cdot P^d(t,t_i)\right) + 1_{\{t<\tau\}} P^d(t,T). \tag{6.58}
$$

6.3.7 Defaultable Interest Rate Swaps

An interest rate swap is an agreement between two counterparties in which each of them promises to make fixed or floating periodic payments to the other in the same currency. Each series of floating payments is based on a specific interest rate (index). In a typical swap contract, a fixed rate payer makes fixed payments every year, whereas a floating rate payer makes several payments a year. The payments due on any date are usually netted against each other before payment is made. The generic interest rate swap refers to a floating to fixed swap in which the payment of one counterparty is based on a floating interest rate index and the payment of the other is based on a predetermined fixed interest rate. If there is no default risk, a generic swap can be characterized as an exchange of two riskless bonds. According to Bicksler & Chen (1986), the swap transaction is identical to an agreement where the counterparty promising to make fixed payments exchanges a fixed coupon default-free bond with coupon-dates and maturity equal to that of the swap for a default-free floating rate note with index, payment dates and maturity equal to that of the swap with the counterparty promising to make floating payments. In practice, the counterparties to swaps are usually not default riskless. But there is only coupon default risk. Netting usually reduces the credit risk even further. There is no exposure to the loss of principal, because the exchange of principals on the two sides of the swaps is avoided by the terms of the swap contract. The price of the default risk in swaps is highly dependent on the default triggering events and the settlement in bankruptcy. Conditions under which swap agreements are terminated are set out in the International Swap Dealers Association (ISDA) Master Agreement. In a swap settled under the "fault" rule, the bankruptcy rules are such that if default occurs, all future payments are null and void. Especially, the non-defaulting counterparty is not obligated to compensate the defaulting counterparty if the remaining market value of the swap contract is positive for the defaulting counterparty. In a swap settled under the "no fault" rule, the counterparty with negative remaining market value is obligated to compensate the other counterparty based on the remaining market value of the swap independent of the fact of the identity of the defaulting counterparty.

Sundaresan (1991) was the first one to consider the pricing of defaultable interest rate swaps. He models both, the non-defaultable short rate and a default premium, as Cox-Ingersoll-Ross processes (see, e.g., Cox et al. (1985)). Cooper & Mello (1991) calculate defaultable swap prices for structural models but only allow for one sided default risk. They don't allow for netting and only consider the "no fault" rule. Sorensen & Bollier (1994) analyze two sided default risk, but not within an advanced credit risk model. Duffie & Huang (1996) are the first to model two sided credit risk for an advanced credit risk model, especially for intensity based models. They apply basic concepts developed by Rendleman (1992). We analyze two sided default risk in interest rate swaps for our hybrid model.

In the following, we model two counterparties A and B, A making fixed payments and B making floating payments. Given is a sequence of reset dates t_i, $i = 1, ..., n$, and a sequence of payment dates $t_p(t_i)$, $i = 1, ..., n$, such that $t_1 \leq t \leq t_p(t_1) \leq t_2 \leq t_p(t_2) \leq ... \leq t_n \leq t_p(t_n) = T$. The periodic floating payments float with an agreed-upon floating index such as the three-month LIBOR. The index yield fixed at time t_i and the time to maturity of the index are denoted by $I(t_i)$ and τ^I, respectively. For the annualized LIBOR, we have

$$I(t_i) = \frac{1}{\tau^I}\left(\frac{1}{P(t_i, t_i + \tau^I)} - 1\right),$$

where the non-defaultable bond price $P(t_i, t_i + \tau^I)$ is given by equation (6.15). The periodic finite fixed payments are denoted by c. A generic interest rate swap is then an exchange of a sequence of fixed payments $\{c\}$ with a sequence of floating payments $\{\tau^I I(t_i)\}$ at times $\{t_p(t_i)\}$. Note that the floating payment $\tau^I I(t_i)$ at time $t_p(t_i)$ depends on the non-defaultable short rate r_{t_i} at time t_i and not on the defaultable short rate $r_{t_p(t_i)}$ at time $t_p(t_i)$. As in the case of floating rate notes we could also consider floating payments that float with $\tau^I I(t_i)$ plus a constant spread S. Because we take into account possible asynchronous swap payments, at time $t_p(t_i)$, there are three payment procedures possible:

- counterparty A receives a floating rate payment of $\tau^I I(t_i)$, i.e. the counterparty A netted payment is given by $C_A(t_p(t_i)) = \tau^I I(t_i)$.
- counterparty A makes a fixed rate payment of c, i.e. the counterparty A netted payment is given by $C_A(t_p(t_i)) = -c$.
- there is a floating rate payment of $\tau^I I(t_i)$ and a fixed rate payment of c, i.e. the counterparty A netted payment is given by $C_A(t_p(t_i)) = \tau^I I(t_i) - c$. If $c - \tau^I I(t_i) > 0$, counterparty B is the receiver of the cash flow, otherwise counterparty A.

We denote the value of the swap contract to counterparty i, $i \in \{A, B\}$, at time t by $F_i(t)$. Obviously, $F_A(t) = -F_B(t)$. Because the standard market practice is the "no fault" rule, we concentrate on that case.

One Sided Default Risk.

Assumption 6.3.1
There is only default risk associated with one of the counterparties, say counterparty i, where $i \in \{A, B\}$.

For counterparty i we assume the dynamics specified by equations (6.5), (6.6) and (6.7). Because there is no default risk associated with the other counterparty, its interest rate dynamics are fully described by equation (6.5). Using the notations of definition 6.3.1, to counterparty A the swap contract is a triple $\left[(Y_A, D_A, T), Z_A, T_A^d\right]$ with $Y_A = 0$, pre-default payments of

$$dD_A(t) = \begin{cases} C_A(t_p(t_j)), & t = t_p(t_j), \; j = 1, ..., n \\ 0, & \text{in all other cases} \end{cases}, \tag{6.59}$$

and predictable settlement payoffs of

$$Z_A(t) = \begin{cases} w_A(t) F_A(t-), & F_A(t-) < 0, \; A \text{ defaults}, \\ F_A(t-), & F_A(t-) \geq 0, \; A \text{ defaults}, \\ w_B(t) F_A(t-), & F_A(t-) \geq 0, \; B \text{ defaults}, \\ F_A(t-), & F_A(t-) < 0, \; B \text{ defaults}, \end{cases} \tag{6.60}$$

$w_i(t)$ is the recovery rate process of counterparty i as defined in assumption 6.2.4. Note, that all characteristics of a defaultable claim are satisfied by our specification of Y_A, D_A, and T_A^d. Hence, we can apply our general theory. For $t < \tau$ the price of the defaultable interest rate swap, $F_A(t)$, is given by

$$F_A(t) = E^{\mathbf{Q}}\left[\int_t^T e^{-\int_t^u r_l dl} Z_A(u)\, dH_i(u) \right. \tag{6.61}$$

$$\left. + \int_t^T e^{-\int_t^u r_l dl} L_i(u)\, dD_A(u) \right| \mathcal{F}_t \Bigg]$$

where $H_i(t)$ is the default indicator function of counterparty i and $L_i(t) = 1 - H_i(t)$. Because a swap contract is not a one-party claim we can't use proposition 6.3.1 and the framework developed in section 6.3.5 to value the contract. We have to consider the default characteristics of both counterparties. But we can extend proposition 6.3.1 to this more general setting:

Proposition 6.3.2.
If we assume

- *the pre-default payments of equation (6.59) and the settlement payoffs of equation (6.60)*
- *the dynamics specified by equations (6.5), (6.6) and (6.7) for the defaultable counterparty i*
- *the dynamics specified by equation (6.5) for the non-defaultable counterparty*

then the price process, $F_A(t)$, given by equation (6.61), satisfies

$$F_A(t) = V_t 1_{\{t<\tau\}},$$

where for $0 \le t < T$ the adapted continuous process V is given by

$$V_t = \begin{cases} E^{\mathbf{Q}}\left[\int_t^T e^{-\int_t^u \left(r_l + s_l \cdot 1_{\{F_A(l-)<0\}}\right) dl} dD_A(u) \Big| \mathcal{F}_t\right], & A \text{ defaults} \\ E^{\mathbf{Q}}\left[\int_t^T e^{-\int_t^u \left(r_l + s_l \cdot 1_{\{F_A(l-)\ge 0\}}\right) dl} dD_A(u) \Big| \mathcal{F}_t\right], & B \text{ defaults} \end{cases} \tag{6.62}$$

and $V_t = 0$ for $t \ge T$. Equation (6.62) has a unique solution in the space consisting of every semimartingale, J, such that $E^{\mathbf{Q}}\left[\sup_t |J_t|^q\right] < \infty$ for some $q > 1$.

Proof.
See proof of proposition 6.3.4.

Remark 6.3.3.
Further pricing techniques are described in section 6.3.7.

Two Sided Symmetric Default Risk.

Assumption 6.3.2
Before default the two counterparties always have the same default risk, but there is no simultaneous default possible.

Let us extend our definition of default from section 6.2.2 to the case of two counterparties A and B. We model the stochastic default times as $\mathbf{F}$–stopping times T_i^d, $i \in \{A, B\}$, valued in $[0, \infty)$. The default time T^d of the swap contract is defined as $T^d = \min\left\{T_A^d, T_B^d\right\}$. If T is the maturity of the swap, we set $\tau = \min\left(T, T^d\right)$. The event $\left\{T^d > \tau\right\}$ is the event of no default. We assume that the interest rate dynamics of both counterparties are driven by identical processes specified by equations (6.5), (6.6) and (6.7). For $t < \tau$ the price of the defaultable swap can be calculated by

$$\begin{aligned} F_A(t) = E^{\mathbf{Q}}\Bigg[&\int_t^T e^{-\int_t^u r_l dl} 1_{\{u \le T^d\}} Z_A(u) \left(dH_A(u) + dH_B(u)\right) \\ &+ \int_t^T e^{-\int_t^u r_l dl} L_A(u) L_B(u) dD_A(u) \Big| \mathcal{F}_t\Bigg], \end{aligned}$$

where D_A and Z_A are given by equations (6.59) and (6.60), respectively, and $H_i(t)$, $i \in \{A, B\}$, are the default indicator functions of counterparties A and B. In addition, $L_i(t) = 1 - H_i(t)$. Given all these assumptions the following proposition holds:

Proposition 6.3.3.
For $0 \leq t < T$,

$$F_A(t) = V_t 1_{\{t<\tau\}},$$

with

$$V_t = E^{\mathbf{Q}}\left[\int_t^T e^{-\int_t^u (r_l+s_l)dl} dD_A(u) \middle| \mathcal{F}_t\right], \tag{6.63}$$

and $V_t = 0$ for $t \geq T$. Equation (6.63) has a unique solution in the space consisting of every semimartingale, J, such that $E^{\mathbf{Q}}[\sup_t |J_t|^q] < \infty$ for some $q > 1$.

Proof.
See proof of proposition 6.3.4.

Remark 6.3.4.
In the case of two sided symmetric default risk, $F_A(t)$ can be calculated with the same techniques as in the pricing of defaultable fixed rate bonds and defaultable floating rate notes.

Two Sided Asymmetric Default Risk. Finally, we generalize the pricing of defaultable swap contracts to the case of two counterparties A and B with individual default risks. Again, we assume that there is no simultaneous default possible[24].

Assumption 6.3.3
Under the measure $\mathbf{Q}$, the dynamics of the non-defaultable short rate are given by equation (6.5).

Assumption 6.3.4
Under the measure $\mathbf{Q}$, the developments of the uncertainty indices of counterparties $i = A$ and B are given by the following stochastic differential equations:

$$du_i(t) = \left[\theta_u^i - \hat{a}_u^i u_i(t)\right] dt + \sigma_u^i \sqrt{u_i(t)} d\hat{W}_u^i(t), \; 0 \leq t \leq T^*, \; i = A, \; B,$$

where $\hat{a}_u^i$, $\sigma_u^i > 0$, $i = A$, B, are positive constants and θ_u^i, $i = A$, B, are non–negative constants.

Assumption 6.3.5
Under the measure $\mathbf{Q}$, the dynamics of the short rate spreads of counterparties $i = A$ and B are given by the following stochastic differential equations:

$$ds_i(t) = \left[b_s^i u_i(t) - \hat{a}_s^i s_i(t)\right] dt + \sigma_s^i \sqrt{s_i(t)} d\hat{W}_s^i(t), \; 0 \leq t \leq T^*, \; i = A, \; B,$$

where $\hat{a}_s^i$, $b_s^i, \sigma_s^i > 0$ are positive constants.

[24] In practice a simultaneous default of two swap contract counterparties with positive probability is very unlikely.

Additionally, we assume that the Brownian Motions $\hat{W}_r, \hat{W}_s^A, \hat{W}_s^B, \hat{W}_u^A$ and $\hat{W}_u^B$ are uncorrelated in pairs.
As in the case of symmetric default risk, $F_A(t)$ is given by

$$F_A(t) = E^{\mathbf{Q}}\left[\int_t^T e^{-\int_t^u r_l dl} 1_{\{u \le \tau\}} Z_A(u) \left(dH_A(u) + dH_B(u)\right) + \int_t^T e^{-\int_t^u r_l dl} L_A(u) L_B(u) \, dD_A(u) \,\middle|\, \mathcal{F}_t\right],$$

where D_A and Z_A are given by equations (6.59) and (6.60), respectively. Under these circumstances the following proposition holds:

Proposition 6.3.4.
For $0 \le t < T$,

$$F_A(t) - V_t 1_{\{t < \tau\}},$$

with

$$V_t = E^{\mathbf{Q}}\left[\int_t^T e^{-\int_t^u \left(r_l + s_A(l) \cdot 1_{\{F_A(l-) < 0\}} + s_B(l) \cdot 1_{\{F_A(l-) \ge 0\}}\right) dl} dD_A(u) \middle| \mathcal{F}_t\right], \tag{6.64}$$

and $V_t = 0$ *for* $t \ge T$.

Proof.
Existence and uniqueness of equation (6.64) follow immediately from lemma 6.2.4. Because $e^{-\int_t^T r_u du} F_A(T) = 0$, for $t < \tau$:

$$\begin{aligned}
F_A(t) &= E^{\mathbf{Q}}\left[\int_t^T e^{-\int_t^l r_u du} \left(L_A(l) L_B(l) \, dD_A(l) + 1_{\{l \le \tau\}} Z_A(l) \left(dH_A(l) + dH_B(l)\right)\right) \middle| \mathcal{F}_t\right] \\
&= E^{\mathbf{Q}}\left[\int_t^T e^{-\int_t^l r_u du} \left(L_A(l) L_B(l) \, dD_A(l) + 1_{\{l \le \tau\}} Z_A(l) \left(dH_A(l) + dH_B(l)\right)\right) + e^{-\int_t^T r_u du} F_A(T) \middle| \mathcal{F}_t\right].
\end{aligned}$$

Applying corollary 6.2.1 to F_A, using equation (6.60), lemma 6.2.3 with $M_A^d(t)$ and $M_B^d(t)$, and $F_A(t) = F_A(t) \cdot 1_{\{t \le \tau\}}$ yields

$$
\begin{aligned}
& dF_A(t) \\
&= r_t F_A(t)\,dt - L_A(t)\,L_B(t)\,dD_A(t) \\
&\quad -1_{\{t\le\tau\}} Z_A(t)\left(dH_A(t)+dH_B(t)\right) + dm_t \\
&= r_t F_A(t)\,dt - L_A(t)\,L_B(t)\,dD_A(t) \\
&\quad -1_{\{t\le\tau\}} F_A(t-)\left(1_{\{F_A(t-)<0\}}\left(w_A(t)-1\right)+1\right) dH_A(t) \\
&\quad -1_{\{t\le\tau\}} F_A(t-)\left(1_{\{F_A(t-)\ge 0\}}\left(w_B(t)-1\right)+1\right) dH_B(t) + dm_t \\
&= r_t F_A(t)\,dt - L_A(t)\,L_B(t)\,dD_A(t) \\
&\quad -1_{\{t\le\tau\}} 1_{\{F_A(t-)<0\}} F_A(t-)\left(w_A(t)-1\right) dH_A(t) \\
&\quad -1_{\{t\le\tau\}} F_A(t-)\,dH_A(t) \\
&\quad -1_{\{t\le\tau\}} 1_{\{F_A(t-)\ge 0\}} F_A(t-)\left(w_B(t)-1\right) dH_B(t) \\
&\quad -1_{\{t\le\tau\}} F_A(t-)\,dH_B(t) + dm_t \\
&= r_t F_A(t)\,dt - L_A(t)\,L_B(t)\,dD_A(t) \\
&\quad -1_{\{t\le\tau\}} 1_{\{F_A(t-)<0\}} F_A(t-)\left(dM_A^d(t) - s_A(t)\,dt\right) \\
&\quad -1_{\{t\le\tau\}} F_A(t-)\,dH_A(t) \\
&\quad -1_{\{t\le\tau\}} 1_{\{F_A(t-)\ge 0\}} F_A(t-)\left(dM_B^d(t) - s_B(t)\,dt\right) \\
&\quad -1_{\{t\le\tau\}} F_A(t-)\,dH_B(t) + dm_t \\
&= 1_{\{t\le\tau\}} r_t F_A(t)\,dt - L_A(t)\,L_B(t)\,dD_A(t) \\
&\quad +1_{\{t\le\tau\}} F_A(t-)\left(s_A(t)\,1_{\{F_A(t-)<0\}} + s_B(t)\,1_{\{F_A(t-)\ge 0\}}\right) dt \\
&\quad -1_{\{t\le\tau\}} F_A(t-)\left(dH_A(t)+dH_B(t)\right) + d\widetilde{m}_t \\
&= 1_{\{t\le\tau\}} F_A(t)\left(r_t + s_A(t)\,1_{\{F_A(t-)<0\}} + s_B(t)\,1_{\{F_A(t-)\ge 0\}}\right) dt \\
&\quad -L_A(t)\,L_B(t)\,dD_A(t) \\
&\quad -1_{\{t\le\tau\}} F_A(t-)\left(dH_A(t)+dH_B(t)\right) + d\widetilde{m}_t.
\end{aligned}
$$

for some (local) $\mathbf{Q}$–martingales m and $\widetilde{m}$, where

$$
\begin{aligned}
d\widetilde{m}_t = {} & -L_A(t-)\,L_B(t-)\,F_A(t-)\,1_{\{F_A(t-)<0\}}\left(dM_A^d(t) - dM_B^d(t)\right) \\
& -L_A(t-)\,L_B(t-)\,F_A(t-)\,dM_B^d(t) + dm_t.
\end{aligned}
$$

Applying lemma 6.2.6 to F_A yields (the prerequisites of the lemma are satisfied by similar arguments as in the proof of propositions 6.2.1 and 6.3.1)

$$
\begin{aligned}
& F_A(t)\cdot 1_{\{t<\tau\}} \\
&= F_A(t) \\
&= E^{\mathbf{Q}}\left[\int_t^T -\left(r_l + s_A(l)\,1_{\{F_A(l-)<0\}} + s_B(l)\,1_{\{F_A(l-)\ge 0\}}\right) 1_{\{l\le\tau\}} F_A(l)\,dl\right. \\
&\quad \left. +L_A(l)\,L_B(l)\,dD_A(l) + 1_{\{l\le\tau\}} F_A(l-)\left(dH_A(l)+dH_B(l)\right) \mid \mathcal{F}_t\right].
\end{aligned}
$$

On the other hand, applying corollary 6.2.1 to V_t yields

$$dV_t = \left(r_t + s_A(t) \cdot 1_{\{F_A(t-)<0\}} + s_B(t) \cdot 1_{\{F_A(t-)\geq 0\}}\right) V_t dt$$
$$-dD_A(t) + d\widehat{m}_t,$$

for some $\mathbf{Q}$-martingale $\widehat{m}$. Using integration by parts,

$$\begin{aligned}
& d\left(V_t 1_{\{t<\tau\}}\right) \\
&= d\left(V_t L_A(t) L_B(t)\right) \\
&= L_A(t-) L_B(t-) dV_t - V_{t-}\left(L_B(t-) dH_A(t) + L_A(t-) dH_B(t)\right) \\
&= L_A(t-) L_B(t-) \left(\left(r_t + s_A(t) 1_{\{F_{t-}^A<0\}} + s_B(t) 1_{\{F_{t-}^A\geq 0\}}\right) V_t dt\right. \\
&\quad \left. -dD_A(t) + d\widehat{m}_t\right) - V_{t-}\left(L_B(t-) dH_A(t) + L_A(t-) dH_B(t)\right) \\
&= -L_A(t-) L_B(t-) dD_A(t) + L_A(t-) L_B(t-) \\
&\quad \left(r_t + s_A(t) 1_{\{F_A(t-)<0\}} + s_B(t) 1_{\{F_A(t-)\geq 0\}}\right) V_t dt \\
&\quad -1_{\{t\leq\tau\}} V_{t-}\left(dH_A(t) + dH_B(t)\right) + L_A(t-) L_B(t-) d\widehat{m}_t.
\end{aligned}$$

Applying lemma 6.2.6 to $V_t 1_{\{t<\tau\}}$ yields (the prerequisites of the lemma are satisfied by similar arguments as in the proof of propositions 6.2.1 and 6.3.1)

$$\begin{aligned}
& V_t 1_{\{t<\tau\}} \\
&= E^{\mathbf{Q}}\left[\int_t^T -\left(r_l + s_A(l) 1_{\{F_A(l-)<0\}} + s_B(l) 1_{\{F_A(l-)\geq 0\}}\right) 1_{\{l\leq\tau\}} V_l dl\right. \\
&\quad \left. + L_A(l) L_B(l) dD_A(l) + 1_{\{l\leq\tau\}} V_{l-}\left(dH_A(l) + dH_B(l)\right) \middle| \mathcal{F}_t\right].
\end{aligned}$$

Because of the uniqueness of a solution we can conclude for $t < \tau$,

$$F_A(t) = V_t 1_{\{t<\tau\}}.$$

A solution for equation (6.64) can be determined, e.g., by choosing its Feynman-Kac representation and solving the appropriate PDE, e.g., by applying a finite-difference algorithm such as the Crank-Nicholson method[25]. Alternatively Monte-Carlo simulations can be used.

6.4 The Pricing of Credit Derivatives

6.4.1 Some Pricing Issues

In the following we develop pricing formulas for credit options, spread options, and default options/swaps based on our three-factor defaultable term structure model. Our work is different from all other results in that we apply our new three-factor model and PDE methods similar to those introduced

[25] See, e.g., Lamberton & Lapeyre (1996).

in section 6.3.5 to the pricing of credit derivatives. Throughout this section we assume that $\beta = \frac{1}{2}$ to make sure that the non-defaultable short rate r always stays positive. We discuss the case $\beta = 0$ in appendix C. To develop pricing formulas in our three-factor framework we have to make some extensions first. Suppose we want to price an option on a defaultable zero-coupon bond (a so called credit option). If the option is knocked out at default of the zero-coupon bond, the buyer of the option receives nothing. Hence, we can interpret the option as a defaultable investment with zero recovery. We show that we can determine the price of the option as the expected value of the promised cash flow at maturity of the option discounted at risky discount rates. But which risky discount rates do we have to use ? As the recovery rate of the option is different from the recovery rate of the reference defaultable zero-coupon bond the risky discount rates are not the same as in the case of the pricing of defaultable zero-coupon bonds. Hence, we have to find a short rate credit spread s^{zero} describing the credit spread process of an obligor which is equivalent to the issuer of the zero-coupon bond (especially of the same quality) but with zero recovery rate. Therefore, for pricing credit derivatives such as credit options we need the following data for all maturities $T > 0$:

- the default-free term structure of bond prices $P(t, T)$,
- the defaultable term structure of bond prices $P^d(t, T)$,
- the defaultable term structure of bond prices under zero recovery $P^{d,zero}(t, T)$.

The first piece of information, the default-free term structure, is easily obtained. Possible choices are government curves or swap curves in developed economies. The second piece is the defaultable term structure of the reference credit. Ideally it is obtained directly from the prices of the reference credit's bonds. Finally, the third piece of input data are the defaultable bond prices under zero recovery. These prices are usually unobservable. But we can derive the zero recovery term structure of bond prices from the default-free and defaultable term structures of bond prices. Therefore we use our three-factor defaultable term structure model. Obviously, the uncertainty index u and the non defaultable short rate r do not depend on the recovery rate, but the recovery rate should implicitly change the short rate spread. Hence, we have to introduce an additional stochastic process to our model, describing the zero recovery short rate spread. We do this consistently to our definition of a defaultable money market account (see definition 6.2.4).

Assumption 6.4.1
The zero recovery short rate spread s^{zero} is given by:

$$s_t^{zero} = \frac{s_t}{1 - w_t}, \quad 0 \leq t \leq T^*, \tag{6.65}$$

where s_t is the short rate spread process defined in equation (6.3) and $0 \leq w_t < 1$ is the recovery rate process of the reference credit asset introduced in assumption 6.2.4.

Definition 6.4.1 (corresponds to definition 6.2.4).
The zero recovery defaultable money market account $B^{d,zero}$ is defined as

$$B_t^{d,zero} = \left(1 - \int_0^t dH_l\right) e^{\int_0^t (r_l + s_l^{zero}) dl},$$

where H_t is the default indicator function of the reference credit asset.

Lemma 6.4.1 (corresponds to lemma 6.2.3).
$M_t^{d,zero} = \int_0^t s_l^{zero} dl - \int_0^t dH_l$ is a $\mathcal{F}_t-$(local) martingale. H_t is the default indicator function of the reference credit asset.

Proof.
See proof of lemma 6.2.3.

Using lemma 6.2.4, lemma 6.2.5, lemma 6.2.6, lemma 6.4.1, and corollary 6.2.1, it can be shown that the following proposition is valid:

Proposition 6.4.1 (corresponds to proposition 6.2.1).
Let $E^{\mathbf{Q}}[|Y|^q] < \infty$ for some $q > 1$. Under the zero recovery assumption (i.e. under the assumption that the contingent claim is knocked out at default of the reference credit asset) and the stochastic processes specified by equations (6.5), (6.6), (6.7), and (6.65), the price process, F_t :

$$F_t = E^{\mathbf{Q}}\left[e^{-\int_t^T r_l dl}\, Y L_T \middle| \mathcal{F}_t\right], \; 0 \leq t < \tau,$$

where $L_T = 1 - H_T$, is given by

$$F_t = V_t 1_{\{t<\tau\}}, \tag{6.66}$$

where the adapted continuous process V is defined by

$$Vt = E^{\mathbf{Q}}\left[e^{-\int_t^T (r_l + s_l^{zero}) dl} Y \mid \mathcal{F}_t\right], \; 0 \leq t < T, \tag{6.67}$$

and $V_t = 0$ for $t \geq T$, i.e. if there has been no default until time t, F_t must equal the expected value of riskless cash flows discounted at zero recovery risky discount rates. Equation (6.67) has a unique solution in the space consisting of every semimartingale, J, such that $E^{\mathbf{Q}}[\sup_t |J_t|^q] < \infty$ for some $q > 1$.

Proof.
See proof of proposition 6.2.1.

Remark 6.4.1.

1. Suppose we want to price a contingent claim that promises to pay off Y at maturity time T of the contingent claim, if the reference credit asset hasn't defaulted until then and zero in case of a default. Then the time t price of the reference credit asset, given there has been no default so far, depends on the stochastic processes r, s, and u. But in addition, the price of the contingent claim depends on s^{zero}, because discounting is done with $e^{-\int_t^T (r_l + s_l^{zero}) dl}$.
2. Similarly to proposition 6.4.1 a proposition corresponding to proposition 6.3.1 can be formulated.

For all the pricing formulas developed in the previous sections we haven't used recovery rates explicitly. Especially, we haven't made any specific assumptions on the dynamics of w_t. In the following we assume that w_t is a known constant, i.e. $w_t = w$ for all $0 \le t \le T^*$. Then the dynamics of the zero recovery short rate spread are given by

$$\begin{aligned} ds_t^{zero} &= \frac{1}{1-w} ds_t \\ &= \left[\frac{b_s}{1-w} u_t - \frac{a_s}{1-w} s_t \right] dt + \frac{\sigma_s}{1-w} \sqrt{s_t} dW_s(t) \\ &= \left[\frac{b_s}{1-w} u_t - a_s s_t^{zero} \right] dt + \frac{\sigma_s}{\sqrt{1-w}} \sqrt{s_t^{zero}} dW_s(t) \\ &= \left[b_s^{zero} u_t - a_s s_t^{zero} \right] dt + \sigma_s^{zero} \sqrt{s_t^{zero}} dW_s(t), \ 0 \le t \le T^*, \end{aligned}$$

where $b_s^{zero} = \frac{b_s}{1-w}$ and $\sigma_s^{zero} = \frac{\sigma_s}{\sqrt{1-w}}$. Under the measure $\mathbf{Q}$, the dynamics of s^{zero} are given by

$$ds_t^{zero} = \left[b_s^{zero} u_t - \widehat{a}_s s_t^{zero} \right] dt + \sigma_s^{zero} \sqrt{s_t^{zero}} d\hat{W}_s(t), \ 0 \le t \le T^*.$$

To summarize, the following steps are necessary to determine the defaultable term structure of bond prices under zero recovery at time t:

1. Determination of the time t default-free and defaultable term structures of bond prices from observable market data.
2. Determination of r_t, s_t, and u_t.
3. Estimation of the parameters of the processes r, s and u (for details see section 6.6).
4. Calculation of the parameters b_s^{zero} and σ_s^{zero}.
5. Calculation of the zero recovery zero-coupon bond prices

$$\begin{aligned} &P^{d,zero}(t,T) \\ &= A^{d,zero}(t,T) \, e^{-B(t,T) r_t - C^{d,zero}(t,T) s_t^{zero} - D^{d,zero}(t,T) u_t}, \end{aligned} \tag{6.68}$$

where $A^{d,zero}(t,T)$, $C^{d,zero}(t,T)$, and $D^{d,zero}(t,T)$ are given by the corresponding formulas for $A^d(t,T)$, $C^d(t,T)$, and $D^d(t,T)$ with b_s and σ_s substituted by b_s^{zero} and σ_s^{zero}, respectively.

So far nothing has been said about the credit quality of the protection seller. The price of a credit derivative should be lower for lower quality protection sellers. The higher the default correlation between the reference credit and the protection seller the less the default swap should be worth. If not otherwise stated, we assume that the protection seller is default-free.

6.4.2 Credit Options

Let us price a European call and put option[26] with maturity $T_O \leq T$ and exercise price K, on a defaultable zero-coupon bond with maturity T, and denote their values by $F^{cc}(z, t, T; T_O, K)$ and $F^{cp}(z, t, T; T_O, K)$, respectively, where $z = (r, s, u)$. As with other options, the credit option premium is highly sensitive to the volatility of the market price of the underlying, and the extent to which the spread between strike and bond price is in or out of the money relative to the applicable current forward yield curve. Depending on the specification, the option can either survive a default or be knocked out by it. In the first case it doesn't suffice to consider P^d alone but we must consider the post-default bond price behavior as well. Basically there are two possibilities: either the obligor survives the default, e.g., some kind of restructuring takes place, and the bonds are still traded with a market value dropped to the recovery rate times the pre-default market value of the debt, or the proceeds from default recovery are reinvested at the default-free short rate and rolled over until expiration. If credit options survive a credit event, default risk and spread risk are transferred between the counterparties. Otherwise only spread risk is transferred. In the following we will assume throughout that options are knocked out at default of the underlying.

Definition 6.4.2.

1. *A European call option with maturity $T_O \leq T$ and exercise price K on a defaultable zero-coupon bond $P^d(t,T)$ pays off*

$$1_{\{T_O < T^d\}} \left(P^d(T_O, T) - K\right)_+ ,$$

if the call is knocked out at default.

2. *A European put option with maturity $T_O \leq T$ and exercise price K on a defaultable zero-coupon bond $P^d(t,T)$ pays off*

$$1_{\{T_O < T^d\}} \left(K - P^d(T_O, T)\right)_+ ,$$

if the put is knocked out at default.

If we assume that there is no default risk with the issuer of the option and the call is knocked out at default, by proposition 6.4.1 the call price at time $t \leq T_O$ is given by

[26] For an introduction to credit options see section 5.3.1.

$$
\begin{aligned}
& F^{cc}(z,t,T;T_O,K) \\
&= E^{\mathbf{Q}}\left[e^{-\int_t^{T_O} r_l dl} 1_{\{T_O<T^d\}} \left(P^d(T_O,T)-K\right)_+ \Big| \mathcal{F}_t \right] \\
&= E^{\mathbf{Q}}\left[e^{-\int_t^{T_O} (r_l+s_l^{zero}) dl} \left(P^d(T_O,T)-K\right) \cdot 1_{\{\mathcal{B}_1\}} \Big| \mathcal{F}_t \right],
\end{aligned}
$$

where $\mathcal{B}_1$ is the range of values[27] of r, s and u over which the option is in the money at the expiration date T_O, that is,

$$
\begin{aligned}
\mathcal{B}_1 &= \left\{ z \in \mathbb{R}_+^3 | P^d(z,T_O,T) - K \geq 0 \right\} \\
&= \left\{ z \in \mathbb{R}_+^3 | B(T_O,T) r + C^d(T_O,T) s + D^d(T_O,T) u \leq K^* \right\}.
\end{aligned}
$$

$K^* = \ln \frac{A^d(T_O,T)}{K}$ and $P^d(T_O,T)$ is the bond price given by equation (6.19). Similarly, by proposition 6.4.1 the put price is given by

$$
\begin{aligned}
& F^{cp}(z,t,T;T_O,K) \\
&= E^{\mathbf{Q}}\left[e^{-\int_t^{T_O} (r_l+s_l^{zero}) dl} \left(K - P^d(T_O,T)\right) \cdot 1_{\{\mathcal{B}_2\}} \Big| \mathcal{F}_t \right],
\end{aligned}
$$

where $\mathcal{B}_2$ is the range of values of r, s and u over which the option is in the money at the expiration date T_O, that is,

$$
\mathcal{B}_2 = \left\{ z \in \mathbb{R}_+^3 | K^* \leq B(T_O,T) r + C^d(T_O,T) s + D^d(T_O,T) u \right\}.
$$

We determine $F^{cc}(z,t,T;T_O,K)$ by using PDE techniques. Under mild regularity conditions and by the Feynman-Kac Formula, $F^{cc}(z,t,T;T_O,K)$ satisfies the PDE

$$
\begin{aligned}
0 &= \frac{1}{2}\sigma_r^2 r^{2\beta} F_{rr}^{cc} + \frac{1}{2}\sigma_s^2 s F_{ss}^{cc} + \frac{1}{2}\sigma_u^2 u F_{uu}^{cc} \\
&\quad + \left[\theta_r(t) - \hat{a}_r r\right] F_r^{cc} + \left[b_s u - \hat{a}_s s\right] F_s^{cc} \\
&\quad + \left[\theta_u - \hat{a}_u u\right] F_u^{cc} + F_t^{cc} - \left(r + \frac{s}{1-w}\right) F^{cc},
\end{aligned}
\tag{6.69}
$$

with boundary condition

$$
F^{cc}(z,T_O,T;T_O,K) = \left(P^d(z,T_O,T) - K\right)_+ . \tag{6.70}
$$

If $\vec{G}\left(y,\check{t},z,t\right)$ is the solution of the problem

$$
\begin{aligned}
0 &= \frac{1}{2}\sigma_r^2 r^{2\beta} \vec{G}_{rr} + \frac{1}{2}\sigma_s^2 s \vec{G}_{ss} + \frac{1}{2}\sigma_u^2 u \vec{G}_{uu} \\
&\quad + \left[\theta_r(t) - \hat{a}_r r\right] \vec{G}_r + \left[b_s u - \hat{a}_s s\right] \vec{G}_s \\
&\quad + \left[\theta_u - \hat{a}_u u\right] \vec{G}_u + \vec{G}_t - \left(r + \frac{s}{1-w}\right) \vec{G},
\end{aligned}
$$

[27] Since we have set β to $\frac{1}{2}$ we can make sure that r always stays positive. In addition, the definitions of s and u guarantee that s and u are positive processes. Hence, throughout section 6.4 $(r,s,u) \in \mathbb{R}_+^3$. In appendix C we discuss the case that $\beta = 0$. This implies that r may become negative.

with boundary condition

$$\overrightarrow{G}\left(y,\check{t},z,\check{t}\right)=\delta\left(r-y_r\right)\delta\left(s-y_s\right)\delta\left(u-y_u\right),$$

where $\delta\left(.\right)$ is the Dirac function, the solution of the PDE (6.69) with boundary condition (6.70) can be written as

$$\begin{aligned}&F^{cc}\left(z,t,T;T_O,K\right)\\&=\iiint\limits_{\mathbf{R}^3}\overrightarrow{G}\left(y,T_O,z,t\right)\left(P^d\left(y,T_O,T\right)-K\right)_+dy.\end{aligned}\tag{6.71}$$

If we know $\overrightarrow{G}$ we can determine $F^{cc}\left(z,t,T;T_O,K\right)$ easily.

Lemma 6.4.2.
$\overrightarrow{G}\left(y,\check{t},z,t\right)$, is given by

$$\overrightarrow{G}\left(y,\check{t},z,t\right)=\frac{1}{\left(2\pi\right)^{3/2}}\iiint\limits_{\mathbf{R}^3}e^{-ixy'}\overleftarrow{G}\left(x,\check{t},z,t\right)dx,\tag{6.72}$$

where

$$\overleftarrow{G}\left(x,\check{t},z,t\right)=A^{\overleftarrow{G}}\left(x,t,\check{t}\right)e^{-B^{\tilde{G}}\left(x,t,\check{t}\right)r-C^{\overleftarrow{G}}\left(x,t,\check{t}\right)s-D^{\overleftarrow{G}}\left(x,t,\check{t}\right)u},$$

with

$$C^{\overleftarrow{G}}\left(x,t,\check{t}\right)=\frac{\kappa_3^{\left(s,\frac{1}{1-w}\right)}\left(x\right)-e^{-\delta_s^{\left(\frac{1}{1-w}\right)}\left(\check{t}-t\right)}}{\kappa_4^{\left(s,\frac{1}{1-w}\right)}\left(x\right)-\kappa_5^{\left(s,\frac{1}{1-w}\right)}\left(x\right)e^{-\delta_s^{\left(\frac{1}{1-w}\right)}\left(\check{t}-t\right)}},$$

$$D^{\overleftarrow{G}}\left(x,t,\check{t}\right)=\frac{-2\left(v^{\overleftarrow{G}}\right)'\left(x,t,\check{t}\right)}{\sigma_u^2v^{\overleftarrow{G}}\left(x,t,\check{t}\right)}.$$

If $\beta=0$:

$$\begin{aligned}&A^{\overleftarrow{G}}\left(x,t,\check{t}\right)\\&=e^{-\int_t^{\check{t}}\left(-\frac{1}{2}\sigma_r^2\left(B^{\tilde{G}}\right)^2\left(x,\tau,\check{t}\right)+\theta_r(\tau)B^{\tilde{G}}\left(x,\tau,\check{t}\right)+\theta_uD^{\overleftarrow{G}}\left(x,\tau,\check{t}\right)\right)d\tau}.\end{aligned}$$

If $\beta=\frac{1}{2}$:

$$A^{\overleftarrow{G}}\left(x,t,\check{t}\right)=e^{-\int_t^{\check{t}}\left(\theta_r(\tau)B^{\tilde{G}}\left(x,\tau,\check{t}\right)+\theta_uD^{\overleftarrow{G}}\left(x,\tau,\check{t}\right)\right)d\tau}.$$

$v^{\overleftarrow{G}}\left(x,\check{t},t\right)$ and $\left(v^{\overleftarrow{G}}\right)'\left(x,\check{t},t\right)$ are complicated functions defined in the proof below. $\kappa_3^{\left(s,\frac{1}{1-w}\right)}\left(x\right)$, $\kappa_4^{\left(s,\frac{1}{1-w}\right)}\left(x\right)$, $\kappa_5^{\left(s,\frac{1}{1-w}\right)}\left(x\right)$, and $\delta_s^{\left(\frac{1}{1-w}\right)}\left(t\right)$ are defined by equations (B.23), (B.24), (B.25), and (B.22), respectively. $B^{\tilde{G}}\left(x,t,\check{t}\right)$ is given by equation (6.52).

Proof.
We apply the proof of appendix B.3 with $\alpha = \frac{1}{1-w}$. Then

$$\begin{aligned}
&\overrightarrow{G}\left(y, \check{t}, z, t\right) = G^{\left(\frac{1}{1-w}\right)}\left(y, \check{t}, z, t\right), \quad \overleftarrow{G}\left(x, \check{t}, z, t\right) = \tilde{G}^{\left(\frac{1}{1-w}\right)}\left(x, \check{t}, z, t\right), \\
&A^{\overleftarrow{G}}\left(x, t, \check{t}\right) = A^{\tilde{G}^{\left(\frac{1}{1-w}\right)}}\left(x, t, \check{t}\right), \quad C^{\overleftarrow{G}}\left(x, t, \check{t}\right) = C^{\tilde{G}^{\left(\frac{1}{1-w}\right)}}\left(x, t, \check{t}\right), \\
&D^{\tilde{G}}\left(x, t, \check{t}\right) = D^{\tilde{G}^{\left(\frac{1}{1-w}\right)}}\left(x, t, \check{t}\right).
\end{aligned}$$

$v^{\overleftarrow{G}}\left(x, \check{t}, t\right) = v^{\tilde{G}^{\left(\frac{1}{1-w}\right)}}\left(x, \check{t}, t\right)$ and $\left(v^{\overleftarrow{G}}\right)'\left(x, \check{t}, t\right)$ are defined as in equations (B.26), (B.27), and (B.28) of appendix B.3.

Equation (6.71) together with lemma (6.4.2) yields the following closed-form solution for $F^{cc}\left(z, t, T_B; T_O, K\right)$:

Theorem 6.4.1.

1. *The price of a European call option (knocked out at default of the underlying) with maturity $T_O \leq T$ and exercise price K on a defaultable zero-coupon bond $P^d\left(t, T\right)$ is given by*

$$\begin{aligned}
&F^{cc}\left(z, t, T; T_O, K\right) \\
&= P^d\left(z, t, T\right) \cdot \\
&\quad \Phi^C\left(\phi_1, \phi_2, \phi_3, \psi_0, \psi_1, \psi_2, \psi_3, \varphi_0, \varphi_1, \varphi_2, \varphi_3, z\right) \\
&\quad - K P^d\left(z, t, T_O\right) \cdot \\
&\quad \Phi^C\left(\phi_4, \phi_5, \phi_6, \psi_4, \psi_5, \psi_6, \psi_7, \varphi_0, \varphi_1, \varphi_2, \varphi_3, z\right),
\end{aligned}$$

where

$$\begin{aligned}
&\Phi^C\left(a_1, a_2, a_3, b_0, b_1, b_2, b_3, c_0, c_1, c_2, c_3, z\right) \\
&= \frac{1}{\left(2\pi\right)^{3/2}} \iiint\limits_{\mathbf{R}^3} \Lambda^C\left(a_1, a_2, a_3, c_0, c_1, c_2, c_3\right) \cdot \\
&\quad b_0\left(x\right) \cdot e^{-b_1(x)r - b_2(x)s - b_3(x)u} dx
\end{aligned} \tag{6.73}$$

with

$$\begin{aligned}
&\Lambda^C\left(a_1, a_2, a_3, c_0, c_1, c_2, c_3\right) \\
&= \iiint\limits_{\mathbf{R}_+^3} e^{-ia_1(x)y_r - ia_2(x)y_s - ia_3(x)y_u} \cdot \\
&\quad 1_{\{c_1 y_r + c_2 y_s + c_3 y_u \leq c_0\}} dy.
\end{aligned} \tag{6.74}$$

$$\begin{array}{ll}
\phi_1 = x_1 - iB(T_O,T), & \phi_2 = x_2 - iC^d(T_O,T), \\
\phi_3 = x_3 - iD^d(T_O,T), & \phi_4 = x_1, \\
\phi_5 = x_2, & \phi_6 = x_3, \\
\psi_0 = A^d(T_O,T)\frac{A^{\overleftarrow{G}}(x,t,T_O)}{A^d(t,T)}, & \psi_1 = B^{\widetilde{G}}(x,t,T_O) - B(t,T), \\
\psi_2 = C^{\overleftarrow{G}}(x,t,T_O) - C^d(t,T), & \psi_3 = D^{\overleftarrow{G}}(x,t,T_O) - D^d(t,T), \\
\psi_4 = \frac{A^{\overleftarrow{G}}(x,t,T_O)}{A^d(t,T_O)}, & \psi_5 = B^{\widetilde{G}}(x,t,T_O) - B(t,T_O), \\
\psi_6 = C^{\overleftarrow{G}}(x,t,T_O) - C^d(t,T_O), & \psi_7 = D^{\overleftarrow{G}}(x,t,T_O) - D^d(t,T_O), \\
\varphi_0 = \ln\frac{A^d(T_O,T)}{K}, & \varphi_1 = B(T_O,T), \\
\varphi_2 = C^d(T_O,T), & \varphi_3 = D^d(T_O,T).
\end{array}$$

$B^{\widetilde{G}}$ is given by equation (6.52), $A^{\overleftarrow{G}}$, $C^{\overleftarrow{G}}$, and $D^{\overleftarrow{G}}$ are defined like in lemma 6.4.2, and A^d, B, C^d, and D^d are defined in theorem 6.3.1. $\Lambda^C(a_1,a_2,a_3,c_0,c_1,c_2,c_3)$ can be computed analytically (lemma 6.4.3).

2. *The price of a European put option (knocked out at default of the underlying) with maturity $T_O \leq T$ and exercise price K on a defaultable zero-coupon bond $P^d(t,T)$ is given by*

$$\begin{aligned}
&F^{cp}(z,t,T;T_O,K) \\
&= KP^d(z,t,T_O)\cdot \\
&\quad \Phi^P(\phi_4,\phi_5,\phi_6,\psi_4,\psi_5,\psi_6,\psi_7,\varphi_0,\varphi_1,\varphi_2,\varphi_3,z) \\
&\quad -P^d(z,t,T)\cdot \\
&\quad \Phi^P(\phi_1,\phi_2,\phi_3,\psi_0,\psi_1,\psi_2,\psi_3,\varphi_0,\varphi_1,\varphi_2,\varphi_3,z),
\end{aligned}$$

where

$$\begin{aligned}
&\Phi^P(a_1,a_2,a_3,b_0,b_1,b_2,b_3,c_0,c_1,c_2,c_3,z) \\
&= \frac{1}{(2\pi)^{3/2}}\iiint_{\mathbf{R}^3}\Lambda^P(a_1,a_2,a_3,c_0,c_1,c_2,c_3)\cdot \\
&\quad b_0(x)\cdot e^{-b_1(x)r-b_2(x)s-b_3(x)u}dx
\end{aligned}$$

with

$$\begin{aligned}
&\Lambda^P(a_1,a_2,a_3,c_0,c_1,c_2,c_3) \\
&= \iiint_{\mathbf{R}^3_+} e^{-ia_1y_r-ia_2y_s-ia_3y_u}\cdot \\
&\quad 1_{\{c_1y_r+c_2y_s+c_3y_u\geq c_0\}}dy.
\end{aligned}$$

Proof.
The value $F^{cc}(z,t,T;T_O,K)$ can be written as

$$
\begin{aligned}
&F^{cc}(z,t,T;T_O,K) \\
&= \iiint\limits_{\mathbf{R}^3} \left(P^d(y,T_O,T)-K\right)\overrightarrow{G}(y,T_O,z,t)\,1_{\{y\in\mathcal{B}_1\}}dy \\
&= \iiint\limits_{\mathbf{R}^3} P^d(y,T_O,T)\,\overrightarrow{G}(y,T_O,z,t)\,1_{\{y\in\mathcal{B}_1\}}dy \\
&\quad -K\iiint\limits_{\mathbf{R}^3} \overrightarrow{G}(y,T_O,z,t)\,1_{\{y\in\mathcal{B}_1\}}dy \\
&= I_1 - I_2
\end{aligned}
$$

where $\mathcal{B}_1$ is defined as above and $\overrightarrow{G}(y,T_O,z,t)$ is given in lemma 6.4.2.
Determination of I_1 :

$$
\begin{aligned}
I_1 &= \iiint\limits_{\mathbf{R}^3} P^d(y,T_O,T)\,\overrightarrow{G}(y,T_O,z,t)\,1_{\{y\in\mathcal{B}_1\}}dy \\
&= \frac{1}{(2\pi)^{3/2}}\iiint\limits_{\mathbf{R}^3}\iiint\limits_{\mathbf{R}^3} P^d(y,T_O,T)\,e^{-iyx'}1_{\{y\in\mathcal{B}_1\}}dy\cdot\overleftarrow{G}(x,T_O,z,t)\,dx \\
&= \frac{P^d(z,t,T)}{(2\pi)^{3/2}}\iiint\limits_{\mathbf{R}^3}\iiint\limits_{\mathbf{R}^3} A^d(T_O,T)\cdot \\
&\quad e^{-iy_r(x_1-iB(T_O,T))-iy_s\left(x_2-iC^d(T_O,T)\right)-iy_u\left(x_3-iD^d(T_O,T)\right)}\cdot 1_{\{y\in\mathcal{B}_1\}}dy\,\cdot \\
&\quad \frac{A^{\overleftarrow{G}}(x,t,T_O)}{A^d(t,T)}e^{-\left(B^{\overleftarrow{G}}(x,t,T_O)-B(t,T)\right)r}\cdot \\
&\quad e^{-\left(C^{\overleftarrow{G}}(x,t,T_O)-C^d(t,T)\right)s-\left(D^{\overleftarrow{G}}(x,t,T_O)-D^d(t,T)\right)u}dx.
\end{aligned}
$$

Determination of I_2 :
Similarly the second integral can be written as

$$
\begin{aligned}
I_2 &= \iiint\limits_{\mathbf{R}^3} K\overrightarrow{G}(y,T_O,z,t)\,1_{\{y\in\mathcal{B}_1\}}dy \\
&= \frac{K}{(2\pi)^{3/2}}\iiint\limits_{\mathbf{R}^3}\iiint\limits_{\mathbf{R}^3} e^{-iyx'}1_{\{y\in\mathcal{B}_1\}}dy \\
&\quad e^{-B^{\overleftarrow{G}}(x,t,T_O)r}A^{\overleftarrow{G}}(x,t,T_O)\,e^{-C^{\overleftarrow{G}}(x,t,T_O)s-D^{\overleftarrow{G}}(x,t,T_O)u}dx \\
&= \frac{KP^d(r,s,u,t,T_O)}{(2\pi)^{3/2}}\iiint\limits_{\mathbf{R}^3}\iiint\limits_{\mathbf{R}^3} e^{-iyx'}1_{\{y\in\mathcal{B}_1\}}dy\,\cdot \\
&\quad \frac{A^{\overleftarrow{G}}(x,t,T_O)}{A^d(t,T_O)}e^{-\left(B^{\overleftarrow{G}}(x,t,T_O)-B(t,T_O)\right)r}\cdot \\
&\quad e^{-\left(C^{\overleftarrow{G}}(x,t,T_O)-C^d(t,T_O)\right)s-\left(D^{\overleftarrow{G}}(x,t,T_O)-D^d(t,T_O)\right)u}dx.
\end{aligned}
$$

The proof for the put works identically.

Lemma 6.4.3.

$$\begin{aligned}
&\Lambda^C\left(a_1, a_2, a_3, c_0, c_1, c_2, c_3\right) \\
&= -\frac{i c_0^3}{c_1 c_2 c_3 \widetilde{a}_1} \cdot \\
&\left(\frac{e^{-i\widetilde{a}_3} - e^{-i\widetilde{a}_2}}{\left(\widetilde{a}_3 - \widetilde{a}_2\right)\left(\widetilde{a}_2 - \widetilde{a}_1\right)} - \frac{e^{-i\widetilde{a}_3} - e^{-i\widetilde{a}_1}}{\left(\widetilde{a}_3 - \widetilde{a}_1\right)\left(\widetilde{a}_2 - \widetilde{a}_1\right)}\right. \\
&\left. + \frac{e^{-i\widetilde{a}_3} - 1}{\widetilde{a}_2 \widetilde{a}_3} - \frac{e^{-i\widetilde{a}_3} - e^{-i\widetilde{a}_2}}{\widetilde{a}_2\left(\widetilde{a}_3 - \widetilde{a}_2\right)}\right),
\end{aligned}$$

where

$$\widetilde{a}_i = \frac{c_0 a_i}{c_i} \text{ for } i = 1, 2, 3.$$

A similar result can be obtained for Λ^P.

Proof.
See appendix B.

Remark 6.4.2.

1. If we assumed that the call option survived a default, and after some restructuring the bonds were still traded with a market value dropped to the bond recovery rate times the pre-default market value of the debt, the price would be given by

$$\begin{aligned}
&F^{cc}\left(z, t, T; T_O, K\right) \\
&= E^{\mathbf{Q}}\left[e^{-\int_t^{T_O} r_l dl} 1_{\{T_O < T^d\}}\left(P^d\left(T_O, T\right) - K\right)_+ \middle| \mathcal{F}_t\right] \\
&\quad + E^{\mathbf{Q}}\left[e^{-\int_t^{T_O} r_l dl} 1_{\{T_O \geq T^d\}}\left(\widetilde{P}^d\left(T_O, T\right) - K\right)_+ \middle| \mathcal{F}_t\right],
\end{aligned}$$

where $\widetilde{P}^d(t, T)$ is the defaultable zero-coupon bond price adjusted for a possible default before t, i.e. it is the price of a zero-coupon bond which hasn't defaulted yet but is equivalent to the defaulted bond with the face value reduced to the bond recovery rate (which we assumed to be constant) times the pre-default market value of the debt.

$$\begin{aligned}
&E^{\mathbf{Q}}\left[e^{-\int_t^{T_O} r_l dl} 1_{\{T_O \geq T^d\}}\left(\widetilde{P}^d\left(T_O, T\right) - K\right)_+ \middle| \mathcal{F}_t\right] \\
&= E^{\mathbf{Q}}\left[e^{-\int_t^{T_O} r_l dl}\left(\widetilde{P}^d\left(T_O, T\right) - K\right)_+ \middle| \mathcal{F}_t\right] \qquad (6.75) \\
&\quad - E^{\mathbf{Q}}\left[e^{-\int_t^{T_O} r_l dl} 1_{\{T_O < T^d\}}\left(\widetilde{P}^d\left(T_O, T\right) - K\right)_+ \middle| \mathcal{F}_t\right] \qquad (6.76)
\end{aligned}$$

Because (6.76) can be calculated similarly to

$$E^{\mathbf{Q}}\left[e^{-\int_t^{T_O} r_l dl} 1_{\{T_O<T^d\}}\left(P^d\left(T_O,T\right)-K\right)_+\Big|\mathcal{F}_t\right],$$

we leave the calculations to the interested reader. For the determination of the value of (6.75) one can use PDE techniques similar to those described above. The necessary mathematical tools are given in lemma B.5.1.

2. Corresponding formulas are valid for put options.
3. The structure of the pricing formulas in the defaultable case is the same as the structure of the bond option pricing formulas in the classical term structure models.

6.4.3 Credit Spread Options

For pricing credit spread options[28] we use the same pricing methodology as in the previous section.

Definition 6.4.3.

1. *A European credit spread call option with maturity $T_O \leq T$ and strike spread K on a defaultable zero-coupon bond $P^d(t,T)$ pays off*

$$1_{\{T_O<T_B^d\}}\left(P^d\left(T_O,T\right)-e^{-(T-T_O)K}P\left(T_O,T\right)\right)_+,$$

if the call is knocked out at default.

2. *A European credit spread put option with maturity $T_O \leq T$ and exercise price K on a defaultable zero-coupon bond $P^d(t,T)$ pays off*

$$1_{\{T_O<T_B^d\}}\left(e^{-(T-T_O)K}P\left(T_O,T\right)-P^d\left(T_O,T\right)\right)_+,$$

if the put is knocked out at default.

If we assume that there is no default risk with the issuer of the option and the call is knocked out at default, the call price at time $t \leq T_O$ is given by

$$\begin{aligned}
&F^{\text{csc}}\left(z,t,T;T_O,K\right)\\
&=E^{\mathbf{Q}}\left[e^{-\int_t^{T_O} r_l dl} 1_{\{T_O<T^d\}}\left(P^d\left(T_O,T\right)-e^{-(T-T_O)K}P\left(T_O,T\right)\right)_+\Big|\mathcal{F}_t\right]\\
&=E^{\mathbf{Q}}\left[e^{-\int_t^{T_O}\left(r_l+s_l^{zero}\right)dl} 1_{\{\mathcal{B}_3\}}\left(P^d\left(T_O,T\right)-e^{-(T-T_O)K}P\left(T_O,T\right)\right)\Big|\mathcal{F}_t\right],
\end{aligned}$$

where $\mathcal{B}_3$ is the range of values of r, s, and u over which the option is in the money at the expiration date T_O, that is,

[28] For an introduction to credit spread options see section 5.3.2.

$$\begin{aligned}
&\mathcal{B}_3 \\
&= \left\{ z \in \mathbb{R}_+^3 \middle| P^d(T_O, T) - e^{-(T-T_O)K} P(T_O, T) \geq 0 \right\} \\
&= \left\{ z \in \mathbb{R}_+^3 \middle| C^d(T_O, T)\, s + D^d(T_O, T)\, u \leq (T - T_O) K + \ln \frac{A^d(T_O, T)}{A(T_O, T)} \right\}.
\end{aligned}$$

Similarly, the put price is given by

$$\begin{aligned}
&F^{csp}(z, t, T; T_O, K) \\
&= E^{\mathbf{Q}} \left[e^{-\int_t^{T_O} (r_l + s_l^{zero}) dl} 1_{\{\mathcal{B}_4\}} \left(e^{-(T-T_O)K} P(T_O, T) - P^d(T_O, T) \right) \middle| \mathcal{F}_t \right],
\end{aligned}$$

where $\mathcal{B}_4$ is the range of values of r, s, and u over which the option is in the money at the expiration date T_O, that is,

$$\begin{aligned}
&\mathcal{B}_4 \\
&= \left\{ z \in \mathbb{R}_+^3 \middle| (T - T_O) K + \ln \frac{A^d(T_O, T)}{A(T_O, T)} \leq C^d(T_O, T)\, s + D^d(T_O, T)\, u \right\}.
\end{aligned}$$

Then the time t value of the call option is

$$\begin{aligned}
&F^{csc}(z, t, T_O, K) \\
&= \int_{\mathcal{B}_3} \left(P^d(y, T_O, T) - e^{-(T-T_O)K} P(y_r, T_O, T) \right) \overrightarrow{G}(y, T_O, z, t)\, dy \\
&= I_3 - I_4,
\end{aligned}$$

where $\overrightarrow{G}(y, T_O, z, t)$ is given by equation (6.72), $\mathcal{B}_3$ is the range of values over which the credit spread option is in the money at the expiration date T. I_3 and I_4 can be calculated by similar methods as I_1 and I_2:

$$\begin{aligned}
I_3 = \frac{P^d(z, t, T)}{(2\pi)^{3/2}} &\iiint_{\mathbf{R}^3} \iiint_{\mathbf{R}^3} \\
&e^{-iy_r(x_1 - iB(T_O, T)) - iy_s\left(x_2 - iC^d(T_O, T)\right) - iy_u\left(x_3 - iD^d(T_O, T)\right)} \cdot 1_{\{y \in \mathcal{B}_3\}} dy \cdot \\
&A^d(T_O, T) \frac{A^{\overleftarrow{G}}(x, t, T_O)}{A^d(t, T)} e^{-\left(B^{\overleftarrow{G}}(x, t, T_O) - B(t, T)\right) r} \cdot \\
&e^{-\left(C^{\overleftarrow{G}}(x, t, T_O) - C^d(t, T)\right) s - \left(D^{\overleftarrow{G}}(x, t, T_O) - D^d(t, T)\right) u} dx,
\end{aligned}$$

and

$$\begin{aligned}
I_4 = \frac{e^{-(T-T_O)K} P^d(z, t, T_O)}{(2\pi)^{3/2}} &\iiint_{\mathbf{R}^3} \\
\iiint_{\mathbf{R}^3} &e^{-iy_r(x_1 - iB(T_O, T)) - iy_s x_2 - iy_u x_3} 1_{\{y \in \mathcal{B}_3\}} dy \cdot \\
&A(T_O, T) \frac{A^{\overleftarrow{G}}(x, t, T_O)}{A^d(t, T_O)} e^{-\left(B^{\overleftarrow{G}}(x, t, T_O) - B(t, T_O)\right) r} \cdot \\
&e^{-\left(C^{\overleftarrow{G}}(x, t, T_O) - C^d(t, T_O)\right) s - \left(D^{\overleftarrow{G}}(x, t, T_O) - D^d(t, T_O)\right) u} dx.
\end{aligned}$$

Therefore the following theorem is obtained.

Theorem 6.4.2.

1. *The price of a European call option (knocked out at default of the underlying) with maturity $T_O \leq T$ and exercise price K on a defaultable zero-coupon bond $P^d(t,T)$ is given by*

$$\begin{aligned} & F^{csc}(z,t,T;T_O,K) \\ = & P^d(z,t,T) \cdot \\ & \Phi^C(\phi_1,\phi_2,\phi_3,\psi_0,\psi_1,\psi_2,\psi_3,\varphi_4,\varphi_5,\varphi_2,\varphi_3,z) \\ & -\overline{K}P^d(z,t,T_O) \cdot \\ & \Phi^C(\phi_1,\phi_5,\phi_6,\psi_8,\psi_5,\psi_6,\psi_7,\varphi_4,\varphi_5,\varphi_2,\varphi_3,z), \end{aligned}$$

where $\overline{K} = e^{-(T-T_O)K}$, Φ^C is given in theorem 6.4.1, and

$$\begin{aligned} \psi_8 &= A(T_O,T)\,\frac{A^{\overleftarrow{G}}(x,t,T_O)}{A^d(t,T_O)}, \\ \varphi_4 &= (T-T_O)K + \ln\frac{A^d(T_O,T)}{A(T_O,T)},\ \varphi_5 = 0. \end{aligned}$$

$A^{\overleftarrow{G}}$, A^d, and A are defined in lemma 6.4.2, theorem 6.3.1, and by equation (6.17), respectively. All other parameters are defined in theorem 6.4.1.

2. *The price of a European put option (knocked out at default of the underlying) with maturity $T_O \leq T$ and exercise price K on a defaultable zero-coupon bond $P^d(t,T)$ is given by*

$$\begin{aligned} & F^{csp}(z,t,T;T_O,K) \\ = & \overline{K}P^d(z,t,T_O) \cdot \\ & \Phi^P(\phi_1,\phi_5,\phi_6,\psi_8,\psi_5,\psi_6,\psi_7,\varphi_4,\varphi_5,\varphi_2,\varphi_3,z) \\ & -P^d(z,t,T) \cdot \\ & \Phi^P(\phi_1,\phi_2,\phi_3,\psi_0,\psi_1,\psi_2,\psi_3,\varphi_4,\varphi_5,\varphi_2,\varphi_3,z). \end{aligned}$$

Remark 6.4.3.

1. Credit spreads are usually very volatile. That is why issuers of credit spread put options often include an additional feature. If the market spread $S(t,T)$ is greater than or equal to some pre-agreed spread $K^* > K$ at any time t before maturity of the option, then the put option immediately pays

$$e^{-(T-t)K}P(t,T) - e^{-(T-t)K^*}P(t,T).$$

2. If we allow the option to survive a default, we must model the post-default bond behavior as well. Similar techniques as for credit options can be applied.

6.4.4 Default Swaps and Default Options

Default Digital Swaps and Default Digital Put Options. We consider a default digital put option[29] that pays off 1 in case of a credit event. First we assume that the payoff takes place at maturity of the contract. If there has been no default until time t, the time t price of the contract is given by

$$\begin{aligned} F_t^{ddp} &= E^{\mathbf{Q}}\left[e^{-\int_t^T r_l dl} 1_{\{T^d \leq T\}} \middle| \mathcal{F}_t\right] \\ &= E^{\mathbf{Q}}\left[e^{-\int_t^T r_l dl}\left(1 - 1_{\{T^d > T\}}\right) \middle| \mathcal{F}_t\right] \\ &= P(t,T) - E^{\mathbf{Q}}\left[e^{-\int_t^T r_l dl} 1_{\{T^d > T\}} \middle| \mathcal{F}_t\right] \\ &= P(t,T) - P^{d,zero}(t,T). \end{aligned}$$

If the payoff takes place at default of the reference credit asset and there has been no default until time t, the time t price of the contract is given by

$$F_t^{ddp} = E^{\mathbf{Q}}\left[\int_t^T e^{-\int_t^{\tilde{t}} r_l dl} dH_{\tilde{t}} \middle| \mathcal{F}_t\right].$$

Hence, since $L_T F_T^{ddp} = 0$,

$$L_t F_t^{ddp} = E^{\mathbf{Q}}\left[\int_t^T e^{-\int_t^{\tilde{t}} r_l dl} dH_{\tilde{t}} + e^{-\int_t^T r_l dl} L_T F_T^{ddp} \middle| \mathcal{F}_t\right]. \tag{6.77}$$

Applying corollary 6.2.1 to equation (6.77) yields

$$d\left(L_t F_t^{ddp}\right) = -dH_t + r_t L_t F_t^{ddp} dt + dm_t \tag{6.78}$$

for some martingale m. Together with lemma 6.4.1 equation (6.78) becomes

$$d\left(L_t F_t^{ddp}\right) = -s_t^{zero} d_t + r_t L_t F_t^{ddp} dt + d\widetilde{m}_t$$

for some martingale $\widetilde{m}$ with $d\widetilde{m}_t = dm_t + dM_t^{d,zero}$. Finally, as

$$\int L_t F_t^{ddp} dM_t^{d,zero} = \int L_t F_t^{ddp} s_t^{zero} dt - \int L_t F_t^{ddp} dH_t$$

is a martingale by the arguments we have used before,

$$\begin{aligned} d\left(L_t F_t^{ddp}\right) &= -L_t F_t^{ddp} dH_t - s_t^{zero} dt + \left(r_t + s_t^{zero}\right) L_t F_t^{ddp} dt + d\widehat{m}_t \\ &= -s_t^{zero} dt + \left(r_t + s_t^{zero}\right) L_t F_t^{ddp} dt + d\widehat{m}_t \end{aligned} \tag{6.79}$$

[29] For an introduction to credit default products see section 5.3.3.

for some martingale $\widehat{m}$ with $d\widehat{m} = d\widehat{m}_t - L_t F_t^{ddp} dM_t^{d,zero}$. Equation (6.79) holds because $L_t F_t^{ddp} dH_t = 0$. By applying corollary 6.2.1 again, we get

$$\begin{aligned} & L_t F_t^{ddp} \\ & = E^{\mathbf{Q}} \left[\int_t^T e^{-\int_t^{\tilde{t}} (r_l + s_l^{zero}) dl} s_{\tilde{t}}^{zero} d\tilde{t} + e^{-\int_t^T (r_l + s_l^{zero}) dl} L_T F_T^{ddp} \middle| \mathcal{F}_t \right]. \end{aligned}$$

Thus,

$$\begin{aligned} L_t F_t^{ddp} &= E^{\mathbf{Q}} \left[\int_t^T e^{-\int_t^{\tilde{t}} (r_l + s_l^{zero}) dl} s_{\tilde{t}}^{zero} d\tilde{t} \middle| \mathcal{F}_t \right] \\ &= \int_t^T E^{\mathbf{Q}} \left[e^{-\int_t^{\tilde{t}} (r_l + s_l^{zero}) dl} s_{\tilde{t}}^{zero} \middle| \mathcal{F}_t \right] d\tilde{t}, \end{aligned} \tag{6.80}$$

as there is sufficient regularity to allow the interchange of expectation and integration[30]. Using the independence of r and s^{zero}, we finally get

$$\begin{aligned} L_t F_t^{ddp} &= \int_t^T P\left(t, \tilde{t}\right) E^{\mathbf{Q}} \left[e^{-\int_t^{\tilde{t}} s_l^{zero} dl} s_{\tilde{t}}^{zero} \middle| \mathcal{F}_t \right] d\tilde{t} \\ &= \int_t^T P^{d,zero}\left(t, \tilde{t}\right) f^{d,zero}\left(t, \tilde{t}\right) d\tilde{t}, \end{aligned} \tag{6.81}$$

which follows immediately from a well-known result (see, e.g., Sandmann & Sondermann (1997)) in non-defaultable interest rate theory, saying basically

$$E^{\mathbf{Q}} \left[e^{-\int_t^{\tilde{t}} r_l dl} r_{\tilde{t}} \middle| \mathcal{F}_t \right] = -E^{\mathbf{Q}} \left[e^{-\int_t^{\tilde{t}} r_l dl} \middle| \mathcal{F}_t \right] \cdot \frac{\partial}{\partial T} \ln P\left(t, T\right) \bigg|_{T=\tilde{t}},$$

and $f^{d,zero}\left(t, \tilde{t}\right)$ is the zero recovery spread forward rate, i.e.

$$f^{d,zero}\left(t, \tilde{t}\right) = -\frac{\partial}{\partial \tilde{t}} \ln \frac{P^{d,zero}\left(t, \tilde{t}\right)}{P\left(t, \tilde{t}\right)}. \tag{6.82}$$

Hence, plugging in equation (6.82) into equation (6.81) yields

[30] $E^{\mathbf{Q}} \left[\int_t^T \left| e^{-\int_t^{\tilde{t}} (r_l + s_l^{zero}) dl} s_{\tilde{t}}^{zero} \right| d\tilde{t} \right] \leq E^{\mathbf{Q}} \left[\int_t^T \left| s_{\tilde{t}}^{zero} \right| d\tilde{t} \right] < \infty$ and $e^{-\int_t^{\tilde{t}} (r_l + s_l^{zero}) dl} s_{\tilde{t}}^{zero}$ is product measurable on $\Omega \times [t, T]$ (see, e.g., Duffie (1992), p. 234).

$$\begin{aligned} L_t F_t^{ddp} &= -\int_t^T P^{d,zero}\left(t,\widetilde{t}\right) \frac{\partial}{\partial \widetilde{t}} \left(\ln \frac{P^{d,zero}\left(t,\widetilde{t}\right)}{P\left(t,\widetilde{t}\right)} \right) d\widetilde{t} \\ &= \int_t^T \frac{P^{d,zero}\left(t,\widetilde{t}\right) \frac{\partial}{\partial \widetilde{t}} \left(P\left(t,\widetilde{t}\right)\right)}{P\left(t,\widetilde{t}\right)} - \frac{\partial}{\partial \widetilde{t}} \left(P^{d,zero}\left(t,\widetilde{t}\right)\right) d\widetilde{t} \\ &= L_t - \int_t^T \frac{P^{d,zero}\left(t,\widetilde{t}\right)}{P\left(t,\widetilde{t}\right)} \frac{\partial}{\partial \widetilde{t}} \left(P\left(t,\widetilde{t}\right)\right) d\widetilde{t} - P^{d,zero}\left(t,T\right) \\ &= L_t - P^{d,zero}\left(t,T\right) \\ &\quad + \int_t^T P^{d,zero}\left(t,\widetilde{t}\right) \left(\frac{A_{\widetilde{t}}\left(t,\widetilde{t}\right)}{A\left(t,\widetilde{t}\right)} - B_{\widetilde{t}}\left(t,\widetilde{t}\right) r_t \right) d\widetilde{t}, \end{aligned} \tag{6.83}$$

We can conclude:

Theorem 6.4.3. *If there has been no default until time t, the price of a default digital put option paying off 1 in case of a credit event, is given by*

$$\begin{aligned} F_t^{ddp} &= 1 - P^{d,zero}\left(t,T\right) \\ &\quad + \int_t^T P^{d,zero}\left(t,\widetilde{t}\right) \left(\frac{A_{\widetilde{t}}\left(t,\widetilde{t}\right)}{A\left(t,\widetilde{t}\right)} - B_{\widetilde{t}}\left(t,\widetilde{t}\right) r_t \right) d\widetilde{t}. \end{aligned}$$

Remark 6.4.4.

1. Equation (6.83) is a closed form solution for the price of a default digital put:
   a) $P^{d,zero}\left(t,\widetilde{t}\right)$ is given by equation (6.68) for all $t \leq \widetilde{t} \leq T$.
   b) $A_{\widetilde{t}}\left(t,\widetilde{t}\right) = \frac{\partial}{\partial \widetilde{t}} A\left(t,\widetilde{t}\right)$ can be easily calculated from $A\left(t,\widetilde{t}\right)$ which is given by equation (6.17).
   c) $B_{\widetilde{t}}\left(t,\widetilde{t}\right) = \frac{\partial}{\partial \widetilde{t}} B\left(t,\widetilde{t}\right)$ can be easily calculated from $B\left(t,\widetilde{t}\right)$ which is given by equation (6.16).
   d) L_t and r_t are known at time t.
2. Alternatively, without demanding any regularity conditions for the interchange of expectation and integration, we can solve equation (6.80) by applying PDE techniques.

Default Swaps and Default Puts. In practice, default swap and default put contracts differ in their specific default payments:

- Replacement of the difference to an equivalent default-free bond. That is, the payoff at the default time T^d is

$$\begin{aligned} &P\left(T^d,T\right) - P^d\left(T^d,T\right) \\ &= P\left(T^d,T\right) - w\left(T^d\right) P^d\left(T^d-,T\right). \end{aligned}$$

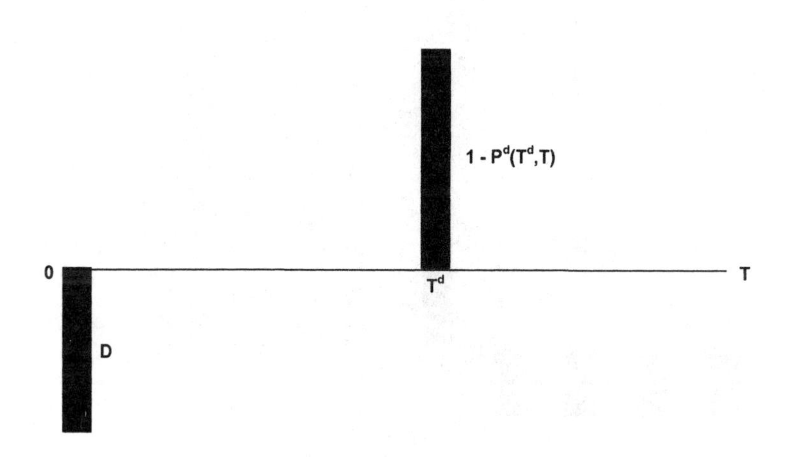

Fig. 6.15. Cash flows of a default put with a default event at time T^d. The default put up-front fee is denoted by D. Replacement of the difference to par: $1 - P^d\left(T^d, T\right)$. The reference credit asset is a defaultable zero coupon bond.

- Replacement of the difference to par: if termination is triggered, there is a payoff that is the difference between the face value and the market value (at default) of a reference credit asset (cash settlement). That is, the payoff at the time T^d of default is

$$\begin{aligned} & 1 - P^d\left(T^d, T\right) \\ & = 1 - w\left(T^d\right) P^d\left(T^d -, T\right). \end{aligned}$$

  A default swap on a defaultable coupon bond only pays off the difference between the post-default coupon bond price and par. There is only principal but not coupon protection. For the case that a default happens, figures 6.15 and 6.16 show the cash flows of a default put and default swap, respectively.

The pricing of a default swap consists of two problems. At origination there is no exchange of cash flows, and we have to determine the default swap spread S that makes the market value of the default swap zero. After origination, the market value of the default swap will change due to changes in the underlying variables. So given the default swap spread S, we have to determine the current market value of the default swap. We assume throughout that the credit swap counterparties (beneficiary and guarantor) are default-free, and

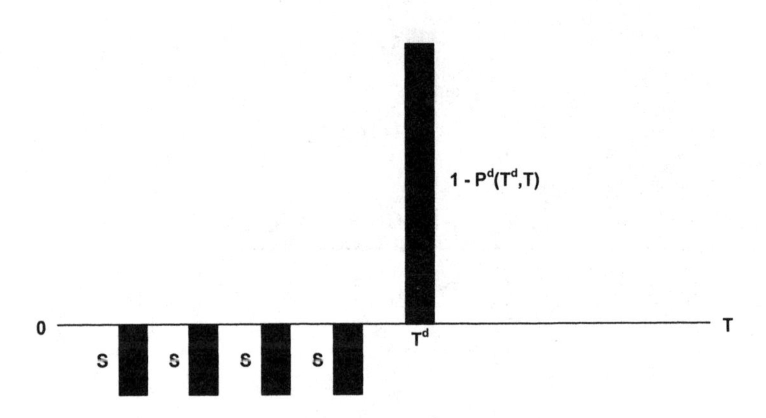

Fig. 6.16. Cash flows of a default swap with a default event at time T^d. The default swap spread is denoted by S. Replacement of the difference to par: $1 - P^d\left(T^d, T\right)$. The underlying reference credit asset is a defaultable zero coupon bond.

that there is a replacement to the difference of par (which is currently market standard).

The underlying reference credit asset is a defaultable zero-coupon bond. We assume that there has been no credit event until time t.

1. Default put options and replacement of the difference to par:
   The time t price is given by

$$\begin{aligned}
F_t^{dp} &= E^{\mathbf{Q}}\left[\int_t^T e^{-\int_t^u r_l dl}\left(1 - Z_u\right) dH_u \middle| \mathcal{F}_t\right] \\
&= F_t^{ddp} - E^{\mathbf{Q}}\left[\int_t^T e^{-\int_t^u r_l dl} Z_u dH_u \middle| \mathcal{F}_t\right] \\
&= F_t^{ddp} - E^{\mathbf{Q}}\left[\int_t^T e^{-\int_t^u r_l dl} Z_u dH_u + e^{-\int_t^T r_l dl} L_T \middle| \mathcal{F}_t\right] \\
&\quad + E^{\mathbf{Q}}\left[e^{-\int_t^T r_l dl} L_T \middle| \mathcal{F}_t\right] \\
&= F_t^{ddp} - P^d\left(t, T\right) + P^{d,zero}\left(t, T\right).
\end{aligned}$$

2. Default put options and replacement of the difference to an equivalent default-free bond:

There is a simple no-arbitrage argument that can be used to determine the value of a credit default option on a defaultable zero-coupon bond that - in case of a credit event - replaces the difference to an equivalent default-free bond . Assume that an investor forms a portfolio consisting of a defaultable zero-coupon bond with maturity T and a credit default option that pays the difference to a default-free zero-coupon bond with maturity T in case of a default. This construction eliminates the entire default risk from the defaultable zero-coupon bond. Hence, the value of the portfolio must always be equal to the value of a non defaultable zero-coupon bond with maturity T, i.e.

$$F_t^{dp} = P(t,T) - P^d(t,T).$$

This equation is valid in the most general assumptions on the processes r, s, and u.

3. Default swaps:
In case of a credit default swap there are regular payments S (the credit swap spread) instead of an up-front fee F_0^{dp}. The value of paying F_0^{dp} at the origination of the credit default put option must be the same as paying S at some predefined times t_i, $i = 1, ..., m$, until a default happens. Note, that t_i, $i = 1, ..., m$, are all possible payment dates. Hence, by using remark 6.4.1 (2.),

$$F_0^{dp} = S \sum_{i=1}^{m} P^{d,zero}(0, t_i).$$

This is equivalent to a credit swap spread of

$$S = \frac{F_0^{dp}}{\sum_{i=1}^{m} P^{d,zero}(0, t_i)}.$$

The underlying reference credit asset is a defaultable fixed rate bond. In the following we want to ignore any accrued premiums that may be outstanding at the time of default. In addition, we assume that there has been no credit event until time t.

1. Default put options and replacement to the difference to par:
The reference credit asset is a coupon bond with discrete coupon payments c_i occurring at dates $t < t_i \leq T$, $i = 1, ..., n$, and maturity T. Then the pricing argument for the default put is exactly the same as in the case of the zero-coupon bond, and we get

$$F_c^{dp}(t) = F_t^{ddp} - P_c^d(t,T) + P_c^{d,zero}(t,T),$$

where

$$P_c^{d,zero}(t,T) = \sum_{i=1}^{n} c_i P^{d,zero}(t, t_i) + P^{d,zero}(t,T).$$

2. Default put options and replacement of the difference to an equivalent default-free bond:
   We can use the same no-arbitrage argument as in the case of zero-coupon bonds, and get
   $$F_c^{dp}(t) = P_c(t,T) - P_c^d(t,T).$$
3. Default swaps:
   The credit swap spread can be calculated by the same argument as in the case of zero-coupon bonds. Hence, if there are regular payments S at some predefined times t_i, $i = 1, ..., m$, until a default happens, S is given by
   $$S = \frac{F_c^{dp}(0)}{\sum_{i=1}^{m} P_0^d(0,t_i)}.$$

The underlying note is a defaultable par floating rate note with the maturity of the default swap. We assume that default is only possible at coupon dates or at maturity and that there exists a default-free floating rate note with floating rate C_t at date t. The coupon payments of the underlying defaultable floating rate note are $C_t + \overline{S}$ at date t, where $\overline{S}$ is a fixed spread. Both floating rate notes have a final payoff of 1 (if there is no credit event). If both floaters trade at par at $t = 0$, we can use an arbitrage argument to price the default swap. We short the par issued defaultable floating rate note and invest the money in a par issued default-free floating rate note and hold this portfolio either until to maturity or a credit event occurs, receiving the coupons on the default-free floating rate note and paying the coupons on the defaultable floating rate note. If there is no credit event before maturity, both floating rate notes mature at par. So there is no net cash flow at maturity. If there is default before maturity, we liquidate the portfolio immediately receiving 1 from the default-free floating rate note, which is par on a coupon date, and paying the remaining market value $P_f^d(T^d,T)$ of the defaulted floating rate note. Hence, the net cash flow equals $1 - P_f^d(T^d,T)$, which is exactly the same as that specified in a default swap contract. The absence of arbitrage implies that the unique arbitrage free default swap spread S is given by $S = \overline{S}$. Table 6.1 compares the cash flows of the default swap to the cash flows of the portfolio at any time until default or maturity.
If we allow for default not only at coupon times we liquidate the portfolio at the coupon date immediately after default. Then the relationship

$$Default\ Swap = Default\text{-}Free\ FRN\ -\ Defaultable\ FRN$$

is only approximately valid.

Conclusion 6.4.1
Because the market value of a default swap on a defaultable floating rate note at origination equals zero, the arbitrage free default swap spread S equals the coupon spread of the reference defaultable floating rate note spread $\overline{S}$ over

Table 6.1. Payoffs of a default swap with a defaultable floating rate note as a reference asset and payoffs of a portfolio consisting of a default free floating rate note (long) and a defaultable floating rate note (short).

	Default Swap	*Default Free FRN - Defaultable FRN*
t=0	0	$1-1=0$
Coupon dates before default including maturity	$-S$	$C_t - \left(C_t + \overline{S}\right) = -\overline{S}$
$t = T^d$ *(without coupons)*	$1 - P_f^d\left(T^d, T\right)$	$1 - P_f^d\left(T^d, T\right)$
$t = T$ *(without coupons)*	0	$1-1=0$

an equivalent default-free floating rate note. The price of the default swap after origination equals the market value of the default-free floating rate note minus the market value of the defaultable floating rate note.

Remark 6.4.5.

1. We can relax the assumption that the underlying floating rate note has the same maturity as the default swap. The relevant par spread $\overline{S}$ for fixing the default swap spread is that of a different floating rate note issued by counterparty C which is of the same maturity as the default swap, and of the same priority as the underlying floating rate note. If there is no available par floating rate note of the same credit quality and seniority with the same maturity as the default swap, often the default swap spread S can be deduced from other spreads.
2. One of the important elements of the arbitrage argument is to short the par issued counterparty C defaultable floating rate note freely. Shorting a floating rate note can be done, e.g., via a reverse repo combined with a cash sale. Through the reverse repo one can receive the floating rate note as a collateral on a loan. Instead of holding the collateral one can sell the note immediately. But there may be some possible additional problems:
   - Transaction costs;
   - Big bid-ask spreads for the counterparty C floating rate note;
   - One may be forced in a special repo, i.e. one may be forced to offer a repo rate that is substantially below the general collateral rate (GCR). Then the default swap spread should be increased by the difference between the GCR and the special repo rate;
   - In practice, for illiquid instruments, the default swap spread can vary from the reference credit counterparty C floating rate note;
3. Sometimes the default swap contract is specified such that the protection buyer must pay the default swap premium that has accrued since the last coupon date.

4. The typical default swap specifies payment at default of the difference between face value without accrued interest and market value of the underlying floating rate note. The arbitrage portfolio described above is worth face value plus accrued interest on the non defaultable floating rate note, less recovery on the underlying defaultable floating rate note.

6.5 A Discrete-Time Version of the Three-Factor Model

6.5.1 Introduction

This section presents a discrete-time version of the model presented in section 6.2. It is a discrete time approximation of the continuous-time model, which enables us to determine prices for default related instruments that don't have any closed form solutions, e.g., caused by callability features or because they are American. In addition, it is a computational tool for extensive numerical investigations of our model. As the defaultable term structure model consists of three stochastic processes, the lattice we have to build involves four dimensions. One dimension in time and the other three in the state space. We assume that the process of the non defaultable short rate is independent of the other two processes. The uncertainty index process affects the process of the short rate credit spread process. So the construction of the four dimensional lattice basically consists of two parts: building the tree for the non defaultable short rate and building the combined tree for the short rate credit spread and the uncertainty index. First, the lattice for the uncertainty index is expanded one step further, than this value is used as an input to extend the lattice for the short rate credit spread.

6.5.2 Constructing the Lattice

In this subsection, we show how credit derivatives can be priced using a four dimensional lattice based on our three-factor model. The building method is a modification of the two-stage procedure proposed by Hull & White (1994) for representing a wide range of one-factor interest rate models. Especially, their procedure is applied for building an interest rate tree corresponding to the non-defaultable short rate model defined by equation (6.5) for $\beta = 0$. All processes are assumed to be considered under the equivalent martingale measure $\mathbf{Q}$. As a first step, let us suppose we had to build up a trinomial tree for the one dimensional stochastic process X_t defined by

$$dX_t = \left(\mu\left(X, t\right) - aX_t\right) dt + \sigma dW_*\left(t\right), \tag{6.84}$$

where $W_*(t)$ denotes a one-dimensional Brownian motion, $a > 0$ and $\sigma > 0$ are constants and $\mu(X, t)$ is a progressively measurable stochastic process which may depend on X_t. Note that $W_*(t)$ may be any of the Brownian motions we consider in our three factor model. Hence, we consider $\mathcal{F}_t$ as defined in the previous sections. Working with time steps of size Δt we know that

$$\Delta X_t = X_{t+\Delta t} - X_t = (\mu(X,t) - aX_t)\,\Delta t + \sigma \Delta W_*(t)\,,$$

which is normally distributed conditional on $\mathcal{F}_t$, with an expected value of

$$E^{\mathbf{Q}}\left[\Delta X_t | \mathcal{F}_t\right] = (\mu(X,t) - aX_t)\,\Delta t, \tag{6.85}$$

and a conditional variance of

$$Var^{\mathbf{Q}}\left[\Delta X_t | \mathcal{F}_t\right] = \sigma^2 \Delta t. \tag{6.86}$$

There are two main conditions for a sequence of discrete-time processes to converge in distribution to the corresponding continuous-time diffusion process (see, e.g., Amin (1995)):

a) The (conditional) mean, variance and covariance terms per time unit Δt of the discrete sequence of processes must converge to that of the continuous-time diffusion process.
b) The limiting process must have continuous sample paths, i.e., as $\Delta t \to 0$, the maximum of the absolute increment to the process (jump size) at any date must converge to zero.

Since we are dealing with the assumption of uncorrelated Brownian motions, i.e. $\rho_{r,s} = \rho_{r,u} = \rho_{s,u} = 0$, these conditions are satisfied for our three factor model if we ensure the mean and variance conditions for each single one factor model. The basic building block to do this is to consider the flat stochastic process X_t^* defined by equation (6.84) with $\mu \equiv 0$. With respect to error minimization (see, e.g., Hull & White (1994)) we choose a constant step size of $\Delta x^* = \sigma \cdot \sqrt{3 \cdot \Delta t}$ for the distance between the nodes of X_t^* at each time step. If we define (i,j) as the tree node with $t = i \cdot \Delta t$ and $X_t^* = X_{i \cdot \Delta t}^* = j \cdot \Delta x^*$, $j_{\max}$ to be the smallest integer greater than $\frac{0.184}{a \cdot \Delta t}$, and $j_{\min} = -j_{\max}$, we can ensure that, for the flat tree, the mean and variance condition holds, and all probabilities are positive if we apply the following branching methods: standard branching for $j_{\min} \le j \le j_{\max}$ from node (i,j) to nodes $u = (i+1, j+1)$, $m = (i+1, j)$ and $d = (i+1, j-1)$ with probabilities

$$p_u = \frac{1}{6} + \frac{a^2 j^2 \Delta t^2 - aj\Delta t}{2}, \; p_m = \frac{2}{3} - a^2 j^2 \Delta t^2, \; p_d = \frac{1}{6} + \frac{a^2 j^2 \Delta t^2 + aj\Delta t}{2},$$

upward branching[31] for $j < j_{\min}$ and downward branching[32] for $j > j_{\max}$. Note that the probabilities at each node only depend on j, a and Δt which makes the procedure very robust with respect to the sensitivity analysis of changing yield or spread curves.
In the second building block we shift each node in the flat tree to ensure that conditions *a)* and *b)* from above hold in the resulting tree. Let us therefore define the shift size α_t at each node by $\alpha_t = X_t - X_t^*$, $t \in [0, T^*]$, which can be described by the differential equation

$$d\alpha_t = (\mu(X,t) - a\alpha_t)\, dt \text{ with } \alpha_0 = X_0.$$

The corresponding discrete time version is given by

$$\begin{aligned} \Delta\alpha_t &= \alpha_{t+\Delta t} - \alpha_t = [\mu(X,t) - a\alpha_t]\, \Delta t \\ &= [\mu(X^* + \alpha, t) - a\alpha_t]\, \Delta t. \end{aligned} \tag{6.87}$$

Hence,

$$\alpha_{t+\Delta t} = \mu(X^* + \alpha, t)\, \Delta t + (1 - a\Delta t)\, \alpha_t.$$

Note, that $\alpha_{t+\Delta t}$ via $\mu(X,t)$ depends on the predecessor node of the specific node in the tree that has to be shifted. Using equations (6.85) with $\mu \equiv 0$ and (6.87), we get

$$\begin{aligned} E^{\mathbf{Q}}[\Delta X_t|\mathcal{F}_t] &= E^{\mathbf{Q}}\left[X_{t+\Delta t}^* + \alpha_{t+\Delta t} - (X_t^* + \alpha_t)\,|\mathcal{F}_t\right] \\ &= [\mu(X,t) - aX_t]\, \Delta t. \end{aligned}$$

Similarly, we get

$$Var^{\mathbf{Q}}[\Delta X_t|\mathcal{F}_t] = Var^{\mathbf{Q}}[\Delta X_t^*|\mathcal{F}_t] = \sigma^2 \Delta t,$$

which ensures that our sequence of discrete-time processes converges in distribution to the corresponding continuous-time diffusion process. For the practical application we start at time $t = 0$ and shift the node $(0,0)$ by $\alpha_0 = X_0$.

[31] Upward branching for $j < j_{\min}$ from node (i,j) to nodes $d = (i+1,j)$, $m = (i+1,j+1)$ and $u = (i+1,j+2)$ with probabilities

$$\begin{aligned} p_u &= \frac{1}{6} + \frac{a^2j^2\Delta t^2 + aj\Delta t}{2}, \quad p_m = -\frac{1}{3} - a^2j^2\Delta t^2 - 2aj\Delta t, \\ p_d &= \frac{7}{6} + \frac{a^2j^2\Delta t^2 + 3aj\Delta t}{2}. \end{aligned}$$

[32] Downward branching for $j > j_{\max}$ from node (i,j) to nodes $d = (i+1,j-2)$, $m = (i+1,j-1)$ and $u = (i+1,j)$ with probabilities

$$\begin{aligned} p_u &= \frac{7}{6} + \frac{a^2j^2\Delta t^2 - 3aj\Delta t}{2}, \quad p_m = -\frac{1}{3} - a^2j^2\Delta t^2 + 2aj\Delta t, \\ p_d &= \frac{1}{6} + \frac{a^2j^2\Delta t^2 - aj\Delta t}{2}. \end{aligned}$$

Suppose we shifted the flat tree up to time $t = i\Delta t$ getting the node values $x_{i,j}$ from $x^*_{i,j}$. Then, the value $x^*_{i+1,j+k}$, $k = -1, 0, 1$, of each node $(i+1, j+k)$, which is reached by node (i, j), is shifted by

$$\alpha_{i+1,j+k} = \mu\left(x_{i,j}, t\right) \Delta t + (1 - a\Delta t)\, \alpha_{i,j}$$

to the value $x_{i+1,j+k}$. Note that the dependence of $\alpha_{i+1,j+k}$ on $x_{i,j}$ can destroy the recombining property of the flat tree.
The third building block is only necessary for Cox-Ingersoll-Ross processes defined by

$$dY_t = \left(\theta\left(Y, t\right) - \hat{a}Y_t\right) dt + \sigma\sqrt{Y_t} dW\left(t\right).$$

Setting $X_t = G\left(y, t\right)$ and using the transformation

$$G\left(y, t\right) = 2\sqrt{y_t},$$

with Itô´s lemma, we get

$$dX_t = \left[\frac{1}{\sqrt{Y_t}}\left(\theta\left(Y, t\right) - \hat{a}Y_t\right) - \frac{1}{4}\frac{1}{\sqrt{Y_t}}\sigma^2\right] dt + \sigma dW\left(t\right).$$

Now, since $Y_t = \frac{1}{4}X_t^2$,

$$\begin{aligned} dX_t &= \left[\frac{2\theta\left(\frac{1}{4}X_t^2, t\right) - \frac{1}{2}\sigma^2}{X_t} - \frac{\hat{a}}{2}X_t\right] dt + \sigma dW\left(t\right) \\ &= \left[\mu\left(X, t\right) - aX_t\right] dt + \sigma dW\left(t\right), \end{aligned}$$

with

$$\mu\left(X, t\right) = \frac{2\theta\left(\frac{1}{4}X_t^2, t\right) - \frac{1}{2}\sigma^2}{X_t} \text{ and } a = \frac{\hat{a}}{2}.$$

Therefore, we can apply the previous building blocks to build up a representing trinomial tree for the stochastic process X_t. We get the corresponding tree for Y_t by setting $Y_t = \frac{1}{4}X_t^2$ in each node of the tree for X_t.
Let us now apply the proposed method for our three factor model. We start with the uncertainty process

$$du_t = \left[\theta_u - \hat{a}_u u_t\right] dt + \sigma_u\sqrt{u_t} d\hat{W}_u\left(t\right).$$

It can be transformed to the process

$$dX_u\left(t\right) = \left[\frac{2\theta_u - \frac{1}{2}\sigma^2}{X_u\left(t\right)} - \frac{\hat{a}_u}{2}X_u\left(t\right)\right] dt + \sigma_u d\hat{W}_u\left(t\right)$$

and

$$\begin{aligned} \alpha_u\left(t + \Delta t\right) &= \alpha_u\left(t\right) + \left[\frac{2\theta_u - \frac{1}{2}\sigma^2}{X_u\left(t\right)} - \frac{\hat{a}_u}{2}\alpha_u\left(t\right)\right] \Delta t \\ &= \frac{2\theta_u - \frac{1}{2}\sigma^2}{X_u^*\left(t\right) + \alpha_u\left(t\right)}\Delta t + \left(1 - \frac{\hat{a}_u}{2}\Delta t\right)\alpha_u\left(t\right) \end{aligned}$$

with $\alpha_u(0) = X_u(0) = 2\sqrt{u_0}$.
Similarly, given u_t at time t, for the spread process

$$ds_t = [b_s u_t - \hat{a}_s s_t]\, dt + \sigma_s \sqrt{s_t} d\hat{W}_s(t)$$

we get the transformations

$$dX_s(t) = \left[\frac{2b_s u_t - \frac{1}{2}\sigma_s^2}{X_s(t)} - \frac{\hat{a}_s}{2} X_s(t)\right] dt + \sigma_s d\hat{W}_s(t)$$

and

$$\alpha_s(t+\Delta t) = \frac{2b_s u_t - \frac{1}{2}\sigma_s^2}{X_s^*(t) + \alpha_s(t)} \Delta t + \left(1 - \frac{\hat{a}_s}{2}\Delta t\right) \alpha_s(t)$$

with $\alpha_s(0) = X_s(0) = 2\sqrt{s_0}$.
The short rate process is given by

$$dr_t = [\theta_r(t) - \hat{a}_r r_t]\, dt + \sigma_s r_t^{\beta} d\hat{W}_r(t)$$

with $\beta \in \left\{0, \frac{1}{2}\right\}$. For $\beta = \frac{1}{2}$ we have to transform the short rate process to

$$dX_r(t) = \left[\frac{2\theta_r(t) - \frac{1}{2}\sigma_r^2}{X_r(t)} - \frac{\hat{a}_r}{2} X_r(t)\right] dt + \sigma_r d\hat{W}_r(t),$$

for $\beta = 0$ the transformation is not necessary and we may consider

$$dX_r(t) = [\theta_r(t) - \hat{a}_r X_r(t)]\, dt + \sigma_r d\hat{W}_r(t),$$

i.e., $X_t = r_t$. The corresponding α's are given by

$$\alpha_r(t+\Delta t) = \frac{2\theta_r(t) - \frac{1}{2}\sigma_r^2}{X_r^*(t) + \alpha_r(t)} \Delta t + \left(1 - \frac{\hat{a}_r}{2}\Delta t\right) \alpha_r(t)$$

with $\alpha_r(0) = X_r(0) = 2\sqrt{r_0}$ for $\beta = \frac{1}{2}$ and

$$\alpha_r(t+\Delta t) = \theta_r(t)\,\Delta t + (1 - \hat{a}_r \Delta t)\, \alpha_r(t)$$

with $\alpha_r(0) = X_r(0) = r_0$ for $\beta = 0$.
Note that the short rate model can be fitted to the yield curve as $\theta_r(t)$ is time-dependent. Also note that for the Hull-White model ($\beta = 0$) the shift size $\alpha_r(t)$ is the same for all nodes (i, j) at time $t = i\Delta t$. To build up an interest rate tree for this case, Hull & White (1994) propose a slightly different method than that just described which they claim to be numerically more efficient.
We are now able to build up a three dimensional trinomial tree for the process (r_t, s_t, u_t). In contrast to the trinomial trees proposed by Chen (1996), Amin (1995), or Boyle (1988) the branching as well as the probabilities don´t change with a changing drift, which makes the tree more efficient, especially under risk management purposes. The probabilities for each node in the three dimensional tree are simply given by the product of the one dimensional processes. Note that despite the fact that we have to consider all possible combinations $(r_{i,j}, s_{i,j}, u_{i,j})$ at time $t = i \cdot \Delta t$ and nodes $j_r \cdot \Delta r$, $j_s \cdot \Delta s$ and $j_u \cdot \Delta u$ the total number is limited by the branching method.

6.5.3 General Interest Rate Dynamics

If we want to allow for correlated Brownian motion, i.e.

$$
\begin{aligned}
Cov\left(dW_r\left(t\right), dW_s\left(t\right)\right) &= \rho_{r,s}dt, \\
Cov\left(dW_r\left(t\right), dW_u\left(t\right)\right) &= \rho_{r,u}dt, \\
Cov\left(dW_s\left(t\right), dW_u\left(t\right)\right) &= \rho_{s,u}dt,
\end{aligned}
$$

the fundamental PDE can be written as

$$
\begin{aligned}
&-\left(\rho_{r,s}\sigma_r\sigma_s r^{\beta}\sqrt{s}F_{rs} + \rho_{r,u}\sigma_r\sigma_u r^{\beta}\sqrt{u}F_{ru} + \rho_{u,s}\sigma_u\sigma_s\sqrt{su}F_{us}\right) \quad (6.88) \\
&= \frac{1}{2}\sigma_r^2 r^{2\beta}F_{rr} + \frac{1}{2}\sigma_s^2 sF_{ss} + \frac{1}{2}\sigma_u^2 uF_{uu} \\
&+ \left[\theta_r\left(t\right) - \hat{a}_r r\right]F_r + \left[b_s u - \hat{a}_s s\right]F_s \\
&+ \left[\theta_u - \hat{a}_u u\right]F_u + F_t - \left(r+s\right)F, \quad \left(r,s,u,t\right) \in \mathbb{R}^3 \times [0,\mathrm{T}),
\end{aligned}
$$

with a boundary condition

$$
F\left(r,s,u,T\right) = h\left(r,s,u\right). \tag{6.89}
$$

It is possible to show that a solution for PDE (6.88) with boundary condition (6.89) can be found by applying a recursive algorithm involving the function G defined in lemma 6.3.1. But as this method is very inconvenient for several reasons[33], we suggest to use tree based methods for the correlated case. Therefore, the procedure proposed in section 6.5.2 should be slightly modified. The extension of tree based methods from the case where the Brownian motions are uncorrelated to the case where they are correlated have been already discussed in depth in the literature. Hence, we omit this discussion here and refer, e.g., to Hull & White (1994).

6.6 Fitting the Model to Market Data

6.6.1 Introduction

Whether we have closed-form solutions for pricing default related instruments or we use numerical methods, in both cases we need estimates of the parameters of the underlying stochastic processes that determine the interest rate dynamics. The ultimate success or failure in implementing pricing formulas is directly related to the ability to collect the necessary information for determining good model parameter values. In the following we suggest two different ways how meaningful values for the parameters of the three

[33] E.g., the algorithm involves the calculation of a large number of integrals all dealing with the complicated function G. In addition, the speed of convergence is not a priori known.

processes r, s, and u can be found. The first one is the method of least squared minimization. Basically, we compare market prices and theoretical prices at one specific point in time and calculate the implied parameters by minimization of the sum of the squared deviations of the market from the theoretical prices. The second one is the Kalman filter method.

6.6.2 Method of Least Squared Minimization

The uncertainty index. Many practitioners use ratings of rating agencies as a proxy for the quality of a firm: Usually the better the rating the smaller the probability of default. We assume that there are 6 rating classes which we denote by $1, 2, ..., 6$. Rating class 6 corresponds to default, rating class 1 consists of the instruments with the highest quality (i.e. the lowest uncertainty) of defaultable bonds. Each firm (or more specific each debt instrument) belongs to exactly one rating class depending on its quality or its "uncertainty" given by the uncertainty index of the firm. There is a one-to-one mapping from the value of the uncertainty index onto the set of rating classes, i.e. there are thresholds ξ_j , $j = 1, ..., 5$, such that a firm is in default if $u > \xi_5$, has rating j if $\xi_{j-1} < u \leq \xi_j$, $j = 2, ..., 5$, or has rating 1 if $u \leq \xi_1$. We call ξ_5 the default boundary and denote it sometimes by ξ^d. Using the non-central χ^2–distribution property of $u(t)$ (for details see, e.g., Lamberton & Lapeyre (1996), pages 129 ff.) the unknown parameters of the process u and the unknown thresholds can be expressed by the transition probabilities of the firm. If we suppose that the transition probabilities are given for the time interval $[0, t]$, the following equations hold:

$$\mathbf{P}\left(u_t \leq \xi_1 | u_0\right) = \mathbf{F}_{\frac{4\theta_u}{\sigma_u^2}, \varsigma}\left(\frac{\xi_1}{L}\right) \tag{6.90}$$

$$\mathbf{P}\left(u_t \leq \xi_1 | u_0\right) + \sum_{k=2}^{j} \mathbf{P}\left(\xi_{k-1} < u_t \leq \xi_k | u_0\right) = \mathbf{F}_{\frac{4\theta_u}{\sigma_u^2}, \varsigma}\left(\frac{\xi_j}{L}\right) \tag{6.91}$$

$$\mathbf{P}\left(u_t > \xi_5 | u_0\right) = 1 - \mathbf{F}_{\frac{4\theta_u}{\sigma_u^2}, \varsigma}\left(\frac{\xi_5}{L}\right)$$

where $L = \frac{\sigma_u^2}{4a_u}\left(1 - e^{-a_u t}\right)$, $\varsigma = \frac{4u_0 a_u}{\sigma_u^2 \left(e^{a_u t} - 1\right)}$ and $\mathbf{F}_{\frac{4\theta_u}{\sigma_u^2}, \varsigma}$ denotes the distribution function of the non-central chi-squared law with $\frac{4\theta_u}{\sigma_u^2}$ degrees of freedom and parameter ς. The transition probabilities are the same within each rating class. To embed this discrete setting in our continuous framework we assume that for each possible initial rating the rating agencies' probabilities correspond to a certain initial value of the uncertainty process within the range of admissible values, i.e., rating 1 corresponds to some uncertainty process value $u_1(0) = \alpha_1 \xi_1$ for some $0 \leq \alpha_1 \leq 1$ and rating k, $2 \leq k \leq 5$, corresponds to some uncertainty process value $u_k(0) = (1 - \alpha_k)\xi_{k-1} + \alpha_k \xi_k$ for some $0 \leq \alpha_k \leq 1$. If we set up equations (6.90) and (6.91) for each rating class (besides rating class 6) we get a system of 25 equations with 25 unknown variables which we must solve under the constraints $\xi_1 < \xi_2 < \xi_3 < \xi_4 < \xi_5$ and $0 \leq \alpha_k \leq 1$, $1 \leq k \leq 5$. One possibility to solve this system of equations is by applying the multidimensional Downhill Simplex method which was introduced by Nelder & Mead (1965). Because we use real world transition probabilities we can only estimate a_u but not $\hat{a}_u$. But for pricing purposes we are rather interested in $\hat{a}_u$ than in a_u. Therefore, we estimate λ_u together with the parameters for the short rate spread from market prices of observed credit spreads.

The non-defaultable short rate. The most intuitive procedure to fit the non-defaultable short rate to market prices (e.g., broker quotes) is to choose the values for $\hat{a}_r$ and σ_r that best fit a series of non-defaultable interest rate options such as caps. The parameter σ_r determines the overall volatility of the non-defaultable short rate. The parameter $\hat{a}_r$ determines the relative volatilities of long and short rates. The procedure is to choose the values of $\hat{a}_r$ and σ_r that minimize

$$\sum_i \left(P_i^* \left(\hat{a}_r, \sigma_r\right) - V_i\right)^2,$$

where P_i^* is the price given by the model for the i-th interest rate option and V_i is the market price of the i-th interest rate option. The procedure is quite easy for the Hull/White model, where cap prices can be calculated analytically. Again, the problem can be solved by applying the multidimensional Downhill Simplex method.

The short rate credit spread. The remaining parameters to be estimated are b_s, $\hat{a}_s$, σ_s and λ_u. We choose the values for these parameters that best fit a series of observed credit spreads. Like with the estimation for the parameters of the non-defaultable short rate the procedure is to choose the values of b_s, $\hat{a}_s$, σ_s and λ_u that minimize

$$\sum_i \left(S_i \left(b_s, \hat{a}_s, \sigma_s, \lambda_u\right) - Y_i\right)^2,$$

where S_i is the model value of the i-th credit spread and Y_i is its observed market value. Again, for the optimization procedure we can use a multidimensional Downhill Simplex method.

Example. We want to apply our theoretical framework and our calibration methods to set up a firm specific model for General Motors. Therefore we use data from Reuters, Bloomberg and S&P to estimate the model parameters.

1. The non-defaultable short rate:
   Using the US Cap-Volas (May 4th, 1999) from Reuters

   $$1 \text{ year } 11.25\ \%, \ 2 \text{ years } 15.25\ \%,$$

   we get the estimations $\hat{a}_r = 1.05 \cdot 10^{-8}$ and $\sigma_r = 1.01949464\ \%$.

Table 6.2. Average one-year transition rates (%) by S&P (July 1998).

	AAA	AA	A	BBB	BB	B	CCC	D
AAA	90.82	8.26	0.74	0.06	0.11	0.00	0.00	0.00
AA	0.65	90.88	7.69	0.58	0.05	0.13	0.02	0.00
A	0.08	2.42	91.30	5.23	0.68	0.23	0.01	0.05
BBB	0.03	0.31	5.87	87.45	4.96	1.08	0.12	0.18
BB	0.02	0.12	0.64	7.70	81.08	8.39	0.98	1.08
B	0.00	0.10	0.24	0.44	6.81	82.90	3.90	5.60
CCC	0.20	0.00	0.40	1.19	2.56	11.24	61.97	22.50

Table 6.3. Modified transition matrix.

	1	2	3	4	5
1	95.31	4.22	0.40	0.07	0.00
2	2.50	91.30	5.91	0.24	0.05
3	0.24	3.26	90.59	5.28	0.63
4	0.15	0.30	5.50	80.00	14.05

2. The uncertainty index:
   We assume that $\alpha_k = \frac{1}{2}\ \forall k$ (i.e. ratings correspond to the average quality within a rating class) and that there are 5 rating classes.
   The average one-year transition rates (%) by S&P (July 1998) are given by the transition matrix in table 6.2. We modify the matrix by aggregating the rating classes AAA and AA to rating class 1, A to rating class 2, BBB and BB to rating class 3, B and CCC to rating class 4. D corresponds to rating class 5. Hence, the modified transition matrix[34] is given by table 6.3.
   In S&P's notation General Motors' rating is A which corresponds to our rating class 2. Using the Downhill Simplex algorithm we get the following parameter estimates for this rating class:

$$\theta_u = 0.0653562131910263,$$
$$\sigma_u = 0.09999999994330677,$$
$$a_u = 0.13499953576965013,$$

   and the threshold values

$$\xi_1 = 0.0436501349994976,$$
$$\xi_2 = 0.2540403487950756,$$
$$\xi_3 = 0.6491354397214147,$$
$$\xi_4 = 1.1806484698650033.$$

[34] We assume that the number of obligors in each rating class is the same.

3. The short rate spread:
   For estimating the remaining parameters we use the following US$ - fixed rate bond data (semiannual coupons, S&P rating class A of General Motors which corresponds to our rating class 2) from Bloomberg (May 4th, 1999):
   a) 9 1/8, 07/18/00, Clean Price: 104.15, Yield: 5.497,
   b) 9.02, 06/07/01, Clean Price: 106.32, Yield: 5.758,
   c) 8 7/8, 06/11/01, Clean Price: 106.07, Yield: 5.760,
   d) 9.2, 07/02/01, Clean Price: 106.86, Yield: 5.769.

   Using $s_0 = 80$ basis points the Downhill Simplex algorithm yields the estimates $b_s = 0.0000997$, $\hat{a}_s = 0.328893$, $\sigma_s = 0.2$ and $\lambda_u = 0.19$.

6.6.3 The Kalman Filtering Methodology

Whereas in the previous section we determined the parameter values for the three-factor defaultable term structure model from market data at one specific day, this section is devoted to estimating parameter values by looking at time series of market values of bonds. Therefore we apply the Kalman filtering technique (see, e.g., Oksendal (1998), pages 79-106, and Harvey (1989)). As the number of parameters in our model is quite big, the idea is to apply the method twice. First, we estimate the parameters $\hat{a}_r$ and σ_r by applying the Kalman filter technique to the non-defaultable bond data corresponding to the non-defaultable short-rate model (see equations (6.1) and (6.5)). As a second step, using the obtained estimators for the non-defaultable model to apply the Kalman filter to different time series of defaultable bonds to get the estimators for the remaining parameters. We apply this method to zero coupon bond data of German and Italian Government bonds and coupon bond data of Greek Government bonds between January 1, 1999 and October 23, 2000. For that time frame the exchange rates between the German and Italian currencies have already been fixed due to the membership of the two countries to the European Monetary Union. The German, Italian, and Greek coupon bonds in our sample are all denominated in Euro. So we don't have to consider any exchange rate risk in our study. We assume that there is no default risk with the German bonds, and assume the Italian and Greek bonds to be risky. We analyze the development of the credit spreads between German and Italian and German and Greek bonds through time, and examine for the ability of our three-factor defaultable term structure model to describe these dynamics realistically. Especially Germany and Italy are appropriate for our analysis because they are one of the most liquid markets for Government bonds within the European Monetary Union. The daily coupon bond prices (from January $1st$, 1999 until October $23rd$, 2000,) are provided by Reuters Information Services.

The application of Kalman filtering methods in the estimation of term structure models using time-series data has been analyzed (among others) by Chen

& Scott (1995), Geyer & Pichler (1996) and Babbs & Nowman (1999). Up to our knowledge we are the first ones to apply this method to defaultable term structure models, especially to our model. The great advantage of the application of the Kalman filter to (defaultable) term structure models is that it allows the underlying state variables (in our case the default free short rate, the short rate credit spread and the uncertainty index) to be completely unobservable. Other approaches usually need to work with proxies for the short rates which induces noise in the parameter estimation process. Which proxy one could use for the values of the uncertainty index would be a serious problem.

All yields in our analysis are adjusted to the continuously compounded yields to maturity assuming 365 days a year. Credit spreads are then calculated as the differences between Italian and German and Greek and German zero coupon bond yields. We analyze the development of these credit spreads through time and analyze the performance of our model to explain these dynamics. Empirical investigation of credit spreads - mostly for corporates - has been done by, e.g., Longstaff & Schwartz (1995a), Duffee (1996b), Wei & Guo (1997), and Düllmann & Windfuhr (2000). Our analysis offers some advantages:

- We use daily data for credit spreads.
- (To our knowledge) we are the first ones to apply Kalman filtering techniques to credit spread investigation.
- We don't use proxies for the unobservable state variables but filter their values from observable data.
- We apply many different in - sample and, most important, out - of - sample tests.

The Data. In the following analysis we use observed data of credit spreads between German and Italian zero coupon Government bonds and German and Greek coupon Government bonds. The S&P ratings of the German, Italian and Greek bonds are AAA, AA, and $A-$, respectively. Hence, all of the bonds are investment grade, and we can assume that the German bonds are default risk free whereas the Italian and Greek bonds are defaultable. The credit spreads are calculated as the differences of the zero rates of the appropriate bonds. We consider a 22 months time series of daily bond prices from January $1st$, 1999, until October $23rd$, 2000, provided by Reuters Information Services[35]. All prices are denominated in Euro. As Germany and Italy are both members of the European Monetary Union, we don't have to take care of possible currency risks involved in these credit spreads. Note further, that our sample for Germany and Italy only consists of pure zero coupon bonds

[35] Most of the previous empirical studies use weekly or monthly data. Besides Düllmann & Windfuhr (2000) we are the only ones to use daily data - at least up to our knowledge.

Table 6.4. Summary statistics for the sample of German, Italian and Greek sovereign bonds.

	Germany	Italy	Greece
Number of bonds	33	36	38
Minimal maturity	0.5 years	0.36 years	2.4 years
Maximal maturity	11.5 years	12.2 years	20.8 years
Average maturity	4.9 years	3.5 years	7.46 years

without any special embedded options such as callability features. The Greek bonds are all pure fixed rate bonds paying annual coupons between 3% p.a. and 9.7% p.a. From table 6.4 it can be seen that we use a universe of 33 German, 36 Italian, and 38 Greek bonds. The ranges of maturities of the bonds span from 0.5 years to 11.5 years, from 0.36 years to 12.2 years, and from 2.4 years to 20.8 years for Germany, Italy, and Greece, respectively.

To determine the parameter values of the three-factor model we need the complete German, Italian and Greek zero curves at each time step (i.e. at each observation day), at least up to some maximal time of maturity. But unfortunately only a finite number of points on these zero curves is observable. To infer the entire German, Italian, and Greek zero curves from the daily prices of the 33 German, 36 Italian, and 38 Greek bonds on a daily basis, we apply the method of Nelson & Siegel (1987). The parsimonious model of Nelson and Siegel is widely used and suggested:

- Litterman & Scheinkman (1991) examine Treasury bonds and find that level, slope, and curvature explain most of the returns on the examined Treasury bonds.
- Jones (1991) shows that the yield curve can be mainly explained by shifts (parallel changes) and twists (changes in the slope).
- Barrett, Gosnell & Heuson (1995) apply it to daily Treasury note and bond data. From the investigation of shifts in the yield curve they draw conclusions for the selection of immunization strategies.
- Willner (1996) uses Nelson and Siegel's approach to extend traditional duration measurement methods to level, slope, and curvature durations.
- The Bundesbank has moved away from its in-house model to the Nelson and Siegel method for estimating the par yield curve (see, e.g., *Estimating the Term Structure of Interest Rates* (1997)).
- Brooks & Yan (1999) show that the Nelson-Siegel functional form can be used to fit the forward curve and the spot curve as well.
- Jordan & Mansi (2000) show that the Nelson and Siegel functional form performs better than other yield-smoothing methods.
- Mansi (2000) compares the use of constant-maturity Treasuries as an alternative to on-the-run yields obtained from the Treasury market. He finds

> that a continuous bootstrapping methodology with the Nelson and Siegel functional form as the interpolation method can reduce pricing errors dramatically.

Nelson and Siegel fit the yield curve using an exponential decay specification and reduce the problem to a four-parameter model: level, slope, curvature, and a time constant. Besides the construction of the zero curves we expect from the Nelson and Siegel methodology a smoothing effect of our data, especially for the Greek data which we expect to include more noise than the German and Italian data. Such an interest rate smoothing model avoids the problem inherent in spline-based models of choosing the optimal knot point specification. Nelson and Siegel take the following four parameter approximation model to determine zero curves from bond prices:

$$R(t,T) = \beta_0(t) + (\beta_1(t) + \beta_2(t)) \frac{1 - e^{-\frac{(T-t)}{\beta_3}}}{\frac{(T-t)}{\beta_3}} - \beta_2(t) e^{-\frac{(T-t)}{\beta_3}}, \qquad (6.92)$$

where

$$P(t,T) = 100e^{-R(t,T)(T-t)}, \; 0 \leq t \leq T.$$

The components in equation (6.92) determine the appropriate choices of weights that can be used to generate zero curves of a variety of shapes. $\beta_0(t)$ stands for the level or long term factor. $-\beta_1(t)$ stands for the slope or the short term factor. $\beta_2(t)$ stands for the curvature or the medium term component. β_3 is the time constant associated with curve hump positioning. It determines the rate of decay towards the long term rate. We estimate the parameters $\beta_i(t)$, $i = 0, 1, 2$, for each day t based on the non-defaultable German zero coupon bond data from a non-linear regression and based on β_3. We choose the specific β_3 value that yields the best results. As table 6.4 shows, the German bonds cover a range of maturities from half a year up to 11.5 years with an average maturity of 4.9 years.

Afterwards we apply the same methodology to the defaultable Italian zero coupon bond data and the defaultable Greece coupon bond data. The Italian bonds cover maturities from three months up to 12.2 years with an average of 3.5 years, the Greece bonds from 2.4 years up to 20.8 years with an average of 7.46 years. For each t we fit the parameters $\beta_i^{d,j}(t)$, $i = 0, 1, 2$, $j \in \{I = Italy, G = Greece\}$, and $\beta_3^{d,j}$ of the Nelson and Siegel model

$$R^{d,j}(t,T) = \beta_0^{d,j}(t) + (\beta_1^{d,j}(t) + \beta_2^{d,j}(t)) \frac{1 - e^{-\frac{(T-t)}{\beta_3^{d,j}}}}{\frac{(T-t)}{\beta_3^{d,j}}} - \beta_2^{d,j}(t) e^{-\frac{(T-t)}{\beta_3^{d,j}}}, \qquad (6.93)$$

where

$$P^{d,I}(t,T) = 100e^{-R^{d,I}(t,T)(T-t)}, \; 0 \leq t \leq T,$$

Table 6.5. Summary statistics for the deviations of the estimated zero rates (continuous compounding) from the observed German and Italian zero rates. The deviation results are in basis points.

	Germany	**Italy**
Average number of bonds per estimated curve	23.8	15.4
Mean of average absolute deviation	18.8	9.0
Standard deviation of average absolute deviation	11.0	10.3
Maximum of average absolute deviation	49.8	39.2

and

$$P_c^{d,G}(t,T) = \sum_{k=1}^{n} c_k e^{-R^{d,G}(t,t_k)(t_k-t)} + 100 e^{-R^{d,G}(t,T)(T-t)},$$

to the available Italian and Greece defaultable bond data. The Greek coupon bonds pay discrete coupons c_i at dates $t < t_k \le T$, $k = 1, ..., n$. To make the three models better comparable we assume that $\beta_3 = \beta_3^{d,j}$ for $j \in \{I,\ G\}$. Finally, based on the estimated parameter values $\beta_i(t)$ and $\beta_i^{d,j}(t)$, $i = 0, 1, 2$, $j \in \{I,\ G\}$, for each t and β_3, we can calculate the estimated zero rates for non-defaultable German and defaultable Italian and Greek Government bonds for each maturity $T \ge t$. It turns out that the best results can be achieved by setting the constant parameter β_3 equal to 3. After having estimated the zero curves we apply a robust outlier test removing all the curves which couldn't be fitted well enough by the Nelson and Siegel methodology[36]. Table 6.5 presents the quality statistics of our zero curve estimations for Germany and Italy, i.e. we summarize the analysis of the deviations of the observed German and Italian bond data from the estimated term structures of interest rates. The mean average absolute errors[37] are 18.8 basis points for Germany and 9.0 basis points for Italy, with standard deviations of 11.0 basis points and 10.3 basis points, correspondingly. Hence, all the curves fit the data fairly well.

[36] Let SSR_t be the sum of the squared deviations of the observed zero rates from the zero rates estimated with the Nelson-Siegel methodology at time t divided by the number of observations. We assume that the curve at time t^* is an outlier if the deviation of SSR_{t^*} from the mean of all SSR_t is more than three times the standard deviation of the values SSR_t, $\forall t$.

[37] By average absolute error of the Nelson-Siegel estimated curve at time t we mean

$$\sqrt{\frac{SSR_t}{\text{Number of bonds used for estimating time } t \text{ curve}}},$$

where SSR_t is the sum of squared deviations of the observed zero rates from the Nelson-Siegel estimated yields at time t.

Tables 6.6, 6.7, and 6.8 show the parameter summary results of the curve fittings for Germany, Italy, and Greece, respectively. The results show that the German long term spot rate is, on average, 5.532%, and the spot curve is, on average, upward-sloping, since the mean of the slope is 2.245%. In addition, on average, the German spot curve shows a curvature of 0.102%. The Italian long term spot rate is, on average, 6.048%, and the spot curve is, on average, upward-sloping, since the mean of the slope is 2.795%. In addition, on average, the Italian spot curve shows a curvature of 0.385%. Hence, on average, the level of the Italian zero rates is higher than the level of the German zero rates. Some of the German and Italian spot curves are downward-sloping, most of them are upward-sloping. The Italian curves tend to be steeper. The Greek long term spot rate is, on average, 7.046%, and the spot curve is, on average, downward-sloping, since the mean of the slope is -1.435%. In addition, on average, the Greek spot curve shows a curvature of -5.684%. On average, the level of the Greece yields is higher than the level of the German and Italian yields. That's exactly what the ratings of the three countries would suggest. In addition, the lower the rating the higher the standard deviation of level, slope and curvature. Most of the Greece curves are downward-sloping and concave. Figures 6.17 and 6.18 show graphically the results of the German and Italian spot curve fittings on October $3rd$, 1999, and August $24th$, 2000. On October $3rd$, 1999, both curves are humped, deeply upward-sloping, convex, and fit the data very well: the level, slope, and curvature parameters of the German and Italian spot curves are 5.345%, 2.989%, and 2.942%, and 6.220%, 3.686%, and 1.532%, respectively. Hence, level and slope parameters of the Italian spot curve are higher than the corresponding quantities of the German spot curve. Obviously, that can be explained by the lower rating of the Italian bonds. On August $24th$, 2000, the situation is quite different. Although the German curve is upward-sloping for short maturities and downward-sloping for long maturities, over all it is very flat. The Italian curve is steeply upward-sloping for short maturities, humped, and downward-sloping for long maturities. For maturities less than a year, the estimated Italian curve lies below the estimated German curve although all observed Italian zero rates are higher than the appropriate German zero rates. Nevertheless, both curves fit the data fairly well. The level, slope, and curvature parameters of the German and Italian spot curves are 5.055%, 0.120%, and 0.602%, and 6.088%, 3.548%, and 2.247%, respectively.

To study the relationship between the German and Italian, and the German and Greek zero curves, we calculate the correlation matrices of the three-, six-, and nine-year German and Italian, and German and Greek zero rates first. The results in table 6.9 show that all German and Italian rates are highly correlated. The correlation of the zero rates of different countries but with the same maturity is almost always higher than the correlation between curves of the same country with different maturities. Overall, we can say, the greater the difference between the maturities the less the correlation of the

Table 6.6. Summary statistics for the Nelson-Siegel parameters, results of fitting the German spot curve. All numbers are given in percent.

	Mean	Std. Dev.	Max.	Min.
Level β_0	5.532	0.257	6.414	4.042
Slope $-\beta_1$	2.245	0.994	3.788	−0.556
Curvature β_2	0.102	1.682	4.343	−3.492

Table 6.7. Summary statistics for the Nelson-Siegel parameters, results of fitting the Italian spot curve. All numbers are given in percent

	Mean	Std. Dev.	Max.	Min.
Level β_0^d	6.048	0.435	9.967	4.305
Slope $-\beta_1^d$	2.795	0.965	7.478	−0.239
Curvature β_2^d	0.385	2.041	8.391	−4.759

Table 6.8. Summary statistics for the Nelson-Siegel parameters, results of fitting the Greek spot curve. All numbers are given in percent.

	Mean	Std. Dev.	Max.	Min.
Level β_0	7.046	0.501	8.400	5.620
Slope $-\beta_1$	−1.435	1.777	2.350	−5.977
Curvature β_2	−5.684	3.533	3.352	−18.905

appropriate curves. For Germany and Greece the situation is quite different. The Greek three year rates show a negative correlation with all German rates and very low correlations with the other Greek rates. There is a high correlation between the six year and nine year Greek rates, and medium correlations of these rates with all German rates. Figures 6.19 for Germany and Italy and 6.20 for Germany and Greece show the co-movements of the rates across time. According to the high correlations between Germany and Italy, the curves mainly differ in their absolute level whereas the shapes are more or less all the same. The longer the maturity of the German and Italian rates the higher the level of the corresponding curves. While the German and Italian rates are mainly increasing over the observation period, the Greek rates are mainly decreasing. The level of all Greek curves is higher than the level of the German and Italian curves. The levels of the Greek curves which correspond to longer maturity rates is lower than the levels of curves which correspond to shorter maturity rates. The German, Italian and Greek rates are converging. The reason for this convergence lies in the announcement that Greece will become a member of the EMU. Because Germany and Italy have belonged to the EMU from the very beginning their rates have already converged prior to the observation period.

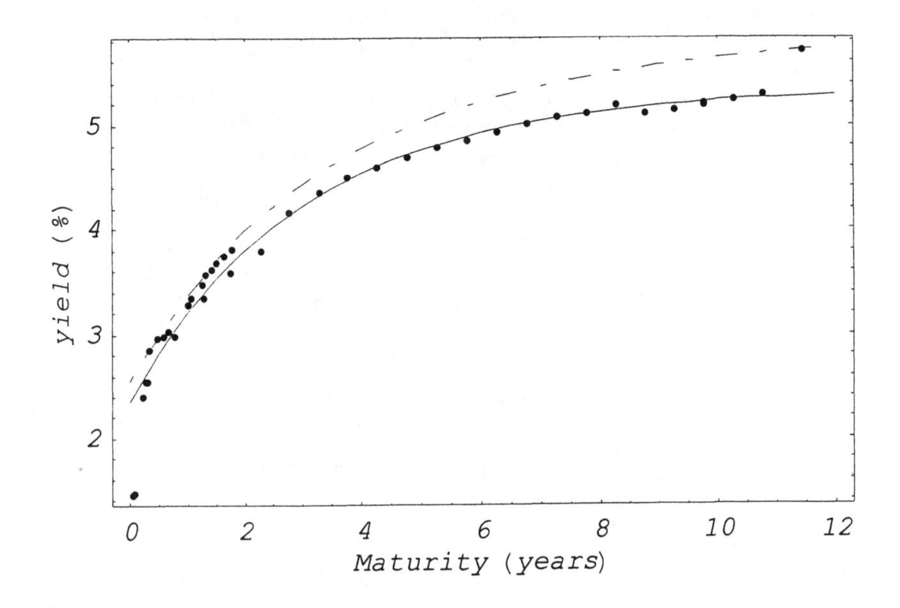

Fig. 6.17. Nelson-Siegel spot curve fitting on October, 3rd, 1999. The line and the dashed line refer to the fitted spot rates of Germany and Italy, respectively. The dots refer to the estimated zero rates of Germany and Italy.

Table 6.9. Correlation matrix of 3Y, 6Y, 9Y German and 3Y, 6Y, 9Y Italian zero rates.

	3Y Ger	**6Y Ger**	**9Y Ger**	**3Y I**	**6Y I**	**9Y I**
3Y Ger	1.000					
6Y Ger	0.967	1.000				
9Y Ger	0.924	0.990	1.000			
3Y I	0.996	0.956	0.911	1.000		
6Y I	0.976	0.991	0.973	0.976	1.000	
9Y I	0.949	0.990	0.978	0.944	0.992	1.000

To further understand the structural differences between the German, Italian and Greek spot curves we investigate the spreads between them. The estimated spread processes for various maturities, calculated as differences between Italian and German yield processes, are presented in figures 6.21, 6.22, and 6.23. The spreads between the Greek and German bonds are shown in figures 6.24, 6.25, and 6.26. The spreads are very volatile for all maturities and all countries. The Greek spreads have fallen from levels as high as 400 basis points to levels below 100 basis points. Because Germany and Italy are one of the most liquid Government bond markets in Europe we can

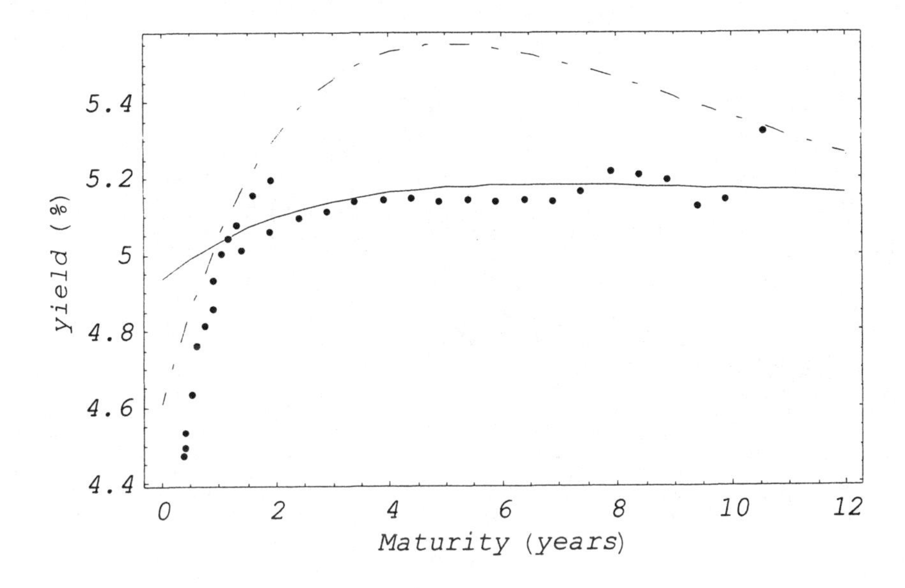

Fig. 6.18. Nelson-Siegel spot curve fitting on August 24th, 2000. The line and the dashed line refer to the fitted spot rates of Germany and Italy, respectively. The dots refer to the estimated zero rates of Germany and Italy.

Table 6.10. Correlation matrix of 3Y, 6Y, 9Y German and 3Y, 6Y, 9Y Greek zero rates.

	3Y Ger	**6Y Ger**	**9Y Ger**	**3Y G**	**6Y G**	**9Y G**
3Y Ger	1.000					
6Y Ger	0.967	1.000				
9Y Ger	0.924	0.990	1.000			
3Y G	−0.486	−0.441	−0.424	1.000		
6Y G	0.482	0.574	0.601	0.391	1.000	
9Y G	0.649	0.746	0.778	0.118	0.951	1.000

assume that these spreads are real credit spreads and that liquidity risk can be neglected.

In the following we partition the credit spread by using the Nelson-Siegel formulas (6.92) and (6.93). First, we observe that we can rewrite equation (6.92) as:

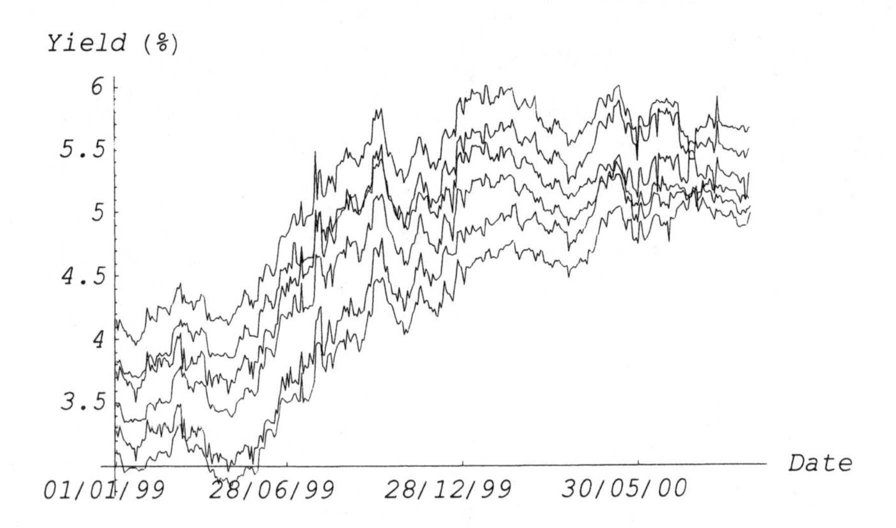

Fig. 6.19. From below: 3Y German, 3Y Italian, 6Y German, 6Y Italian, 9Y German, 9Y Italian zero rates.

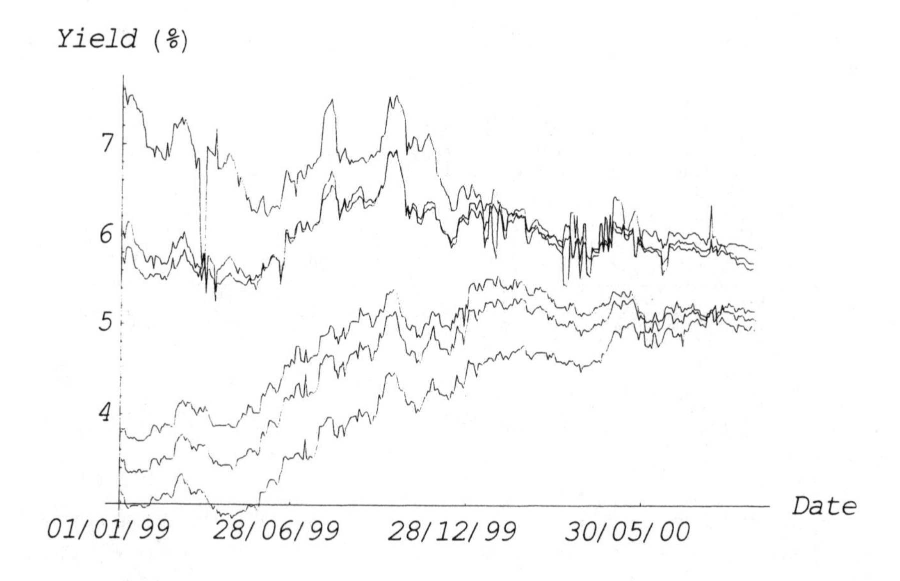

Fig. 6.20. From below: 3Y German, 6Y German, 9Y German, 9Y Greek, 6Y Greek, 3Y Greek zero rates.

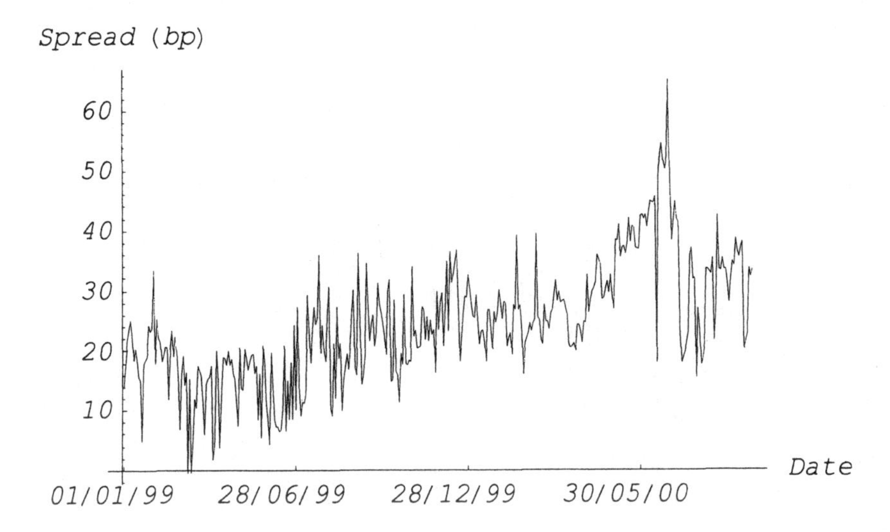

Fig. 6.21. Estimated credit spreads between Italian and German Government zero coupon bonds with maturity of 3 years. Time period: from January 1*st*, 1999, until October 23*rd*, 2000.

$$R(t,T) = \beta_0(t) + \beta_1(t)\frac{\beta_3}{(T-t)}\left(1 - e^{-\frac{(T-t)}{\beta_3}}\right)$$
$$+\beta_2(t)\frac{\beta_3}{(T-t)}\left(1 - e^{-\frac{(T-t)}{\beta_3}}\left(\frac{(T-t)}{\beta_3}+1\right)\right).$$

We can obtain corresponding formulas for $R^{d,j}(t,T)$, $j \in \{I,G\}$. Then the credit spread $S^j(t,T) = R^{d,j}(t,T) - R(t,T)$ can be partitioned into three parts:

$$S^j(t,T) = \left(\beta_0^{d,j}(t) - \beta_0(t)\right) + \left(\beta_1^{d,j}(t) - \beta_1(t)\right)\frac{\beta_3}{(T-t)}\left(1 - e^{-\frac{(T-t)}{\beta_3}}\right)$$
$$+\left(\beta_2^{d,j}(t) - \beta_2(t)\right)\frac{\beta_3}{(T-t)}\left(1 - e^{-\frac{(T-t)}{\beta_3}}\left(\frac{(T-t)}{\beta_3}+1\right)\right) \quad (6.94)$$

The first term on the right-hand side of equation (6.94) is the level component, the second term is the slope component and the third term is the curvature component for the credit spread at time t with maturity T. In our sample set, $\left(\beta_0^{d,j}(t) - \beta_0(t)\right) > 0$ for all t. In case of Italy, the mean level component contributes to most of the mean credit spread. The mean slope component offsets the mean level component. The mean credit spread increases with maturity. The mean curvature component plays a less important role. In

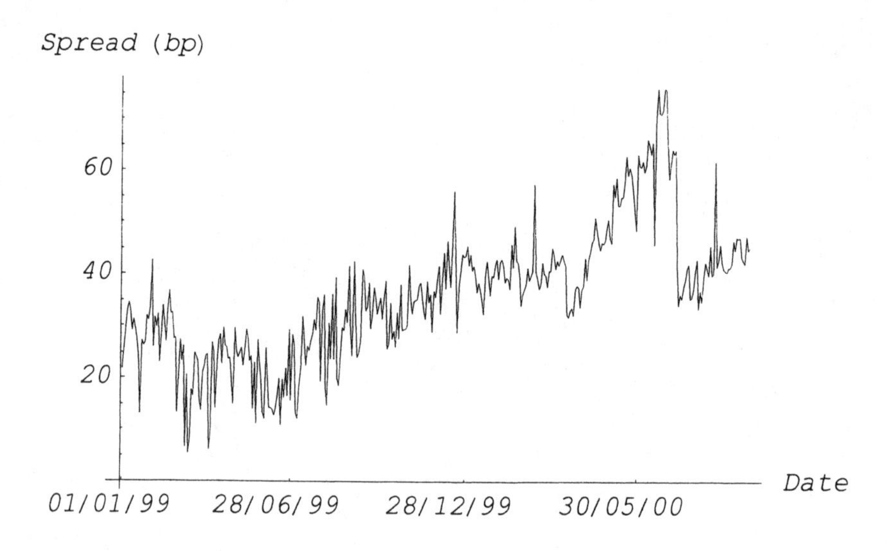

Fig. 6.22. Estimated credit spreads between Italian and German Government zero coupon bonds with maturity of 6 years. Time period: from January $1st$, 1999, until October $23rd$, 2000.

case of Greece, the mean level and the mean slope components contribute to most of the mean credit spread. The mean curvature component offsets the mean level component. The credit spreads coincide almost with the slope component.

All together we get Italian and Greek credit spread curves for 370 days between January $1st$, 1999 and October $23rd$, 2000. Figure 6.29 shows the plot of the Italian curves: over the observed time period, we get a range of different shapes: monotonic, humped and S shaped. Figure 6.30 shows the tightening of the Greek spreads over time.

Table 6.11 shows the summary statistics for the credit spreads between the German and Italian zero coupon bonds for different maturities. Again, the statistics are based on 370 observation days. The range of expected spreads between the two markets spans from 24 to 41 basis points depending on the time to maturity. Spreads are usually higher for longer maturities. The standard deviation of the credit spreads is more or less the same for all maturities (around 12 basis points). The skewness parameters[38] are all positive and suggest that the credit spreads are more likely to be above the mean. The

[38] See, e.g., Larsen & Marx (2001), p. 233: Let X be a random variable with mean μ and variance σ^2. The skewness of a probability density function (pdf) can be

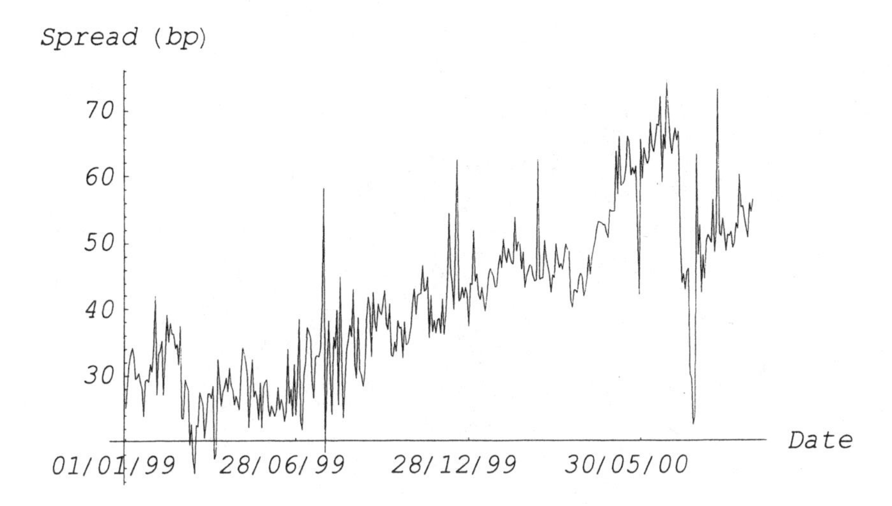

Fig. 6.23. Estimated credit spreads between Italian and German Government zero coupon bonds with maturity of 9 years. Time period: from January 1*st*, 1999, until October 23*rd*, 2000.

excess kurtosis[39] parameters suggest that the probability density functions are platykurtic for all maturities. For short term debt (maturities of 1 and

measured in terms of its third moment about the mean. If a pdf is symmetric, $E\left[(X-\mu)^3\right]$ will obviously be 0; for pdf's not symmetric $E\left[(X-\mu)^3\right]$ will not be zero. In practice, the symmetry (or lack of symmetry) of a pdf is often measured by the coefficient of skewness, γ_1, where

$$\gamma_1 = \frac{E\left[(X-\mu)^3\right]}{\sigma^3}.$$

A variable with a negative skew is more likely to be far below the mean than it is to be far above the mean.

[39] See, e.g., Larsen & Marx (2001), p. 233: Let X be a random variable with mean μ and variance σ^2. The excess kurtosis of a probability density function (pdf) can be measured in terms of its fourth moment about the mean. The coefficient of kurtosis, γ_2, is defined by

$$\gamma_2 = \frac{E\left[(X-\mu)^4\right]}{\sigma^4} - 3.$$

A pdf whose excess kurtosis is positive has more mass in the tails than a Gaussian pdf with the same variance. Hence, γ_2, is a useful measure of peakedness: relatively "flat" pdf's are said to be platykurtic; more peaked pdf's are called leptokurtic.

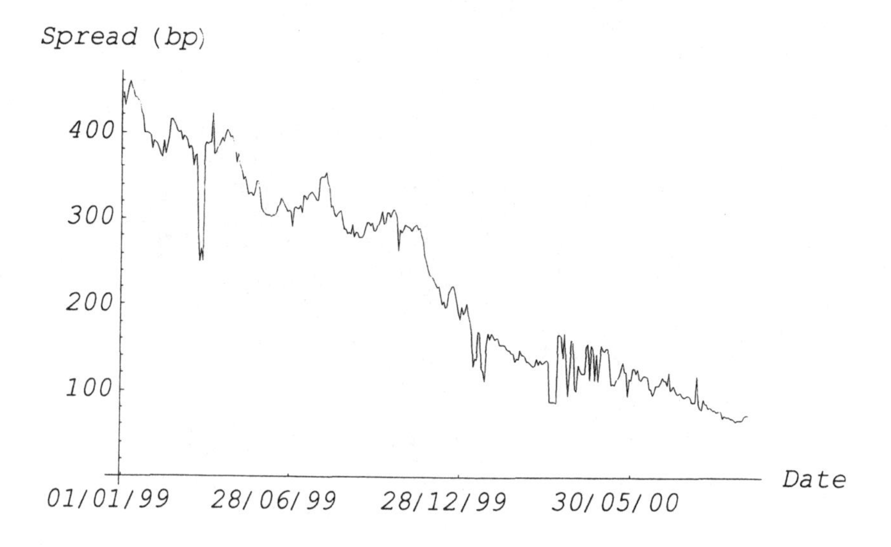

Fig. 6.24. Estimated credit spreads between Greek and German Government zero coupon bonds with maturity of 3 years. Time period: from January 1*st*, 1999, until October 23*rd*, 2000.

2 years) we find very high standard deviations of the spreads. That's why we exclude these maturities from the further analysis. For Greece we have summarized the credit spread statistics in table 6.12. The range of expected spreads between Greece and Germany spans from 113 to 232 basis points and is much higher than the Italian spread range. In opposite to the Italian case the spreads tend to be higher for shorter maturities. The standard deviations of the Greece spreads decrease with the maturity and are much higher than for Italy. The Greece credit spreads are more likely to be above than below the mean, and the probability density functions are leptokurtic for all maturities. One method that has been suggested for testing whether the distribution underlying a sample of n elements is normal, is to refer the statistic

$$L = n\left(\frac{skewness^2}{6} + \frac{excess\ kurtosis^2}{24}\right) \tag{6.95}$$

to the chi-squared distribution with two degrees of freedom[40]. The critical values for 5% and 1% are 5.99 and 9.21, respectively. We carry out the test using our credit spread data for the maturities between 3 and 9 years. According to the Italian values of L given in table 6.11 and the Greece values of L given in table 6.12, we can reject the normality hypothesis for the credit

[40] See, e.g., Greene (2000), p. 161.

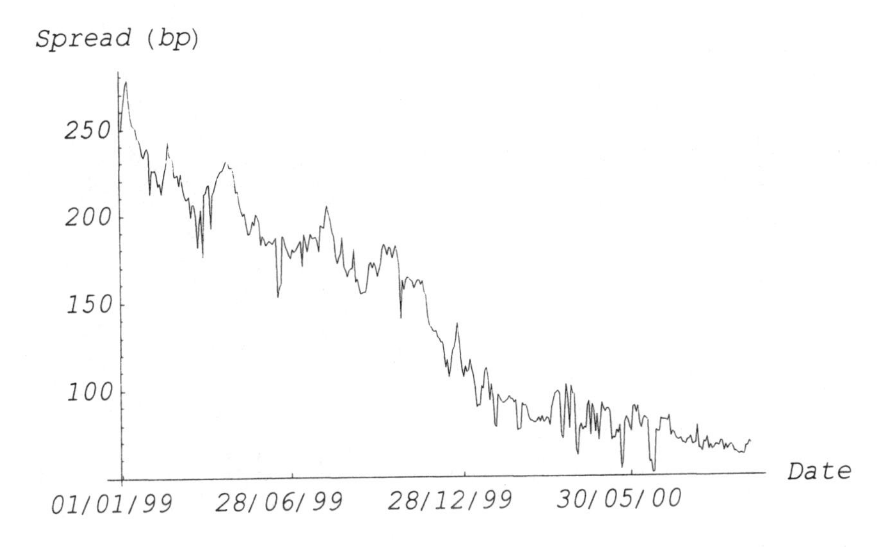

Fig. 6.25. Estimated credit spreads between Greek and German Government zero coupon bonds with maturity of 6 years. Time period: from January 1*st*, 1999, until October 23*rd*, 2000.

Table 6.11. Summary statistics for German - Italian credit spreads for different maturities. Mean, standard deviation and maximum values are given in basis points.

Mat.	Mean	Std. Dev.	Max.	Skewness	Excess Kurtosis	L
3 years	24.35	10.00	65.33	0.46	0.73	21.02
4 years	29.30	11.97	72.40	0.47	0.89	25.29
5 years	33.05	12.88	75.48	0.47	0.79	22.74
6 years	35.89	13.08	76.24	0.45	0.56	17.33
7 years	38.06	12.91	77.05	0.43	0.23	12.12
8 years	39.75	12.62	76.38	0.40	−0.14	10.01
9 years	41.06	12.39	74.73	0.36	−0.46	11.33

spread samples (for all maturities between 3 and 9 years) on a 1% level. In summary, we find that the distributions of the Italian and Greek spreads are completely different, and that both we cannot accept a normal distribution hypothesis.

In addition to the analysis of the credit spread levels we investigate the first-differences in the credit spread levels. The results in tables 6.13 and 6.14 suggest that credit spread changes are far away from being normally distrib-

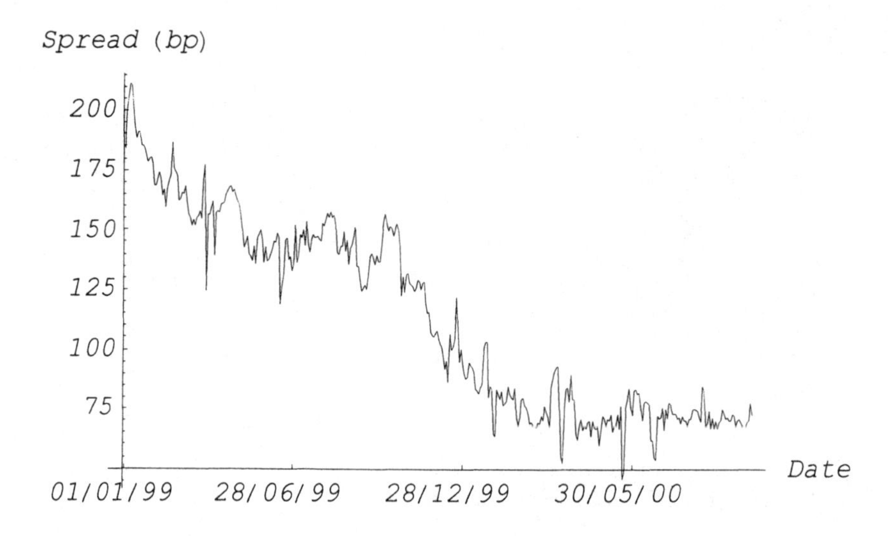

Fig. 6.26. Estimated credit spreads between Greek and German Government zero coupon bonds with maturity of 9 years. Time period: from January 1*st*, 1999, until October 23*rd*, 2000.

Table 6.12. Summary statistics for German - Greek credit spreads for different maturities. Mean, standard deviation and maximum values are given in basis points.

Mat.	Mean	Std. Dev.	Max.	Skewness	Excess Kurtosis	L
3 years	232.16	114.70	460.16	0.15	−1.40	31.28
4 years	188.20	90.12	378.75	0.19	−1.37	31.52
5 years	158.51	72.70	320.30	0.23	−1.36	31.53
6 years	138.80	60.29	278.31	0.26	−1.34	31.43
7 years	126.04	51.40	248.12	0.26	−1.32	31.26
8 years	118.10	45.02	226.41	0.26	−1.32	31.05
9 years	113.46	40.46	211.17	0.24	−1.33	30.92

uted. As, e.g., Duffie & Pan (1997) and Pedrosa & Roll (1998) point out, this implies troubling implications for risk management. The large values of the excess kurtosis are caused by fat tails in the probability density function. There are much more extreme events than would have been in the case of a normal distribution. The skewness is negative for all maturities but the 3−year Greece credit spread changes.

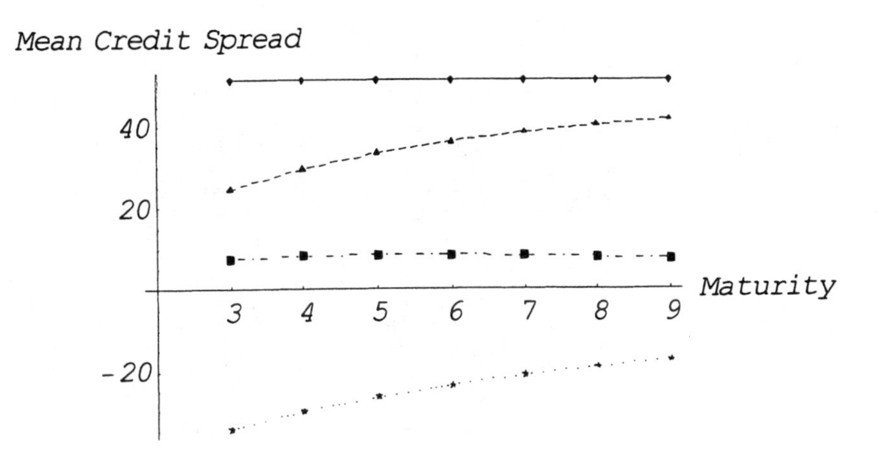

Fig. 6.27. Components of Italian mean credit spreads for different maturities. From below: mean slope component, mean curvature component, mean credit spread, mean level component. Maturity in years, mean credit spreads and credit spread components in basis points.

Table 6.13. Summary statistics for the credit spread changes between Germany and Italy for different maturities. Mean, standard deviation and maximum are given in basis points.

Maturity	**Skewness**	**Excess Kurtosis**
3 years	−0.60	16.17
4 years	−1.05	23.71
5 years	−0.93	11.14
6 years	−0.54	6.31
7 years	−0.35	4.08
8 years	−0.31	4.17
9 years	−0.26	8.18

A prerequisite for producing reliable statistical results is that the credit spread time series is inter-temporal stationary, i.e. that there is no unit root. We use the augmented Dickey-Fuller regression (Dickey & Fuller 1981) to test for unit roots:

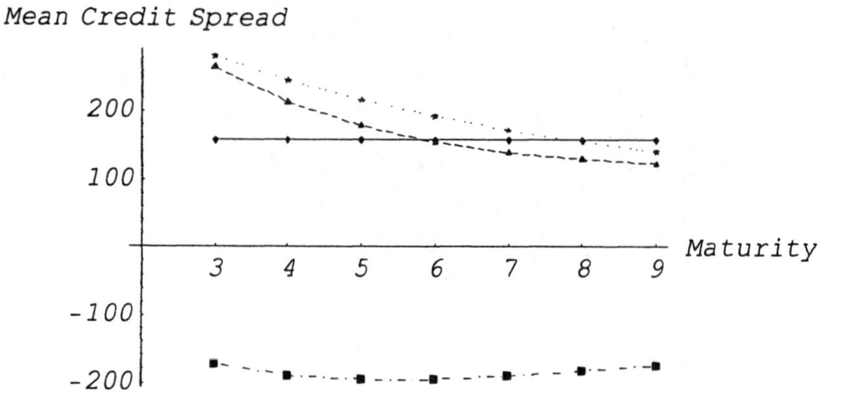

Fig. 6.28. Components of Greek mean credit spreads for different maturities. From below: mean curvature component, mean level component, mean credit spread, mean slope component. Maturity in years, mean credit spreads and credit spread components in basis points.

Table 6.14. Summary statistics for the credit spread changes between Germany and Greece for different maturities. Mean, standard deviation and maximum are given in basis points.

Maturity	Skewness	Excess Kurtosis
3 years	0.69	11.74
4 years	−0.05	7.36
5 years	−0.26	6.89
6 years	−0.22	6.95
7 years	−0.21	8.07
8 years	−0.24	9.18
9 years	−0.26	9.21

$$\begin{aligned}\Delta S_t^j(\tau) &= S_t^j(\tau) - S_{t-1}^j(\tau) \\ &= a_0^j + \gamma^j S_{t-1}^j(\tau) + a_2^j t + \sum_{i=2}^{p} \beta_i^j \Delta S_{t-i+1}^j(\tau) + \varepsilon_t^j,\ j \in \{I,\ G\}\end{aligned}$$

where $S_t^j(\tau) = S^j(t, t+\tau)$ denotes the time t credit spread between $\tau-$year Italian and German zero coupon government bonds. The drift term parameter a_0^j, the linear time trend parameter a_2^j, γ^j, and β_i^j are parameters which have

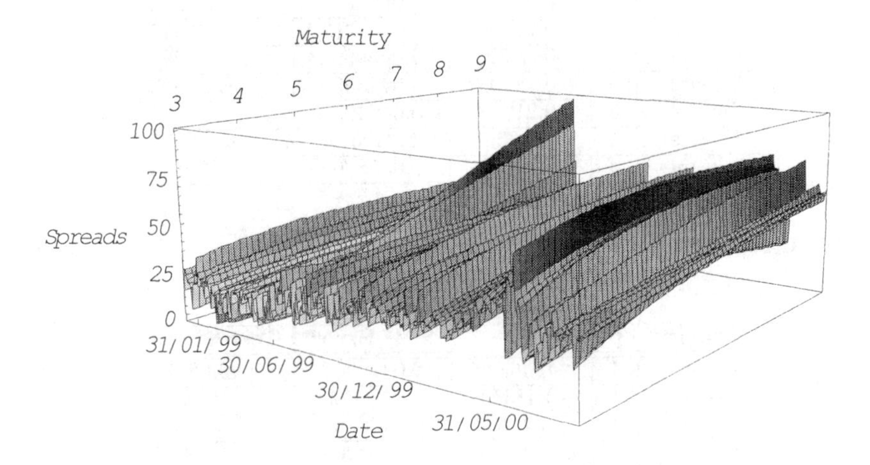

Fig. 6.29. Evolution of Italian credit spread curves. Credit spreads (bps), maturity (years).

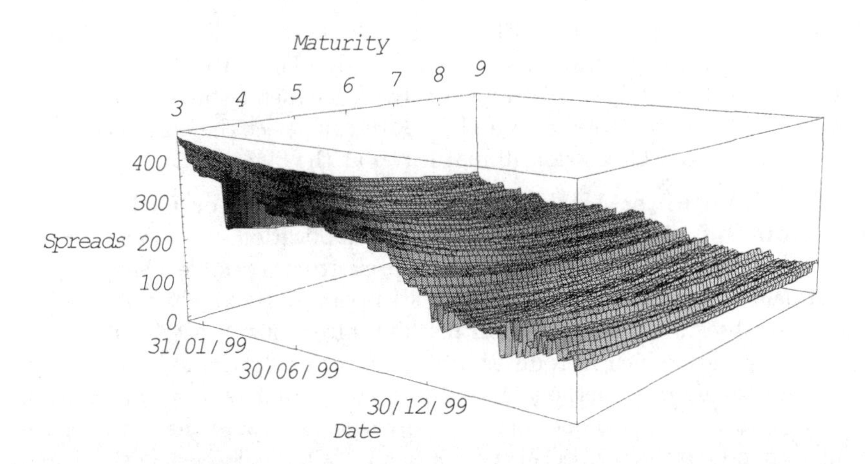

Fig. 6.30. Evolution of Greek credit spread curves. Credit spreads (bps), maturity (years).

Table 6.15. Unit root tests for credit spreads between Italian and German zero coupon government bonds.

Maturity	p	γ	Std. Error	Test Statistic
3 years	5	−0.283617	0.0561638	−5.04982
4 years	5	−0.211004	0.048394	−4.36013
5 years	5	−0.187264	0.0450996	−4.15224
6 years	5	−0.187931	0.0443555	−4.23693
7 years	5	−0.212585	0.0461903	−4.60237
8 years	4	−0.266	0.049571	−5.36603
9 years	4	−0.346507	0.0558706	−6.20196

Table 6.16. Unit root tests for credit spreads between Greece and German zero rates.

Maturity	p	γ	Std. Error	Test Statistic
3 years	1	−0.46146	0.0479452	−9.62474
4 years	1	−0.483547	0.0479009	−10.0948
5 years	1	−0.49087	0.0480069	−10.225
6 years	1	−0.484757	0.0481268	−10.0725
7 years	1	−0.473419	0.0481574	−9.83066
8 years	1	−0.46307	0.0481434	−9.61856
9 years	1	−0.455755	0.0481921	−9.45706

to be estimated by regression. ε_t^j is the spherical disturbance. We choose p to be the highest significant lag from the autocorrelation function of S^j. The coefficient of interest is γ^j. If $\gamma^j = 0$, the sequence $\left\{S_t^j(\tau)\right\}$ contains a unit root. Tables 6.15 and 6.16 report the Dickey-Fuller t-test results of $H_0 : \gamma^j = 0$ for Italy and Greece, respectively. The critical 95% and 99% values are −3.43 and −3.99, respectively. A smaller t-statistic rejects H_0. Hence, according to tables 6.15 and 6.16 we can reject the existence of unit roots for Italy and Greece for all maturities at the 99% level.

The Kalman Filter Methodology for the Parameter Estimation of the Non-Defaultable Short Rate. The application of Kalman filtering methods to the parameter estimation of term structure models has the great advantage that it allows the underlying state variables to be handled as completely unobservable, whereas other methods often must use some short term rates as proxies, which introduces additional noise. In the following we use the Kalman filtering method to estimate the parameters of the Hull and White term structure model defined in section 6.2 using the German zero coupon bond data described in section 6.6.3. First we derive the state space model formulation[41] of the non-defaultable term structure model defined by equation (6.1) and then present the Kalman filter algorithm. This is used in

[41] For a general definition of state space models see, e.g., Harvey (1989), chapter 3.

the calculation of the likelihood function of the observed zero rates and the computation of the unobserved state variables and parameters of the model. Because we use real world zero rate observations for the Kalman filter, we must describe the zero rate dynamics under the "real world" measure $\mathbf{P}$. We follow the assumptions of section 6.2 with the additional assumption that $\beta = 0$, and model the dynamics of r_t under $\mathbf{P}$ the following way:

$$dr_t = [\theta_r(t) - a_r r_t]\, dt + \sigma_r dW_r(t), \ 0 \le t \le T^*.$$

In addition,

$$a_r = \hat{a}_r - \lambda_r \sigma_r^2, \text{ and } \hat{W}_r(t) = W_r(t) + \int_0^t \gamma_r(l)\, dl,$$

with

$$\gamma_r(t) = \lambda_r \sigma_r r_t,$$

for some constant λ_r. We denote the parameter vector consisting of the unknown non-defaultable short rate process parameters by $\Phi_r = (a_r, \sigma_r, \lambda_r)$. The German zero coupon bond prices follow the non-defaultable Hull-White model (see equations (6.15), (6.16), and (6.17)), and the continuous zero rates of the non-defaultable discount bonds are given by

$$\begin{aligned} R(t,T) &= -\frac{1}{T-t} \ln P(t,T) \\ &= a_1(t,\tau) + b_1(\tau) r_t, \end{aligned}$$

where $\tau = T - t$, $a_1(t,\tau) = -\frac{\ln A(t,t+\tau)}{\tau}$, and $b_1(\tau) = \frac{B(t,t+\tau)}{\tau}$. Note, that $B(t,T)$ only depends on the time to maturity τ. We fix N different times to maturity $\tau_1, \ldots, \tau_N$, and denote the vector of the zero rates of the corresponding N different bonds by $RV(t)$:

$$RV(t) = \begin{pmatrix} R(t, t+\tau_1) \\ R(t, t+\tau_2) \\ \vdots \\ R(t, t+\tau_N) \end{pmatrix}.$$

Let

$$A_1(t) = A_1(t, \Phi_r) = \begin{pmatrix} a_1(t,\tau_1) \\ a_1(t,\tau_2) \\ \vdots \\ a_1(t,\tau_N) \end{pmatrix}, \ B_1 = B_1(\Phi_r) = \begin{pmatrix} b_1(\tau_1) \\ b_1(\tau_2) \\ \vdots \\ b_1(\tau_N) \end{pmatrix}.$$

Then the observable vector $RV(t)$ of zero rates and the unobservable non-defaultable short rates are related through the measurement equation:

$$RV(t) = A_1(t) + B_1 r_t.$$

We consider n time instances $t_1, \ldots, t_n$ on the time interval $(t, T]$ with $t_0 = t$ and $t_n = T$, and denote the time increments by $\Delta t_k = t_k - t_{k-1}$, $k = 1, \ldots, n$. At each time step t_k, $k = 1, \ldots, n$, we observe a vector consisting of N different zero rates (of N different maturities) :

$$RV_k = RV(t_k) = \begin{pmatrix} R(t_k, t_k + \tau_1) \\ R(t_k, t_k + \tau_2) \\ \vdots \\ R(t_k, t_k + \tau_N) \end{pmatrix}.$$

We allow for noise in the sampling process of the data. Hence, we assume that the zero rates are observed with measurement errors ε_k, that are jointly normally distributed with zero mean and $(N \times N)$–dimensional covariance matrix H. We choose the covariance matrix to be diagonal but allow for maturity-specific diagonal elements of H, i.e.

$$H = \begin{pmatrix} h_1^2 & 0 & \cdots & 0 \\ 0 & h_2^2 & \cdots & 0 \\ \vdots & \vdots & \ddots & \vdots \\ 0 & 0 & \cdots & h_N^2 \end{pmatrix}.$$

These diagonal elements h_i, $1 \leq i \leq N$, must be estimated together with Φ_r, and the vector consisting of these diagonal elements is denoted by $\Psi_r = (h_1, ..., h_N)$. Let $r_k = r_{t_k}$ and $A_{1,k} = A_1(t_k)$. Then the measurement equation is given by

$$RV_k = A_{1,k} + B_1 r_k + \varepsilon_k, \; \varepsilon_k \sim N(0, H). \tag{6.96}$$

Applying Itó's formula to $v_t = e^{a_r t} r_t$ yields:

$$dv_t = e^{a_r t}\theta_r(t)dt + \sigma_r e^{a_r t} dW_r(t).$$

The solution of this SDE is given by

$$v_t = v_0 + \int_0^t e^{a_r x}\theta_r(x)dx + \sigma_r \int_0^t e^{a_r x} dW_r(x).$$

Hence, r_t is given by

$$r_t = r_0 e^{-a_r t} + \int_0^t e^{a_r(x-t)}\theta_r(x)dx + \sigma_r \int_0^t e^{a_r(x-t)} dW_r(x),$$

which can also be written as

$$\begin{aligned} r_t &= r_s e^{-a_r(t-s)} + \int_s^t e^{a_r(x-t)}\theta_r(x)dx + \sigma_r \int_s^t e^{a_r(x-t)} dW_r(x) \\ &= r_s e^{-a_r(t-s)} + \int_0^{t-s} e^{a_r(x-(t-s))}\theta_r(x+s)dx \\ &\quad + \sigma_r \int_0^{t-s} e^{a_r(x-(t-s))} dW_r(x). \end{aligned}$$

Thus, with respect to the time increments,

$$r_{k+1} = r_k e^{-a_r \Delta t_{k+1}} + \int_0^{\Delta t_{k+1}} e^{-a_r(\Delta t_{k+1}-x)} \theta_r(x+t_k)dx + \sigma_r \int_0^{\Delta t_{k+1}} e^{-a_r(\Delta t_{k+1}-x)} dW_r(x),$$

which can be approximated by

$$r_{k+1} = r_k e^{-a_r \Delta t_{k+1}} + \int_0^{\Delta t_{k+1}} e^{-a_r(\Delta t_{k+1}-x)} \theta_r(t_k)dx + \sigma_r \int_0^{\Delta t_{k+1}} e^{-a_r(\Delta t_{k+1}-x)} dW_r(x).$$

Let $\widetilde{r}_{k|k-1}$ and $\widetilde{r}_k$denote the optimal estimators[42] of the unknown short rate r_k based upon the observations up to time t_{k-1} and t_k, i.e. given $RV_1, ..., RV_{k-1}$, and $RV_1, ..., RV_k$, respectively. Then,

$$\begin{aligned} \widetilde{r}_{k+1|k} &= E_k\left[r_{k+1}\right] \\ &= \widetilde{r}_k e^{-a_r \Delta t_{k+1}} + \int_0^{\Delta t_{k+1}} e^{-a_r(\Delta t_{k+1}-x)} \theta_r(t_k)dx, \end{aligned} \tag{6.97}$$

where $E_k[\cdot]$ denotes the expected value operator conditional on all information up to time t_k, i.e. given the observations $RV_1, ..., RV_k$. The mean squared error between the optimal estimator $\widetilde{r}_{k|k-1}$ and the observation r_{k+1}, conditioned on all observations up to time t_k, and denoted by $Var_{k+1|k}$, is given by[43]

$$\begin{aligned} &Var_{k+1|k} \\ &= E_k\left[r_{k+1} - \widetilde{r}_{k+1|k}\right]^2 \\ &= E_k\left[r_k e^{-a_r \Delta t_{k+1}} + \sigma_r \int_0^{\Delta t_{k+1}} e^{-a_r(\Delta t_{k+1}-x)} dW_r(x) - \widetilde{r}_k e^{-a_r \Delta t_{k+1}}\right]^2 \\ &= e^{-2a_r \Delta t_{k+1}} E_k\left[r_k - \widetilde{r}_k\right]^2 + \sigma_r^2 \int_0^{\Delta t_{k+1}} e^{-2a_r x} dx. \end{aligned}$$

If we denote $E_k\left[r_k - \widetilde{r}_k\right]^2$ by Var_k, then

$$Var_{k+1|k} = e^{-2a_r \Delta t_{k+1}} Var_k + \sigma_r^2 \int_0^{\Delta t_{k+1}} e^{-2a_r x} dx. \tag{6.98}$$

[42] By optimal estimators we mean optimal in the mean squared error sense.

[43] Note, that r_k and the time increments of the standard Brownian motion W are independent for $t \geq t_k$.

Now we can rewrite the measurement equation (6.96) as

$$RV_{k+1} = A_{1,k+1} + B_1 \left(r_{k+1} - \widetilde{r}_{k+1|k} \right) + B_1 \widetilde{r}_{k+1|k} + \varepsilon_{k+1}.$$

Then,

$$\widetilde{RV}_{k+1|k} = A_{1,k+1} + B_1 \widetilde{r}_{k+1|k}.$$

Let

$$v_{k+1} = RV_{k+1} - \widetilde{RV}_{k+1|k} = RV_{k+1} - A_{1,k+1} - B_1 \widetilde{r}_{k+1|k}, \tag{6.99}$$

and consider the vector $\begin{pmatrix} r_{k+1} \\ RV_{k+1} \end{pmatrix}$, which is normally distributed conditional on $RV_1, \ldots, RV_k$. The conditional mean is given by

$$E_k \begin{bmatrix} r_{k+1} \\ RV_{k+1} \end{bmatrix} = \begin{pmatrix} \widetilde{r}_{k+1|k} \\ \widetilde{Y}_{k+1|k} \end{pmatrix},$$

and the conditional covariance matrix by

$$\begin{aligned} & Cov_k \begin{bmatrix} r_{k+1} \\ RV_{k+1} \end{bmatrix} \\ &= \begin{pmatrix} Var_{k+1|k} & Cov_k(r_{k+1}, A_{1,k+1} + B_1 r_{k+1}) \\ Cov_k(A_{1,k+1} + B_1 r_{k+1}, r_{k+1}) & Cov_k(RV_{k+1}, RV_{k+1}) \end{pmatrix} \\ &= \begin{pmatrix} Var_{k+1|k} & Var_{k+1|k} B_1' \\ B_1 Var_{k+1|k} & F_{k+1} \end{pmatrix}, \end{aligned}$$

where

$$\begin{aligned} & F_{k+1} \\ &= E_k \left(v_{k+1} v_{k+1}' \right) \qquad (6.100) \\ &= E_k \left[\left(B_1 \left(r_{k+1} - \widetilde{r}_{k+1|k} \right) + \varepsilon_{k+1} \right) \left(B_1 \left(r_{k+1} - \widetilde{r}_{k+1|k} \right) + \varepsilon_{k+1} \right)' \right] \\ &= B_1 Var_{k+1|k} B_1' + H. \end{aligned}$$

If we apply results from Harvey (1989), chapter 3, we get the updating equations

$$\widetilde{r}_{k+1} = \widetilde{r}_{k+1|k} + K_{k+1} v_{k+1}, \tag{6.101}$$

and

$$Var_{k+1} = Var_{k+1|k} - K_{k+1} B_1 Var_{k+1|k}, \tag{6.102}$$

where

$$K_{k+1} = Var_{k+1|k} B_1^T F_{k+1}^{-1} \tag{6.103}$$

is the Kalman gain vector. Equations (6.97) and (6.98) are prediction equations for r_{k+1} and Var_{k+1}. When the new observation Y_{k+1} becomes available we can obtain more precise estimators using the updating equations (6.101) and (6.102). The Kalman filter determined by the equations (6.96)-(6.103) enables us to evaluate the likelihood function L for the set of parameters Φ_r and Ψ_r :

$$L(\Phi_r, \Psi_r | Y_1, \ldots, Y_n) = \prod_{k=1}^{n} f_{Y_k|Y_1,\ldots,Y_{k-1}}$$
$$= \prod_{k=1}^{n} \frac{1}{\left(\sqrt{2\pi}\right)^N} \frac{1}{\sqrt{\det F_k}}$$
$$\cdot \exp\left\{ -\frac{1}{2} \left(Y_k - \widetilde{Y}_{k|k-1} \right)' F_k^{-1} \left(Y_k - \widetilde{Y}_{k|k-1} \right) \right\},$$

where $f_{Y_k|Y_1,\ldots,Y_{k-1}}$ is the probability density function of Y_k given the observations $Y_1, \ldots, Y_{k-1}$. Thus, apart from a constant, the log-likelihood function is given by

$$\ln L(\Phi_r, \Psi_r | Y_1, \ldots, Y_n) = -\frac{1}{2} \sum_{k=1}^{n} \ln(\det F_k) - \frac{1}{2} \sum_{k=1}^{n} v_k' F_k^{-1} v_k. \tag{6.104}$$

Parameter Estimation Results for $r(t)$. For the parameter estimation of the non-defaultable short rate model we use two time series of different length:

1. Estimation 1: Seven samples of data, each of length 150, i.e. in the Kalman filter algorithm we set $N = 7$ and $n = 150$. We use a time series with daily German zero rate data from September 30*th*, 1999, until May 30*th*, 2000, excluding several days for which there is no bond data available, the Nelson-Siegel estimates of the credit spreads appear to be negative or the bond data is identified as an outlier. We consider the times to maturity $\tau_i = i + 2$, $i = 1, \ldots, 7$ (expressed in years). Thus, the input data consists of 150 observed vectors of zero rates
$$RV_k = \begin{pmatrix} R(t_k, t_k + \tau_1) \\ R(t_k, t_k + \tau_2) \\ \vdots \\ R(t_k, t_k + \tau_7) \end{pmatrix}, \quad 1 \leq k \leq 150.$$
2. Estimation 2: The same time series as in the previous numbered list item but with daily data of German zero rates from January 1, 1999, until May 30, 2000, i.e. with $N = 7$ and $n = 300$.

After initialization we perform the 150 (300) Kalman filter iterations to calculate the log-likelihood function (6.104). Finally, we maximize this function with respect to the two parameter vectors $\Phi_r = (a_r, \sigma_r, \lambda_r)$ and $\Psi_r = (h_1, \ldots., h_7)$. Therefore, we apply the multi-dimensional downhill simplex method of Nelder & Mead (1965). Tables 6.17 and 6.18 summarize the

Table 6.17. Kalman filter German maximum likelihood parameter estimates. Estimation 1: Sample period: 150 days from September 30th, 1999, until May 30th, 2000. Estimation 2: Sample period: 300 days from January 1st, 1999, until May 30th, 2000.

Param.	h_1	h_2	h_3	h_4	h_5	h_6	h_7
Estim. 1	0.00227	0.00119	0.00059	0.00025	0	0.00023	0.00046
Estim. 2	0.00246	0.00139	0.00073	0.00031	0	0.00027	0.00053

Table 6.18. Kalman filter maximum likelihood parameter estimates of the non-defaultable short rate. Estimation 1: Sample period: 150 days from September 30th, 1999, until May 30th, 2000. Estimation 2: Sample period: 300 days from January 1st, 1999, until May 30th, 2000.

Parameter	**Estimation 1**	**Estimation 2**
a_r	0.04481038	0.03398197
σ_r	0.00997727	0.01030943
λ_r	−3.6167735	−2.9502785
Log-Likelihood	7280.2	14219.4

parameter estimates. The average mean reversion levels of the non-defaultable short rate process implied by our parameter estimates for the first and second estimation equal 4.46% and 4.41%, respectively, which nicely fit to figures 6.31 and 6.32, respectively.

Testing the Assumptions on the Residuals. For the Kalman filter algorithm it is essential that we assume that the standardized residuals (the so called innovations) $\frac{v_k}{\sqrt{F_k}}$ are i.i.d. random variables. To verify this assumption we

1. apply a robust test for normal distribution;
2. calculate the Box-Ljung-statistic (Q-statistic) to test for serial correlation;
3. calculate the H-statistic to test for homoscedasticity;

We refer to the statistic (6.95) to test the sample of standardized innovations for normal distribution. The critical values for significance levels of 5% and 1% are 5.99 and 9.21, respectively.

- Estimation 1: According to table 6.19, we reject the normal distribution hypothesis on a 1%-significance level only for the maturity of 3 years. In case of a maturity of 4 years we reject the normal distribution hypothesis on the 5%-significance level. In all other cases we don't reject the hypothesis on 1% and 5% significance levels.

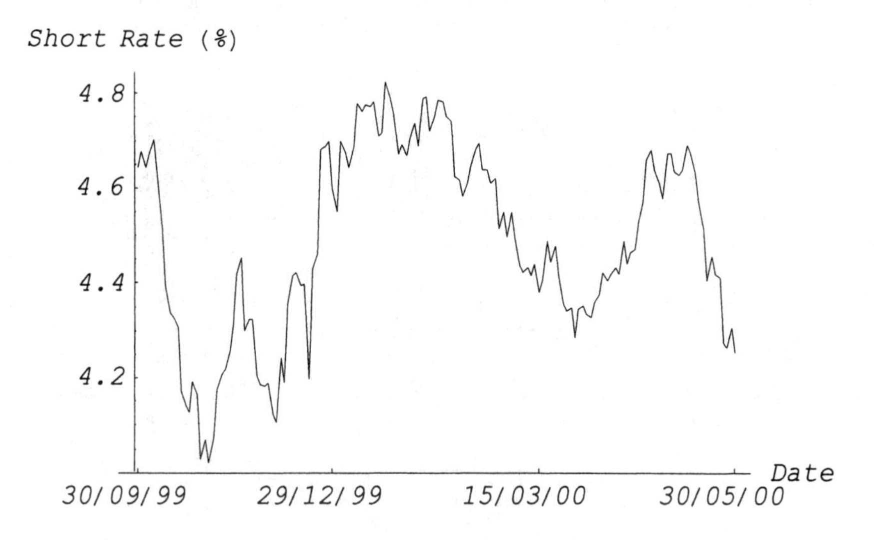

Fig. 6.31. Kalman filtered German short rate process. Time period: 150 days from September 30*th*, 1999, until May 30*th*, 2000.

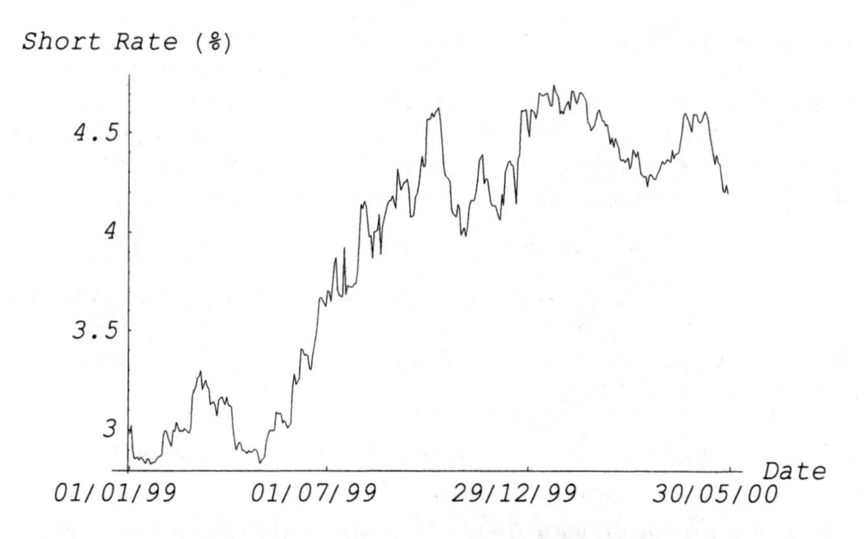

Fig. 6.32. Kalman filtered German short rate process. Time period: 300 days from January 1*st*, 1999, until May 30*th*, 2000.

Table 6.19. Results of the test for normal distribution of the German standardized Kalman filter residuals. Sample period 1: 150 days from September 30th, 1999, until May 30th, 2000. Sample Period 2: 300 days from January 1st, 1999, until May 30th, 2000.

Maturity	L (Estimation 1)	L (Estimation 2)
3 years	9.53294	7.51221
4 years	7.90222	11.198
5 years	5.20366	9.8423
6 years	1.78387	4.41243
7 years	2.43089	26.1632
8 years	5.0033	10.3134
9 years	4.16506	8.5817

Table 6.20. Critical values of the Box-Ljung test with 150 and 300 data points and 10 estimated parameters.

	χ^2 (2)	χ^2 (7)
95% quantile	5.99	14.07
99% quantile	9.21	18.48

- Estimation 2: For the maturities of 4, 5, 7, and 8 years we reject the normal distribution hypothesis on the 1%–significance level. On the 5%–significance level we have to reject the maturity of 3 years as well.

To test the hypothesis that there is no serial correlation of the standardized residuals (i.e. we consider the standardized residual sample autocorrelations), we apply the Box-Ljung test (see, e.g., Harvey (1989), page 259) on 1% and 5%- significance levels to different maturities. Under the null hypothesis that there is no serial correlation of the standardized Kalman filter residuals, the Box-Ljung Q–statistic has a χ^2–distribution (asymptotically) with

$$\sqrt{n} - number\ of\ estimated\ parameters$$

degrees of freedom, i.e. 2 and 7 for estimations 1 and 2, respectively. We reject the null hypothesis on the α–level only if the test statistic is greater than the $(1-\alpha)$%-quantile of the $\chi^2(\sqrt{n}-10)$ distribution. The critical values are given in table 6.20.

Table 6.21 summarizes the results of the Q–statistics for several maturities.

- Estimation 1: We reject the null hypothesis that there is no serial correlation on a 1%–significance level only for the maturity of 3 years and on a 5%–significance level for the maturities of 3, 4, 5, and 8 years.
- Estimation 2: We don't reject the null hypothesis that there is no serial correlation on a 1%–significance level for all maturities. On a 5%–significance level we reject the null hypothesis for the maturities of 4, 5, 8, and 9 years.

Table 6.21. Box-Ljung tests for German standardized Kalman filter residuals. Sample period 1: 150 days from September 30th, 1999, until May 30th, 2000. Sample Period 2: 300 days from January 1st, 1999, until May 30th, 2000.

Maturity	Q-statistic (Estimation 1)	Q-statistic (Estimation 2)
3 years	10.13	10.40
4 years	9.20	18.40
5 years	7.57	15.53
6 years	4.93	6.10
7 years	0.40	4.57
8 years	6.09	16.14
9 years	5.85	17.39

Table 6.22. Critical values of the test for homoscedasticity.

	F (50, 50)	F (100, 100)
0.5% quantile	0.476938	0.594923
2.5% quantile	0.570791	0.674195
97.5% quantile	1.75195	1.48325
99.5% quantile	2.09671	1.68089

Finally, we apply a test for homoscedasticity described in Harvey (1981), page 157. If there is homoscedasticity then the H-statistic has a F-Ratio distribution with n_1 numerator and n_2 denominator degrees of freedom where $n_1 = n_2$ is the integer that is closest to $\frac{n}{3}$. The critical values are summarized in table 6.22. If the value of the test statistic (H-statistic) is greater than the $\left(1 - \frac{\alpha}{2}\right)$ −quantile of the $F\left(n_1, n_2\right)$ −distribution or smaller than the $\frac{\alpha}{2}$−quantile of the $F\left(n_1, n_2\right)$ − distribution, than we reject the null hypothesis. If the value of the test statistic is smaller than the $\left(1 - \frac{\alpha}{2}\right)$ −quantile of the $F\left(n_1, n_2\right)$ −distribution and greater than the $\frac{\alpha}{2}$−quantile of the $F\left(n_1, n_2\right)$ −distribution, than we don't reject the null hypothesis.

- Estimation 1 : $n_1 = n_2 = 50$. According to table 6.23, we don't reject the hypothesis of homoscedasticity on a 5%−level for the maturities of 4, 5, and 7 years. On the 1%−significance level we don't reject the hypothesis for the 3, 4, 5, and 7 year zero rates.
- Estimation 2 : $n_1 = n_2 = 100$. We don't reject the hypothesis of homoscedasticity on a 5%-level for all maturities but 4 and 9 years. But we don't reject homoscedasticity on a 1%−significance level for these maturities as well.

Conclusion 6.6.1

Overall, based on the various tests we can conclude that the assumptions on the Kalman filter algorithm are sufficiently satisfied by the German data sets.

Table 6.23. Tests for homoscedasticity of German standardized Kalman filter residuals. Sample Period 1: 150 days from September 30th, 1999, until May 30th, 2000. Sample period 2: 300 days from January 1st, 1999, until May 30th, 2000.

Maturity	H-statistic:H (50) (Estimation 1)	H-statistic:H (100) (Estimation 2)
3 years	0.548653	0.730148
4 years	0.743375	0.605777
5 years	1.51961	0.741434
6 years	2.45601	1.24844
7 years	0.600204	1.32904
8 years	3.11734	1.19796
9 years	3.3072	1.6434

Testing the Model Performance of the Non-Defaultable Short Rate Model. In the following we test the non-defaultable short rate model performance by applying a test[44] used, e.g., by Titman & Torous (1989). We test the power of the model to explain the changes of the observed German zero rates. If the results are satisfactory we can conclude that the model can price non-defaultable bonds accurately. Let $\Delta R_k(\tau) = R(t_k, t_k + \tau) - R(t_{k-1}, t_{k-1} + \tau)$, $2 \leq k \leq 150$ (300), denote the changes of the observed zero rates with time to maturity τ, and let $\widehat{\Delta R_k}(\tau)$, $2 \leq k \leq 150$ (300), denote the changes of the corresponding fitted German zero rates. The test equation is given by the following linear regression:

$$\Delta R_k(\tau) = a + b\widehat{\Delta R_k}(\tau) + \varepsilon_k^*, \text{ where } \varepsilon_k^* \sim N\left(0, h^2\right). \tag{6.105}$$

We apply the test to the maturities $\tau_i = i + 2$, $i = 1, ..., 7$ (expressed in years). If $\widehat{\Delta R_k}(\tau)$ is an unbiased estimate for $\Delta R_k(\tau)$, then a and b should be close to 0 and 1, respectively. The adjusted $\overline{R}^2$ statistics are an additional measure for the quality of the model. Table 6.24 and 6.25 show the results of the regression for estimation 1 and 2, respectively.

- Estimation 1:
  - The test of the hypothesis that $a = 0$ is not rejected for all maturities on a 5%−significance level.
  - Except for maturities of 3 and 4 years the parameter values for b are close to 1.
  - The test of the joint hypothesis that $a = 0$ and $b = 1$ is not rejected on a 1%−significance level for all maturities but 3 and 4 years. It is not

[44] Because the test is well-known and widely used we refer for any details on the test statistics to the article of Titman & Torous (1989). E.g., testing for $a = 0$ uses the standard t−statistics. The p-values are calculated by comparing the obtained statistic to the t−distribution with $n - p$ degrees of freedom, where n is the sample size and p is the number of predictors.

Table 6.24. Test of model explanatory power for Germany: Linear regression results of estimation 1.

Mat.	a	b	$\overline{R}^2$	p-Value (a = 0)	p-Value (a = 0, b = 1)
3 years	−0.0000130245	0.71069	0.677014	0.599483	0
4 years	$-7.53239 \cdot 10^{-6}$	0.835213	0.777075	0.736346	0.00007395
5 years	$-6.10068 \cdot 10^{-6}$	0.920887	0.872794	0.722183	0.0242764
6 years	$-5.71593 \cdot 10^{-6}$	0.96917	0.960037	0.544857	0.143196
7 years	$-5.55063 \cdot 10^{-6}$	0.991991	0.999794	0.577653	0.142372
8 years	$-5.34673 \cdot 10^{-6}$	0.997844	0.954494	0.591318	0.139627
9 years	$-5.04453 \cdot 10^{-6}$	0.992707	0.828923	0.802432	0.0482509

rejected on a 5%− significance level for 6, 7, and 8 years. Hence, the model prices these zero coupon bonds with acceptable accuracy.
- The adjusted $\overline{R}^2$ values vary between 67.70 % and 99.98 %.

• Estimation 2:
The results for the longer time horizon of 300 days are not as good as the results for estimation 1.
- The test of the hypothesis that $a = 0$ is only rejected for a maturity of 3 years on the 1%−significance level, and, in addition, on the 5%−significance level for a maturity of 4 years.
- Except for maturities of 3 and 4 years the parameter values for b are close to 1.
- The test of the joint hypothesis that $a = 0$ and $b = 1$ is not rejected on a 1%−significance level for the maturities of 5, 6, and 7 years. It is not rejected on a 5%− significance level for maturities of 6 and 7 years. Hence, we can assume that the model prices zero coupon bonds with maturities of 5 − 7 years with acceptable accuracy.

Conclusion 6.6.2
Overall, estimation 1 *yields better results than estimation* 2 *does. This is not very surprising, as figure 6.32 shows that there has been a tremendous change in the level of the filtered non-defaultable German short rate between July* 1999 *and September* 1999. *But even for estimation* 1 *the model performs worse for* 3 *year zero rates. Hence, we can conclude that there may be a structural difference between the* 3 *year zero rates and the zero rates for longer maturities.*

Out-of-Sample Testing Based on the German Data. So far we have done some in-sample-testing to examine the performance of the one-factor non-defaultable term structure model. To complete the analysis of the model we add out-of-sample-tests for estimation 1 and estimation 2. We analyze if

Table 6.25. Test of model explanatory power for Germany: Linear regression results of estimation 2.

Mat.	a	b	$\overline{R}^2$	p-Value (a = 0)	p-Value (a = 0, b = 1)
3 years	−0.0000551972	0.724096	0.75725	0.001371	0
4 years	−0.0000286776	0.885319	0.85426	0.049230	0
5 years	−0.0000168865	0.965857	0.91886	0.129549	0.027711
6 years	−0.0000103833	0.993544	0.97415	0.091742	0.168942
7 years	$-6.22198 \cdot 10^{-6}$	0.986557	0.99988	0.734251	0.145742
8 years	$-3.25004 \cdot 10^{-6}$	0.965154	0.96515	0.625646	0.000266
9 years	$-9.52978 \cdot 10^{-6}$	0.914719	0.85560	0.943822	0.000540

we can apply the Kalman filter with the parameters in tables 6.17 and 6.18 to the pricing of zero rates later than May 30*th*, 2000. Therefore we consider 70 additional time steps $t_k^* = 1, ..., 70$, for $k = 1, ..., 70$, from May 30*th*, 2000, until October 23*rd*, 2000, and apply the Kalman filter to the observed zero rates for this time frame. Note, that we keep the parameters of tables 6.17 and 6.18 fixed, i.e. we filter the values for r_t and determine the non-defaultable short rate predictors with given values for the vectors Φ_r and Ψ_r. To test the performance of the out-of-sample model we apply the linear regression (6.105) to the new data. We test the maturities $\tau_i = i + 2$, $i = 1, ..., 7$ (expressed in years). Tables 6.26 and 6.27 show the results of the regression:

- Estimation 1:
  - The test of the hypothesis that $a = 0$ is not rejected for all maturities on a 5%–significance level.
  - The test of the joint hypothesis that $a = 0$ and $b = 1$ is rejected on a 5%–significance level only for the maturity of 8 and 9 years, on a 1%-significance level it is rejected only for a maturity of 9 years. Hence, the model prices these zero coupon bonds with acceptable accuracy.
  - The adjusted $\overline{R}^2$ values vary between 44.6 % and 99.9 %.

  Overall, the test suggests that the out-of-sample behavior of the model is very encouraging.
- Estimation 2:
  - The test of the hypothesis that $a = 0$ is not rejected for all maturities on a 5%–significance level.
  - The test of the joint hypothesis that $a = 0$ and $b = 1$ is rejected on a 5%–significance level only for the maturity of 9 years. It is not rejected on a 1%-significance level for all maturities. Hence, the model prices these zero coupon bonds with acceptable accuracy.
  - The adjusted $\overline{R}^2$ values vary between 44.6 % and 99.9 %.

Table 6.26. Test of out-of-sample model explanatory power for Germany: Linear regression results for estimation 1.

Mat.	a	b	$\overline{R}^2$	p-Value (a = 0)	p-Value (a = 0, b = 1)
3 years	−0.0000550418	0.864412	0.44628	0.301913	0.245901
4 years	−0.0000314571	0.938133	0.53581	0.484341	0.360516
5 years	−0.0000197269	0.970996	0.68472	0.551017	0.221816
6 years	−0.0000126984	0.985759	0.88139	0.471629	0.270834
7 years	$-7.73987 \cdot 10^{-6}$	0.990779	0.99931	0.567842	0.154642
8 years	$-3.82299 \cdot 10^{-6}$	0.990284	0.85459	0.838539	0.039232
9 years	$-5.15315 \cdot 10^{-7}$	0.986782	0.58180	0.989008	0.008135

Table 6.27. Test of out-of-sample model explanatory power for Germany: Linear regression results for estimation 2.

Mat.	a	b	$\overline{R}^2$	p-Value (a = 0)	p-Value (a = 0, b = 1)
3 years	−0.0000561881	0.881174	0.44618	0.292339	0.2902
4 years	−0.0000318168	0.952851	0.53589	0.479379	0.311749
5 years	−0.0000196989	0.981572	0.67975	0.551555	0.189156
6 years	−0.0000124838	0.991632	0.88141	0.479081	0.238614
7 years	$-7.43832 \cdot 10^{-6}$	0.991836	0.99937	0.653212	0.183240
8 years	$-3.48693 \cdot 10^{-6}$	0.986573	0.85699	0.852494	0.051134
9 years	$-1.73726 \cdot 10^{-7}$	0.978417	0.58197	0.996293	0.021683

Overall, the test suggests that the out-of-sample behavior of the model is very encouraging.

Conclusion 6.6.3
The out-of-sample tests yield good results for both time series. Only for the 9 *year maturity zero rates in estimation* 1*, the* $p - value(a = 0, b = 1) < 0.01$. *All other p-values are much greater than* 0.01. *The out-of-sample results are even better than the in-sample results. Though, we have considered a shorter time horizon than in the in-sample case.*

Spread Parameter Estimation by Using Kalman Filtering. In the following we apply the Kalman filtering method to estimate the parameters of the three-factor defaultable term structure model described in section 6.2 using the German, Italian, and Greece bond data described in section 6.6.3. First we derive the state space model formulation, and then present the Kalman filter algorithm. This is used in the calculation of the likelihood function of the observed credit spreads and the computation of the unobserved state variables and parameters of the model. We have already shown how we can determine the parameters for the non-defaultable one-factor model, i.e. how to find good estimators for the process r_t. So it suffices to show how to determine the remaining parameters of the Italian and Greece short rate credit spread and uncertainty index processes. Because we use real world credit spread observations, we must describe the credit spread dynamics under the "real world" measure **P**. We follow the assumptions of section 6.6.3 and model the dynamics of s_t^j and u_t^j, $j \in \{I, G\}$, under the measure **P** the following way:

$$du_t^j = \left[\theta_u^j - a_u^j u_t^j\right] dt + \sigma_u^j \sqrt{u_t^j} dW_u^j(t),$$

$$ds_t^j = \left[b_s^j u_t^j - a_s^j s_t^j\right] dt + \sigma_s^j \sqrt{s_t^j} dW_s^j(t),$$

where

$$a_i^j = \hat{a}_i^j - \lambda_i^j \left(\sigma_i^j\right)^2 \text{ and } \hat{W}_i^j(t) = W_i^j(t) + \int_0^t \gamma_i^j(l)\, dl, \; i = s, u,$$

with

$$\gamma_s^j(t) = \lambda_s^j \sigma_s^j \sqrt{s_t^j} \text{ and } \gamma_u^j(t) = \lambda_u^j \sigma_u^j \sqrt{u_t^j},$$

for some constants λ_s^j and λ_u^j. Note, that we assume that $dW_s^I(t)$, $dW_s^G(t)$, $dW_u^I(t)$, $dW_u^G(t)$, and $dW_r(t)$ are pairwise uncorrelated. We denote the parameter vectors consisting of the unknown parameters by

$$\Phi_{s,u}^j = (a_s^j, a_u^j, \sigma_s^j, \sigma_u^j, b_s^j, \theta_u^j, \lambda_s^j, \lambda_u^j), j \in \{I, G\}.$$

The theoretical formulas for the observed credit spreads $S^j(t, t+\tau)$, $j \in \{I, G\}$, (compare to equation (6.94)) are given as the differences between the defaultable and non-defaultable continuous zero rates. For country $j \in \{I, G\}$ the zero rate of a defaultable bond $P^{d,j}$ at time $t \in [0, T]$ for a maturity time $T \in [t, T^*]$ is given by

$$\begin{aligned}
& R^{d,j}(t,T) \\
&= -\frac{1}{T-t} \ln P^{d,j}(t,T) \\
&= -\frac{\ln A^{d,j}(t,T)}{T-t} + \frac{B(t,T)}{T-t} r_t + \frac{C^{d,j}(t,T)}{T-t} s_t^j + \frac{D^{d,j}(t,T)}{T-t} u_t^j \\
&= -\frac{\ln A(t,T)}{T-t} - \frac{2\theta_u^j}{\left(\sigma_u^j\right)^2 (T-t)} \ln \left| \frac{v^j(T,T)}{v^j(t,T)} \right| + \frac{B(t,T)}{T-t} r_t \\
&\quad + \frac{C^{d,j}(t,T)}{T-t} s_t^j + \frac{D^{d,j}(t,T)}{T-t} u_t^j.
\end{aligned}$$

Hence,

$$\begin{aligned} S^j(t,t+\tau) &= R^{d,j}(t,T) - R(t,T) \\ &= a^j(\tau) + c^j(\tau)\, s_t^j + d^j(\tau)\, u_t^j,\ \tau \geq 0, \end{aligned}$$

with

$$a^j(\tau) = -\frac{2\theta_u^j}{\left(\sigma_u^j\right)^2 \tau} \ln\left|\frac{v^j(T,T)}{v^j(\tau)}\right|,\ c^j(\tau) = \frac{C^{d,j}(\tau)}{\tau},\ \text{and}\ d^j(\tau) = \frac{D^{d,j}(\tau)}{\tau},$$

where $v^j(\tau) = v^j(t,t+\tau)$, $C^{d,j}(\tau) = C^{d,j}(t,t+\tau)$, and $D^{d,j}(\tau) = D^{d,j}(t,t+\tau)$. Note that a^j, c^j, and d^j do not explicitly depend on t. Rather they depend on the time to maturity of the defaultable bond. We consider $N-$dimensional vectors of credit spreads with different maturities $\tau_1, ..., \tau_N \geq 0$ at times $t_1 < t_2 < ... < t_n$:

$$SV^j(t_k) = SV_k^j = \begin{pmatrix} S^j(t_k, t_k+\tau_1) \\ S^j(t_k, t_k+\tau_2) \\ \vdots \\ S^j(t_k, t_k+\tau_N) \end{pmatrix},\ 1 \leq k \leq n,\ j \in \{I, G\}.$$

If we denote

$$A^j = A^j(\Phi_{s,u}^j) = \begin{pmatrix} a^j(\tau_1) \\ a^j(\tau_2) \\ \vdots \\ a^j(\tau_N) \end{pmatrix},\ C^j = C^j(\Phi_{s,u}^j) = \begin{pmatrix} c^j(\tau_1) & d^j(\tau_1) \\ c^j(\tau_2) & d^j(\tau_2) \\ \vdots & \vdots \\ c^j(\tau_N) & d^j(\tau_N) \end{pmatrix},$$

and

$$X_t^j = X^j(t, \Phi_{s,u}^j) = \begin{pmatrix} s_t^j \\ u_t^j \end{pmatrix},$$

where $\Phi_{s,u}^j$ is a subset of the parameter set which we want to estimate, we can define the measurement equation as

$$SV^j(t) = A^j + C^j X_t^j,\ t \geq 0.$$

The system of stochastic differential equations (6.2) and (6.3) with $X_t^j = \left(s_t^j, u_t^j\right)'$ can be written in matrix form as

$$dX_t^j = \left[H^j X_t^j + J^j\right] dt + V^j(X_t^j) dW_t^j, \tag{6.106}$$

where

$$H^j = H^j(\Phi_{s,u}^j) = \begin{pmatrix} -a_s^j & b_s^j \\ 0 & -a_u^j \end{pmatrix},\ J^j = J^j(\Phi_{s,u}^j) = \begin{pmatrix} 0 \\ \theta_u^j \end{pmatrix},$$

$$V^j(X_t^j) = V^j\left(X_t^j, \Phi_{s,u}^j\right) = \begin{pmatrix} \sigma_s^j\sqrt{s_t^j} & 0 \\ 0 & \sigma_u^j\sqrt{u_t^j} \end{pmatrix},$$

and

$$dW^j = \begin{pmatrix} dW_s^j \\ dW_u^j \end{pmatrix}.$$

We allow for noise in the sampling process of the data. Hence, we assume that for $j \in \{I, G\}$ the spreads SV_k^j are observed with measurement errors ε_k^j that are jointly normally distributed with zero mean and $(N \times N)-$dimensional covariance matrix G^j. Theoretically the matrix G^j can take various forms. But to simplify the calculations of the Kalman filter we choose the covariance matrix to be diagonal. Because trading activity and therefore bid-ask spreads usually vary across maturities (see, e.g., Geyer & Pichler (1996)) we allow for maturity-specific diagonal elements of G^j, i.e.

$$G^j = \begin{pmatrix} \left(g_1^j\right)^2 & 0 & \cdots & 0 \\ 0 & \left(g_2^j\right)^2 & \cdots & 0 \\ \vdots & \vdots & \ddots & \vdots \\ 0 & 0 & \cdots & \left(g_N^j\right)^2 \end{pmatrix}.$$

The diagonal elements g_i^j, $1 \leq i \leq N$, must be estimated together with $\Phi_{s,u}^j$, and the vector consisting of these diagonal elements is denoted by

$$\Psi_{s,u}^j = \left(g_1^j, ..., g_N^j\right).$$

Thus, the measurement equation is given by

$$SV_k^j = A^j + C^j X_k^j + \varepsilon_k^j, \tag{6.107}$$

where

$$X_k^j = X_{t_k}^j \text{ and } \varepsilon_k^j \sim N\left(0, G^j\right).$$

ε_k^j is assumed to be a sequence of i.i.d. random vectors. Now, let

$$\Delta t_{k+1} = t_{k+1} - t_k,\ k = 1, ..., n-1,$$

and approximate the variance vector $V^j\left(X_t^j, \Phi_{s,u}^j\right)$ over a subinterval $(t_k, t_{k+1}]$ by its value at the beginning of the subinterval, i.e. by $V_k^j = V^j\left(X_k^j, \Phi_{s,u}^j\right)$. Then the transition equation (6.106) can be approximated by

$$dX^j(t) = \left(H^j X^j(t) + J^j\right) dt + V_k^j dW_t^j,\ t \in (t_k, t_{k+1}]. \tag{6.108}$$

For $t \in (t_k, t_{k+1}]$ we rewrite equation (6.108) as

$$e^{\left(-H^j t\right)} dX_t^j - e^{\left(-H^j t\right)} H^j X_t^j dt = e^{\left(-H^j t\right)} \left[J^j dt + V_k^j dW_t^j \right]. \tag{6.109}$$

By applying the two-dimensional version of the Itô formula[45] to $e^{\left(-H^j t\right)} X_t^j$, we get

$$d\left(e^{\left(-H^j t\right)} X_t^j\right) = e^{\left(-H^j t\right)} dX_t^j - e^{\left(-H^j t\right)} H^j X_t^j dt. \tag{6.110}$$

Plugging equation (6.110) into equation (6.109) yields

$$\begin{aligned} & e^{\left(-H^j t_k\right)} X_k^j - e^{\left(-H^j t_{k+1}\right)} X_{k+1}^j \\ &= \int_{t_{k+1}}^{t_k} e^{\left(-H^j l\right)} J^j dl + \int_{t_{k+1}}^{t_k} e^{\left(-H^j l\right)} V_k^j dW_l^j, \end{aligned}$$

which is equivalent to

$$X_{k+1}^j = e^{H^j \Delta t_{k+1}} X_k^j - \int_{t_{k+1}}^{t_k} e^{H^j (t_{k+1} - l)} J^j dl - \int_{t_{k+1}}^{t_k} e^{H^j (t_{k+1} - l)} V_k^j dW_l^j.$$

Easy calculations yield our final transition equation

$$X_{k+1}^j = e^{H^j \Delta t_{k+1}} X_k^j + \int_0^{\Delta t_{k+1}} e^{H^j l} J^j dl + \int_0^{\Delta t_{k+1}} e^{H^j (\Delta t_{k+1} - l)} V_k^j dW_l^j. \tag{6.111}$$

The measurement equation (6.107) and transition equation (6.111) represent the state space formulation of our model. Now let $\widetilde{X}_{k+1|k}^j$ and $\widetilde{X}_{k+1}^j$ denote the optimal estimators[46] of the unknown state vector X_{k+1}^j based on the available information (i.e. the observed credit spreads) up to and including time t_k and t_{k+1}, respectively. The optimal estimators are the conditional means of X_{k+1}^j which are denoted by $E_k\left[X_{k+1}^j\right]$ and $E_{k+1}\left[X_{k+1}^j\right]$, respectively. Let Cov_k^j denote the covariance matrix of the estimation error, then

$$Cov_k^j = E_k\left[\left(X_k^j - \widetilde{X}_k^j\right)\left(X_k^j - \widetilde{X}_k^j\right)'\right].$$

Then the prediction equations are

$$\widetilde{X}_{k+1|k}^j = e^{H^j \Delta t_{k+1}} \widetilde{X}_k^j + \int_0^{\Delta t_{k+1}} e^{H^j l} J^j dl \tag{6.112}$$

and

[45] See, e.g., Oksendal (1998), p. 48.

[46] Here we mean optimal in the mean squared error sense.

$$\begin{aligned}
&Cov^j_{k+1|k} \\
&= E_k\left[(X^j_{k+1} - \widetilde{X}^j_{k+1|k})(X^j_{k+1} - \widetilde{X}^j_{k+1|k})'\right] \\
&= E_k\left[\left(e^{H^j \Delta t_{k+1}}\left(X^j_k - \widetilde{X}^j_k\right) + \int_0^{\Delta t_{k+1}} e^{H^j(\Delta t_{k+1}-l)} V^j_k dW^j_l\right)\right. \\
&\quad \left.\cdot\left(e^{H^j \Delta t_{k+1}}\left(X^j_k - \widetilde{X}^j_k\right) + \int_0^{\Delta t_{k+1}} e^{H^j(\Delta t_{k+1}-l)} V^j_k dW^j_l\right)'\right] \\
&= e^{H^j \Delta t_{k+1}} \cdot Cov^j_k \cdot e^{(H^j)' \Delta t_{k+1}} + \int_0^{\Delta t_{k+1}} e^{H^j l}\widetilde{V}^j_k \widetilde{V}'_k e^{(H^j)' l} dl, \quad (6.113)
\end{aligned}$$

where

$$\widetilde{V}^j_k = E_k\left[V^j_k\right] = \begin{pmatrix} \sigma^j_s\sqrt{\widetilde{s}^j_k} & 0 \\ 0 & \sigma^j_u\sqrt{\widetilde{u}^j_k} \end{pmatrix}.$$

Note, that X^j_k and the time increments of the standard Brownian motion are independent for $t \geq t_k$. If we rewrite the credit spread vector as

$$SV^j_{k+1} = A^j + C^j\left(X^j_{k+1} - \widetilde{X}^j_{k+1|k}\right) + C^j \widetilde{X}^j_{k+1|k} + \varepsilon^j_{k+1},$$

then the optimal estimator of the credit spread vector at time t_{k+1}, given all credit spreads up to time t_k, is given by

$$\widetilde{SV}^j_{k+1|k} = E_k\left[SV^j_{k+1}\right] = A^j + C^j \widetilde{X}^j_{k+1|k}.$$

If we denote the prediction error vector of the credit spreads by υ^j_{k+1}, then

$$\upsilon^j_{k+1} = SV^j_{k+1} - \widetilde{SV}^j_{k+1|k} = SV^j_{k+1} - A^j - C^j \widetilde{X}^j_{k+1|k}. \quad (6.114)$$

Obviously, given $SV^j_1, ..., SV^j_k$ the vector $\begin{pmatrix} X^j_{k+1} \\ SV^j_{k+1} \end{pmatrix}$ is conditionally normally distributed with conditional mean

$$E_k\left[\begin{pmatrix} X^j_{k+1} \\ SV^j_{k+1} \end{pmatrix}\right] = \begin{pmatrix} \widetilde{X}^j_{k+1|k} \\ \widetilde{SV}^j_{k+1|k} \end{pmatrix}$$

and conditional covariance matrix

$$\begin{aligned}
&Cov_k\left[\begin{pmatrix} X^j_{k+1} \\ SV^j_{k+1} \end{pmatrix}, \begin{pmatrix} X^j_{k+1} \\ SV^j_{k+1} \end{pmatrix}\right] \\
&= \begin{pmatrix} Cov^j_{k+1|k} & Cov_k\left[X^j_{k+1}, A^j + C^j X^j_{k+1}\right] \\ Cov_k\left[A^j + C^j X^j_{k+1}, X^j_{k+1}\right] & Cov_k\left[SV^j_{k+1}, SV^j_{k+1}\right] \end{pmatrix} \\
&= \begin{pmatrix} Cov^j_{k+1|k} & Cov^j_{k+1|k}\left(C^j\right)' \\ C^j Cov^j_{k+1|k} & F^j_{k+1} \end{pmatrix},
\end{aligned}$$

where - according to equation (6.114) -

$$
\begin{aligned}
F_{k+1}^{j} &= E_k\left[v_{k+1}^{j}\left(v_{k+1}^{j}\right)'\right] \\
&= E_k\left[C^j\left(X_{k+1}^{j}-\widetilde{X}_{k+1|k}^{j}\right)\left(X_{k+1}^{j}-\widetilde{X}_{k+1|k}^{j}\right)'\left(C^j\right)'\right]+G^j \\
&= C^j Cov_{k+1|k}^{j}\left(C^j\right)'+G^j.
\end{aligned}
$$

If we apply results from Harvey (1989), chapter 3, we get the updating equations

$$\widetilde{X}_{k+1}^{j} = \widetilde{X}_{k+1|k}^{j} + K_{k+1}^{j} v_{k+1}^{j} \tag{6.115}$$

and

$$Cov_{k+1}^{j} = Cov_{k+1|k}^{j} - K_{k+1}^{j} C^j Cov_{k+1|k}^{j}, \tag{6.116}$$

with the so called Kalman gain matrix

$$K_{k+1}^{j} = Cov_{k+1|k}^{j}\left(C^j\right)'\left(F_{k+1}^{j}\right)^{-1}. \tag{6.117}$$

In the update step, the additional information given by SV_{k+1}^{j} is used to get a better estimate of X_{k+1}^{j}, the so called filtered estimate. Hence, the Kalman filter is capable of estimating the unobservable short rates s^j and u^j from observable credit spreads. In addition, for parameter estimation, one can define the likelihood function L^j, evaluated with the Kalman filter (6.112)-(6.117), by

$$
\begin{aligned}
& L^j\left(\Phi_{s,u}^{j},\Psi_{s,u}^{j}|SV_1^j,\ldots,SV_n^j\right) \\
&= \prod_{k=1}^{n} f_{SV_k^j|SV_1^j,\ldots,SV_{k-1}^j}^{j}\left(\Phi_{s,u}^{j},\Psi_{s,u}^{j}\right) \\
&= \prod_{k=1}^{n} \frac{1}{\left(\sqrt{2\pi}\right)^N}\frac{1}{\sqrt{\det F_k^j}} \\
&\quad \cdot \exp\left\{-\frac{1}{2}\left(SV_k^j-\widetilde{SV}_{k|k-1}^{j}\right)'\left(F_k^j\right)^{-1}\left(SV_k^j-\widetilde{SV}_{k|k-1}^{j}\right)\right\},
\end{aligned}
$$

where $f_{SV_k^j|SV_1^j,\ldots,SV_{k-1}^j}^{j}\left(\Phi_{s,u}^{j},\Psi_{s,u}^{j}\right)$ denotes the conditional probability density function of SV_k^j given all information up to and including t_{k-1}. Up to a constant, the log-likelihood function can be expressed as

$$
\begin{aligned}
& \ln L^j\left(\Phi_{s,u}^{j},\Psi_{s,u}^{j}|SV_1^j,\ldots,SV_n^j\right) \\
&= -\frac{1}{2}\sum_{k=1}^{n}\ln(\det F_k^j) - \frac{1}{2}\sum_{k=1}^{n}\left(v_k^j\right)'\left(F_k^j\right)^{-1}v_k^j.
\end{aligned} \tag{6.118}
$$

The parameter vectors $\Phi_{s,u}^{j}$ and $\Psi_{s,u}^{j}$ can be estimated by maximizing the log-likelihood function (6.118) with respect to $\Phi_{s,u}^{j}$ and $\Psi_{s,u}^{j}$.

Parameter Estimation Results for s^j and $u^j, j \in \{\mathbf{I}, \mathbf{G}\}$. For the parameter estimation of the two-factor credit spread model, we use seven samples of data, each of length 150 for each country, i.e. in the Kalman filter algorithm we set $N = 7$ and $n = 150$. We don't consider a time series of length 300 as in the case of non-defaultable parameter estimation because the functions involved in the credit spread parameter estimation are far more complex. We use time series with daily data from September 30*th*, 1999, until May 30*th*, 2000, excluding several days for which there is no bond data available, the Nelson-Siegel estimates of the credit spreads appear to be negative or the bond data is identified as an outlier. We consider the times to maturity $\tau_i = i+2$, $i = 1, ..., 7$ (expressed in years). Thus, for each $j \in \{I, G\}$ the input data consists of 150 observed spread vectors

$$SV_k^j = \begin{pmatrix} S^j(t_k, t_k + \tau_1) \\ S^j(t_k, t_k + \tau_2) \\ \vdots \\ S^j(t_k, t_k + \tau_7) \end{pmatrix}, \quad 1 \leq k \leq 150.$$

For the iterative calculation of the log-likelihood function (6.118) we have to initialize the Kalman filter first, i.e. we have to guess some good starting values for $\widetilde{X}^j$ and Cov_0^j. Often they are set to their unconditional expectation and unconditional covariance matrix, respectively. In our case - as the state vector consisting of the short rate spread and the uncertainty index is not observable - we use starting values that are implied by the observable data. Solving the following system of two equations

$$S^j(t_0, t_0 + \tau_1) = a^j(\tau_1) + c^j(\tau_1)\widetilde{s}_0^j + d^j(\tau_1)\widetilde{u}_0^j,$$
$$S^j(t_0, t_0 + \tau_2) = a^j(\tau_2) + c^j(\tau_2)\widetilde{s}_0^j + d^j(\tau_2)\widetilde{u}_0^j,$$

yields $\widetilde{X}_0^j = \begin{pmatrix} \widetilde{s}_0^j \\ \widetilde{u}_0^j \end{pmatrix}$. We initialize the matrix Cov_0^j simply by setting it equal to $\begin{pmatrix} \frac{1}{2} & \frac{1}{2} \\ \frac{1}{2} & \frac{1}{2} \end{pmatrix}$. After initialization we perform the 150 Kalman filter iterations to calculate the log-likelihood function (6.118). Finally, we maximize this function with respect to the two parameter vectors $\Phi_{s,u}^j = \left(a_s^j, a_u^j, \sigma_s^j, \sigma_u^j, b_s^j, \theta_u^j, \lambda_s^j, \lambda_u^j\right)$ and $\Psi_{s,u}^j = \left(g_1^j, ..., g_7^j\right)$. To find the maximum we apply the multi-dimensional Downhill Simplex method of Nelder & Mead (1965). Tables 6.28 and 6.29 summarize the parameter estimates of our model derived from the observed credit spreads between Germany and Italy. The average mean reversion level of the Italian short rate credit spread and the mean reversion level of the Italian uncertainty index implied by our parameter estimates equal 25 basis points and 0.049%, respectively, which nicely fit to figures 6.33 and 6.34. In case of Greece the average mean reversion level of the Greek short rate credit spread and the mean reversion level of the Greek uncertainty index are given by 164 basis points and 0.65%.

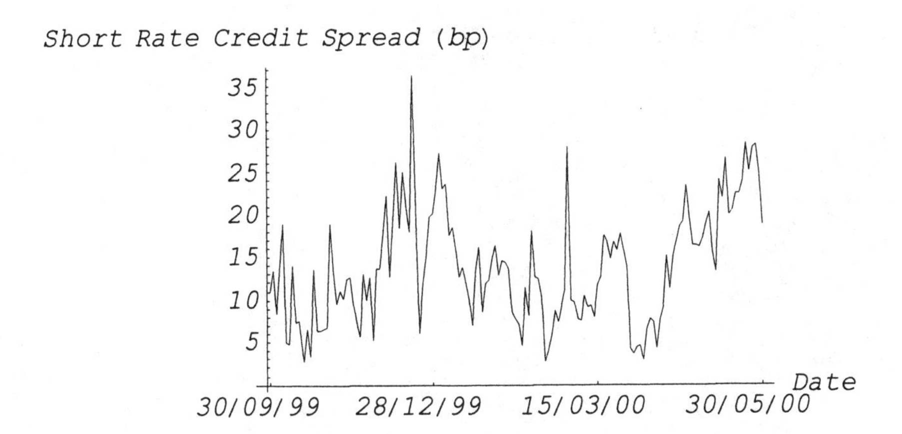

Fig. 6.33. Kalman filtered Italian short rate credit spread process. Time period: 150 days from September $30th$, 1999, until May $30th$, 2000.

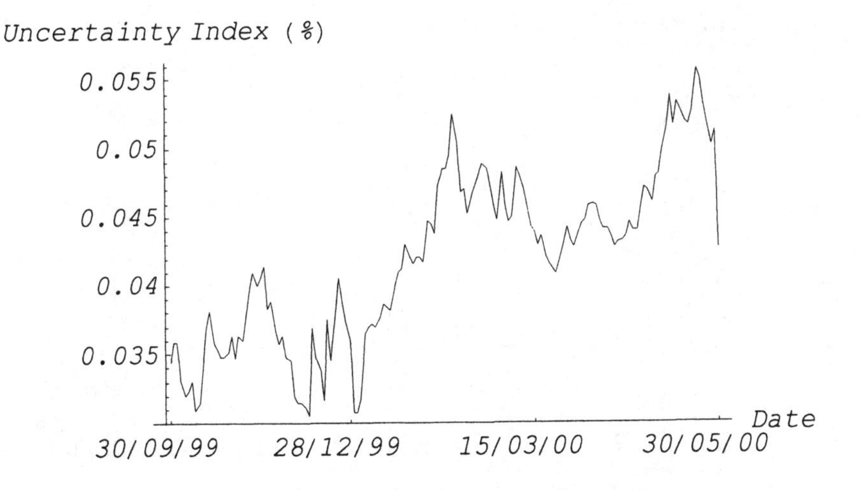

Fig. 6.34. Kalman filtered Italian uncertainty index process. Time period: 150 days from September $30th$, 1999, until May $30th$, 2000.

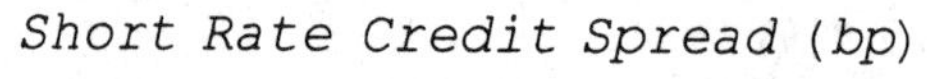

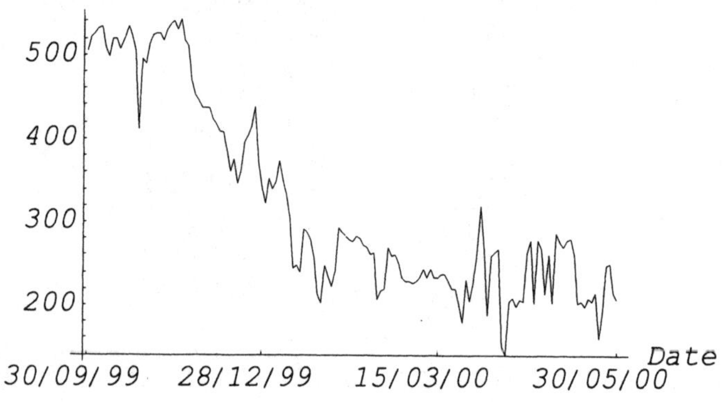

Fig. 6.35. Kalman filtered Greek short rate credit spread process. Time period: 150 days from September 30*th*, 1999, until May 30*th*, 2000.

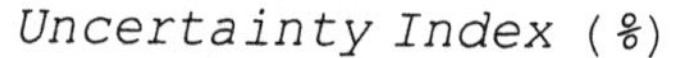

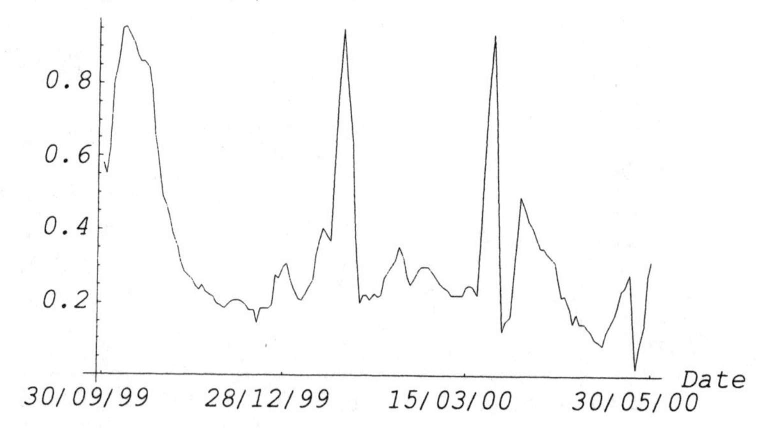

Fig. 6.36. Kalman filtered Greek uncertainty index process. Time period: 150 days from September 30*th*, 1999, until May 30*th*, 2000.

Table 6.28. Kalman filter maximum likelihood parameter estimates for Italy: Part I. Sample period: 150 days from September 30th, 1999, until May 30th, 2000.

Parameter	g_1^I	g_2^I	g_3^I
Estimation	0.000418245	0.000289569	0.000236120
g_4^I	g_5^I	g_6^I	g_7^I
0.000225986	0.000057738	0.000143889	0.000241359

Table 6.29. Kalman filter maximum likelihood parameter estimates for Italy. Part II: Parameters of the short rate credit spread and uncertainty index processes. Sample period: 150 days from September 30th, 1999, until May 30th, 2000.

Parameter	**Estimation**	**Parameter**	**Estimation**
b_s^I	0.273903738	θ_u^I	0.000033835
a_s^I	0.053199919	a_u^I	0.068884441
σ_s^I	0.158772058	σ_u^I	0.030100786
λ_s^I	−1.868477669	λ_u^I	−1.190989999

Table 6.30. Kalman filter maximum likelihood parameter estimates for Greece: Part I. Sample period: 150 days from September 30th, 1999, until May 30th, 2000.

Parameter	g_1^G	g_2^G	g_3^G
Estimation	0.000889889	0.000557789	0.000302059
g_4^G	g_5^G	g_6^G	g_7^G
0.000240028	0.000374741	0.000452236	0.000513706

Table 6.31. Kalman filter maximum likelihood parameter estimates for Greece. Part II: Parameters of the short rate credit spread and uncertainty index processes. Sample period: 150 days from September 30th, 1999, until May 30th, 2000.

Parameter	**Estimation**	**Parameter**	**Estimation**
b_s^G	0.373241211	θ_u^G	0.00041234
a_s^G	0.147542315	a_u^G	0.06342136
σ_s^G	0.456662213	σ_u^G	0.07121165
λ_s^G	−0.23211112	λ_u^G	−1.1235121

Testing the Assumptions on the Italian and Greek Kalman Filter Residuals. First we test the assumption that the standardized Kalman filter residuals are normally distributed. We apply the test statistic (6.95) to the residuals with the critical values given in table 6.20 (column $\chi^2(2)$) and find:

- For Italy: According to table 6.32 we don't reject the normal distribution hypothesis on the 5%−significance level for maturities of 3, 4, 5, and 6 years. In addition to 3, 4, 5, and 6 years, we don't reject the normal dis-

Table 6.32. Test for normal distribution of the Italian standardized Kalman filter residuals. Sample period: 150 days from September 30th, 1999, until May 30th, 2000.

Mat.	3 years	4 years	5 years	6 years	7 years	8 years	9 years
L	1.79604	1.9342	1.0273	0.86506	7.9013	12.113	10.911

Table 6.33. Test for normal distribution of the Greek standardized Kalman filter residuals. Sample period: 150 days from September 30th, 1999, until May 30th, 2000.

Mat.	3 years	4 years	5 years	6 years	7 years	8 years	9 years
L	5.88917	3.16618	0.187345	4.91219	4.94	2.7143	1.0266

tribution hypothesis on a 1%−significance level for 7 years. We reject the hypothesis on the 1%−significance level for maturities of 8 and 9 years.
- For Greece: According to table 6.33 we don't reject the normal distribution hypothesis on the 5%−significance level for all maturities under consideration.

To test the standardized residuals for serial correlation we apply the Box-Ljung test on 1% and 5%- significance levels to different maturities. Under the null hypothesis that the standardized Kalman filter residuals for the specific maturities are not correlated, the Box-Ljung Q−statistic $Q(L)$ is asymptotically χ^2−distributed with

$$L - number\ of\ estimated\ parameters$$

degrees of freedom. As it is necessary, that the lag L grows with the number of observations, we choose $L = 2\sqrt{n} = 2\sqrt{150} \approx 24$. We reject the null hypothesis on the 1% significance level only if the test statistic is greater than the 99%-quantile of 21.67, or on the 5% significance level if the test statistic is greater than the 95%-quantile of 16.92. Table 6.34 summarizes the results of the Italian Q−statistics for several maturities: for the maturities of 4, 5, 8, and 9 years we find that the correlation is not significantly different from zero on the 5%-significance level, and for the maturities of 3, 6, and 7 years the correlation of the residuals is not significantly different from zero on the 1%-significance level. Table 6.35 shows the results of the Greek Q−statistics for several maturities: for the maturities of 4, 5, 8, and 9 years we find that the correlation is not significantly different from zero on the 1%-significance level.

Table 6.34. Box-Ljung tests for Italian standardized Kalman filter residuals. Sample period: 150 days from September 30th, 1999, until May 30th, 2000.

Mat. (years)	3	4	5	6	7	8	9
Q-statistic (Box-Ljung)	21.52	14.95	15.96	18.38	21.16	14.24	13.88

Table 6.35. Box-Ljung Tests for Greek standardized Kalman filter residuals. Sample period: 150 days from September 30th, 1999, until May 30th, 2000.

Mat. (years)	3	4	5	6	7	8	9
Q-statistic (Box-Ljung)	26.93	20.79	18.96	29.95	21.87	18.12	18.95

Table 6.36. Test for homoscedasticity of the Italian standardized Kalman filter residuals. Sample period: 150 days from September 30th, 1999, until May 30th, 2000.

Maturity	H-statistic: H (50)
3 years	0.81934
4 years	0.459982
5 years	0.526177
6 years	0.611612
7 years	0.700424
8 years	0.791193
9 years	0.846494

Finally, we use the test for homoscedasticity as described in Harvey (1981), page 157. The 1% and 5% critical values are given in table 6.22. According to table 6.36, for Italy we don't reject the hypothesis on the 5%−significance level for the maturities of 3, 6, 7, 8 and 9 years. In addition, we don't reject the hypothesis of homoscedasticity on the 1%−level for the maturity of 5 years. According to table 6.37, for Greece we don't reject the hypothesis on the 5%−significance level for the maturities of 3, 6, 7, 8 and 9 years. In addition, we don't reject the hypothesis of homoscedasticity on the 1%−level for the maturities of 4 and 5 years.

Conclusion 6.6.4
Overall, based on the various tests we can conclude that the assumptions on the Kalman filter algorithm are sufficiently satisfied by our Italian and Greek data sets.

Testing of the Model Performance Based on the Italian and Greek Spread Data. In the following, we test the model performance by applying

Table 6.37. Test for homoscedasticity of the Greek standardized Kalman filter residuals. Sample period: 150 days from September 30th, 1999, until May 30th, 2000.

Maturity	H-statistic: H (50)
3 years	0.923415
4 years	0.489632
5 years	0.538179
6 years	0.937615
7 years	0.887592
8 years	1.191873
9 years	0.976572

a test, e.g., used by (Titman & Torous 1989). We test the power of the model to explain the changes of observed credit spreads. If the results are satisfactory we can conclude that the model can price defaultable bonds and credit spreads accurately. Let

$$\Delta S_k^j(\tau) = S^j(t_k, t_k + \tau) - S^j(t_{k-1}, t_{k-1} + \tau), \; 2 \leq k \leq 150, \; j \in \{I, G\},$$

denote the changes of the observed credit spreads with time to maturity τ, and let $\widehat{\Delta S_k^j}(\tau)$, $2 \leq k \leq 150$, $j \in \{I, G\}$ denote the changes of the corresponding fitted credit spreads. The test equation is given by the following linear regression:

$$\Delta S_k^j(\tau) = a^j + b^j \widehat{\Delta S_k^j}(\tau) + \varepsilon_k^{j,*}, \text{ where } \varepsilon_k^{j,*} \sim N\left(0, \left(h^{j,*}\right)^2\right), \; j \in \{I, G\}. \tag{6.119}$$

We apply the test to the maturities $\tau = 3, ..., 9$ (all in years). If $\widehat{\Delta S_k^j}(\tau)$ is an unbiased estimate for $\Delta S_k^j(\tau)$, then a^j and b^j should be close to 0 and 1, respectively. In addition, the adjusted $\overline{R}^2$ statistics are a measure for the quality of the model. Table 6.38 shows the results of the regression for the Italian data:

- The test of the hypothesis that $a^I = 0$ is not rejected for all maturities on a 5%−significance level.
- The parameter values for b^I are always close to 1.
- The test of the joint hypothesis that $a^I = 0$ and $b^I = 1$ is not rejected on a 5%−significance level for all maturities but 6 years. Hence, the model prices these zero coupon bonds with acceptable accuracy.
- The adjusted $\overline{R}^2$ values vary between 71.7 % and 97.3 %.

Table 6.39 shows the results of the regression for the Greek data:

- The test of the hypothesis that $a^G = 0$ is not rejected for all maturities on a 5%−significance level.

Table 6.38. Test of in-sample model explanatory power for Italy: Linear regression results.

Mat.	a^I	b^I	$\overline{R}^2$	p-Value ($a^I = 0$)	p-Value ($a^I = 0, b^I = 1$)
3 years	$6.11333 \cdot 10^{-6}$	0.938757	0.71706	0.814845	0.440907
4 years	$7.63096 \cdot 10^{-6}$	1.03916	0.83112	0.713915	0.452307
5 years	$7.73356 \cdot 10^{-6}$	1.06541	0.90744	0.610017	0.055909
6 years	$6.96232 \cdot 10^{-6}$	1.05084	0.96796	0.414893	0.004006
7 years	$5.64142 \cdot 10^{-6}$	1.01643	0.97338	0.452594	0.355994
8 years	$3.9653 \cdot 10^{-6}$	0.975102	0.88281	0.80251	0.320922
9 years	$2.05174 \cdot 10^{-6}$	0.934458	0.71720	0.937439	0.396085

Table 6.39. Test of in-sample model explanatory power for Greece: Linear regression results.

Mat.	a^G	b^G	$\overline{R}^2$	p-Value ($a^G = 0$)	p-Value ($a^G = 0, b^G = 1$)
3 years	$2.4842 \cdot 10^{-6}$	0.73553	0.82381	0.980181	0
4 years	$2.1248 \cdot 10^{-6}$	0.89199	0.94441	0.960677	0.01273
5 years	$3.1223 \cdot 10^{-6}$	0.86653	0.99282	0.807109	0.01196
6 years	$4.8976 \cdot 10^{-6}$	0.89471	0.97768	0.805628	0.03453
7 years	$7.0636 \cdot 10^{-6}$	0.90389	0.93128	0.825408	0.05342
8 years	$9.3651 \cdot 10^{-6}$	0.86207	0.87587	0.818706	0.01627
9 years	0.0000116379	0.8776	0.81765	0.809207	0.02481

- The parameter values for b^G are not very close to 1. Besides for the 7 year credit spreads the distance of all b^G values from 1 is greater than 0.1.
- The test of the joint hypothesis that $a^G = 0$ and $b^G = 1$ is not rejected on a 5%−significance level for all maturities but 6 years. Hence, the model prices these zero coupon bonds with acceptable accuracy.

Conclusion 6.6.5
The out-of-sample behavior is very different for Italy and Greece. For Italy the model behaves very well. For Greece the joint hypothesis that $a^G = 0$ and $b^G = 1$ is not rejected only for the 7 year credit spreads on a 5%−significance level. The reason is quite obvious: for Italy we use zero coupon bonds whereas for Greece we have only coupon bonds available which induces additional noise. Furthermore, the Greek bonds are far less liquid than the Italian bonds.

Out-of-Sample Testing Based on the Italian and Greek Spread Data. In the previous section we did some in-sample-testing to examine the performance of the defaultable term structure model. To complete the analysis of the model we add an out-of-sample-test. We analyze if we can apply the model with the estimated parameters to the pricing of credit spreads after May 30*th*, 2000. Therefore we consider 70 additional time steps $t_k^* = 1, ..., 70$, for $k = 1, ..., 70$, from May 30*th*, 2000, until October 23*rd*, 2000, and apply the Kalman filter to the observed credit spreads for this time period. Note, that we keep the parameters of tables 6.28, 6.29, 6.30, and 6.31 fixed, i.e. we filter the values for $s^j(t)$ and $u^j(t)$, $j \in \{I, G\}$, and determine the credit spread predictors with given values for the vectors $\Phi^j_{s,u}$, $\Psi^j_{s,u}$, $j \in \{I, G\}$. To test the performance of the out-of-sample model we apply the linear regression (6.119) to the new data. We test the maturities $\tau_i = i + 2$, $i = 1, ..., 7$ (expressed in years). Table 6.40 shows the results of the regression for Italy:

- The test of the hypothesis that $a^I = 0$ is not rejected for all maturities on a 5%-significance level.
- The test of the joint hypothesis that $a^I = 0$ and $b^I = 1$ is not rejected on a 1%−significance level for all maturities, on a 5%-significance level for all maturities but 6 and 9 years. Hence, the model prices these zero coupon bonds with acceptable accuracy.
- The adjusted $\overline{R}^2$ values vary between 68.9 % and 95.2 %.

Overall the test suggests that the out-of-sample behavior of the model is very encouraging for the Italian data set.
Table 6.41 shows the results of the regression for Greece:

- The test of the hypothesis that $a^G = 0$ is not rejected on a 5%-significance level for all maturities but the 5 year credit spreads.
- The test of the joint hypothesis that $a^G = 0$ and $b^G = 1$ is not rejected on a 1%−significance level for all maturities but the 5 year credit spreads, on a 5%-significance level for all maturities but 3, 4, and 5 years. Hence, the model prices these zero coupon bonds with acceptable accuracy.
- The adjusted $\overline{R}^2$ values vary between 78.4 % and 99.3 %.

Conclusion 6.6.6
As with the in-sample tests, the model behaves better for the Italian than for the Greek data. But overall we find that the three-factor defaultable term structure model is able to describe real world credit spreads realistically.

6.7 Portfolio Optimization under Credit Risk

6.7.1 Introduction

The process of performing an optimal asset allocation basically deals with the problem of finding a portfolio that maximizes the expected utility of the

Table 6.40. Test of out-of-sample model explanatory power for Italy: Linear regression results.

Mat.	a^I	b^I	$\overline{R}^2$	p-Value ($a^I = 0$)	p-Value ($a^I = 0, b^I = 1$)
3 years	0.0000150071	0.90796	0.689351	0.828829	0.471914
4 years	0.0000109933	1.03144	0.802159	0.852410	0.134735
5 years	0.0000104826	1.07602	0.888956	0.813476	0.265945
6 years	0.0000122546	1.07674	0.952238	0.667574	0.037469
7 years	0.0000153419	1.05566	0.951507	0.590850	0.145532
8 years	0.0000190276	1.02615	0.862036	0.701499	0.219308
9 years	0.000022813	0.995921	0.714741	0.766278	0.044077

Table 6.41. Test of out-of-sample model explanatory power for Greece: Linear regression results.

Mat.	a^G	b^G	$\overline{R}^2$	p-Value ($a^G = 0$)	p-Value ($a^G = 0, b^G = 1$)
3 years	0.000038583	0.89947	0.899371	0.546014	0.028564
4 years	0.000031957	0.89833	0.969776	0.283150	0.015647
5 years	0.000029042	0.90245	0.992684	0.027074	0.009645
6 years	0.000028369	0.91061	0.987908	0.064070	0.054732
7 years	0.000028880	0.92303	0.953614	0.303275	0.214875
8 years	0.000029837	0.94031	0.883109	0.493757	0.382732
9 years	0.000030760	0.96279	0.783838	0.610533	0.198473

portfolio manager. As long as it is supposed that the returns of the portfolio assets follow a normal distribution, the return distribution of any portfolio will be normal, too. In this case, as it is done throughout the traditional portfolio theory introduced by Markowitz (1952) and Sharpe (1964), the problem of finding an expected utility maximizing portfolio for a risk averse portfolio manager, represented by a concave utility function, can be restricted to finding an optimal combination of the two parameters mean and variance. This dramatically simplifies the whole asset allocation process and is known as mean-variance analysis. It is the aim of the portfolio manager to find a portfolio that maximizes the expected return given a specific risk level or a portfolio that minimizes the risk given a specific return level. In this case risk is measured by the variance of the portfolio return. Unfortunately, selection rules based on the two parameters mean and variance are of limited generality. They are optimal only if the utility function is quadratic or the return distribution is normal. Furthermore, the use of variance as a measure of risk for asymmetric return distributions has already been questioned by financial theorists like, e.g., Markowitz (1991), pages 188-201. He states that portfolio analyses based on semi-variance tend to produce better portfolios than those based on variance. Unfortunately, the computing cost involved with the use of semi-variance is much higher than with the use of variance. We need the entire joint distribution for the first technique while we only need the covariance matrix for the latter.

Extensive research has been done to derive concepts for ordering uncertain prospects resulting in principles like the stochastic dominance of order 1, 2 or 3 (see, e.g., Bawa (1975) or Martin, Cox & MacMinn (1988)). Bawa argues that it is quite reasonable to assume that the average investor decides consistent with a finite, increasing and concave utility function with decreasing absolute risk aversion, and that there is a strong rationale for using mean-lower partial variance rules as an approximation for the portfolio selection of this kind of investor. Under this rule one portfolio is better than the other if its mean is not lower and its lower partial variance is not higher than that of the other portfolio at every possible benchmark. As the lower partial variance is a generalization of the semi-variance, this follows the suggestion of Markowitz, still leaving the computational problems, because we have to calculate the lower partial variances for each distribution at every possible benchmark. If we can not assume special distributions to simplify that problem, the solution is often approximated by optimizing with a finite set of representative benchmarks. The so-called lower partial moments can be considered as a generalization of the lower partial variance (see, e.g., Harlow & Rao (1989) or Harlow (1991)). Therefore, they are consistent with the previous concept. Because they measure the risk of falling below a given benchmark, they have become very popular measures for the risk of a portfolio, especially when the return distribution is asymmetric. Consequently, we will use lower partial moments to control the downside risk of the portfolio exposure. Because we do not have any information on the distribution of the possible or optimal portfolios we approximate the distribution function based on a simulation model. The asset universe in our sample are sovereign bonds with different times to maturity and qualities of the issuers. Some of these bonds may default and thus stop paying interest. Or the value of a bond may fall only due to a downgrading of the issuing country.

We use our three-factor defaultable term structure model with $\beta = 0$ introduced in section 6.2 for pricing sovereign bonds. Based on parameter estimates, determined by Kalman filter techniques, we simulate the prices of the bonds for a future time horizon. For each future time step and for each given portfolio composition these scenarios yield distributions of future cash flows and portfolio values. In section 6.7.2 we show how the portfolio composition can be optimized by maximizing the expected final value or return of the portfolio under a set of constraints. One set of restrictions is due to a minimum required cash flow per period to cover the liabilities of a company. The second set limits the tolerated risk. As discussed above, both risks are measured by lower partial moments to account for the downside potential we have to consider. To visualize our methodology, in section 6.7.3 we present a case study for a portfolio consisting of various German, Italian, and Greek sovereign bonds.

6.7.2 Optimization

This section describes the problem of maximizing the final value of a portfolio under limited risk. We assume that there is a planning horizon $T \in [0, T^*]$. The universe of assets consists of n different types of sovereign bonds $i \in \{1, ..., n\}$ from Germany, Italy (I), and Greece (G) with maturities $0 \leq T_1, ..., T_n \leq T^*$. The prices of the bonds at any time $t \in [0, T]$ are given by $P_1(t), ..., P_n(t)$. Note, that in this context $P_i(t)$ is a general notation for the price of a bond - not a priori distinguishing between defaultable and non-defaultable bonds, or zero coupon and coupon bonds, i.e. $P_i(t) \in \{P(t, T_i), P^{d,I}(t, T_i), P^{d,G}(t, T_i), P_c(t, T_i), P_c^{d,I}(t, T_i), P_c^{d,G}(t, T_i)\}$. $\varphi = (\varphi_1, ..., \varphi_n)$ denotes the vector of the portfolio weights. In addition to the bond positions there is a cash account whose value at time $t \in [0, T]$ is denoted by $C_0(\varphi, t)$.

We consider three sources of risk:

- Market Risk caused by changing German interest rates over time.
- Spread Risk caused by the possibility of changes of the qualities of the Greek and Italian sovereign bonds.
- Default Risk caused by the possibility of default. As in the previous sections we assume Germany to be without any default risk. If we decide to invest in default risky sovereign bonds we may lack in coupon payments which we do not receive in case of default[47].

We want to find a portfolio allocation that satisfies certain limit conditions. More precisely, we set the following restrictions on our portfolio:

- A restriction on the portfolio value: We don't want the portfolio value fall below some given benchmark B at specific times $t \in \mathcal{T}_{\mathsf{B}} = \{T_1^{\mathsf{B}}, ..., T_{m_{\mathsf{B}}}^{\mathsf{B}}\} \subset [0, T]$, $m_{\mathsf{B}} \in \mathbb{N}$.
- A restriction on the cash flows: We assume a stream of liabilities L_t at times $t \in \mathcal{T}_{\mathsf{L}} = \{T_1^{\mathsf{L}}, ..., T_{m_{\mathsf{L}}}^{\mathsf{L}}\} \subset [0, T]$, $m_{\mathsf{L}} \in \mathbb{N}$, which have to be covered by the coupon payments of the bond portfolio. Because these coupon payments are under default risk we also set a restriction on the cash flows. We don't want them fall below the corresponding liability.

In addition, to the definition of the risk sources and the limits we make the following assumptions:

[47] Although we have to admit that in practice the default risk of Italy and Greece is almost zero.

Assumption 6.7.1

1. *All coupon payments, received between liability payment dates are given to a cash account earning continuous interest of* R_{ca}*, i.e. a cash flow* $C_i(t)$ *received from bond* $i \in \{1, ..., n\}$ *at time* $t \in \left(T_{l-1}^{\mathsf{L}}, T_l^{\mathsf{L}}\right]$, $l \in \{1, ..., m_{\mathsf{L}}\}$, $T_0^{\mathsf{L}} = 0$*, will have a cash value of*
$$C_i\left(t, T_l^{\mathsf{L}}\right) = C_i(t)\, e^{R_{ca}\left(t, T_l^{\mathsf{L}}\right)\left(T_l^{\mathsf{L}} - t\right)}$$
on the cash account at time T_l^{L}.
2. *Country* $j \in \{I, G\}$ *defaults as soon as the corresponding uncertainty process* u^j *crosses some default boundary* $\xi^{d,j}$*. The default boundary is implicitly given and can be evaluated from the default probability of country* j*. Therefore, the time of the default event can be expressed as* $T^{d,j} = \inf\left\{t \in [0, T] : u_t^j > \xi^{d,j}\right\}$*. As usual, we denote the default indicator function of country* $j \in \{I, G\}$ *by* H^j.
3. *In case of a default of a coupon bond* i *with price* $P_i(t) = P_c^{d,j}(t, T_i)$ *and maturity* $T_i \in \left[T^{d,j}, T^*\right]$ *before time* T *there are no more coupon payments, but a recovery payment of* $wP_c^{d,j}\left(T^{d,j}-, T_i\right)$ *fixed at time* $T^{d,j}$*. As usual,* $w \in [0, 1]$ *denotes the recovery rate and* $P_c^{d,j}\left(T^{d,j}-, T_i\right)$ *is the price of the bond just before default.*
4. *The recovery payment is not available before the original maturity time* T_i *of the defaulted bond. Then the recovery payment causes a cash flow of*
$$Z\left(T^{d,j}, T_i\right) = e^{R_{ca}^{d,j}\left(T^{d,j}, T_i\right)\left(T_i - T^{d,j}\right)} wP_c^{d,j}\left(T^{d,j}-, T_i\right)$$
on the cash account. $R_{ca}^{d,j}\left(T^{d,j}, T_i\right)$ *denotes the interest rate earned on the recovery payment for the time period* $\left[T^{d,j}, T_i\right]$.

Hence, at each coupon payment date $0 \leq t_{i1} < \cdots < t_{in_i} = T_i$, if $j \in \{1, ..., n_i - 1\}$, the cash flow of a defaultable bond is

$$C_i^d(t_{ij}) = C_i(t_{ij}) \cdot \left(1 - H_{t_{ij}}^j\right)$$

and

$$C_i^d(T_i) = C_i(T_i)\left(1 - H_{T_i}\right) + H_{T_i} Z\left(T^{d,j}, T_i\right).$$

The cash flows within each liability period $\left(T_{l-1}^{\mathsf{L}}, T_l^{\mathsf{L}}\right]$, $l \in \{1, ..., m_{\mathsf{L}}\}$, are then put in the cash account accruing to a cash value of

$$C_{i0}\left(T_{l-1}^{\mathsf{L}}, T_l^{\mathsf{L}}\right) = \sum_{\substack{j \in \{1, ..., n_i\} \\ t_{ij} \in \left(T_{l-1}^{\mathsf{L}}, T_l^{\mathsf{L}}\right]}} C_i^d(t_{ij})\, e^{R_{ca}\left(t_{ij}, T_l^{\mathsf{L}}\right)\left(T_l^{\mathsf{L}} - t_{ij}\right)}$$

at the liability payment dates $T_l^{\mathsf{L}} \in \mathcal{T}_{\mathsf{L}}$. In case of a non-defaultable bond this value is simply

$$C_{i0}\left(T_{l-1}^{\mathsf{L}},T_l^{\mathsf{L}}\right)=\sum_{\substack{j\in\{1,...,n_i\}\\ t_{ij}\in\left(T_{l-1}^{\mathsf{L}},T_l^{\mathsf{L}}\right]}} C_i\left(t_{ij}\right)e^{R_{ca}\left(t_{ij},T_l^{\mathsf{L}}\right)\left(T_l^{\mathsf{L}}-t_{ij}\right)}.$$

Hence, the time $T_l^{\mathsf{L}}\in\mathcal{T}_{\mathsf{L}}$ cash value of the portfolio cash flows within period $\left(T_{l-1}^{\mathsf{L}},T_l^{\mathsf{L}}\right]$, $l\in\{1,...,m_{\mathsf{L}}\}$, is given by

$$C_0\left(\varphi,T_{l-1}^{\mathsf{L}},T_l^{\mathsf{L}}\right)=\sum_{i=1}^{n}\varphi_i C_{i0}\left(T_{l-1}^{\mathsf{L}},T_l^{\mathsf{L}}\right)$$

leading to a total value of the cash account at time T_l^{L} of

$$C_0\left(\varphi,T_l^{\mathsf{L}}\right)=C_0\left(\varphi,T_{l-1}^{\mathsf{L}}\right)e^{R_{ca}\left(T_{l-1}^{\mathsf{L}},T_l^{\mathsf{L}}\right)\left(T_l^{\mathsf{L}}-T_{l-1}^{\mathsf{L}}\right)}+C_0\left(\varphi,T_{l-1}^{\mathsf{L}},T_l^{\mathsf{L}}\right)-L\left(T_l^{L}\right)$$

for all $l\in\{1,...,m_{\mathsf{L}}\}$. We set

$$C_0\left(\varphi,0\right)=\varphi_0\left(0\right)=C_0-\sum_{i=1}^{n}\varphi_i P_i\left(0\right)$$

with C_0 denoting the initial budget to be invested at time $t=0$.

Having estimated the parameters of the stochastic processes r, s^I, u^I, s^G, and u^G we now consider the vector of risk factors

$$\mathsf{F}=\left(\mathsf{F}_1,...,\mathsf{F}_5\right)=\left(r,s^I,u^I,s^G,u^G\right).$$

The optimization problem is based on a simulation of the vector F over time. For this simulation, let $V_i\left(0\right)$ and $V_i\left(t\right)$ denote the value of a given stochastic variable V_i at times 0 and t, respectively. In our case, this may be either the cash account ($i=0$) or the dirty price of bond i ($i=1,...,n$) plus potential coupon payments between times 0 and t, everything seen from time 0. According to our model we know that the values $V_i\left(t\right)$ are dependent on the vector F of risk factors and that the functional relation between the risk vector and the future value of V_i is known, i.e.

$$V_i\left(t\right)=V_i\left(\mathsf{F},t\right),\ i=1,...,n.$$

Under this assumption we simulate the risk vector F and get values

$$\mathsf{F}^k=\left(\mathsf{F}_1^k,...,\mathsf{F}_5^k\right)$$

with probabilities $p_k>0$, $k=1,...,K$. Dependent on these simulated values of risk factors we get values

$$V_i^k\left(t\right)=V_i\left(\mathsf{F}^k,t\right),\ k=1,...,K,$$

for each $i = 1, .., n$. For each possible bond portfolio $\varphi = (\varphi_1, ..., \varphi_n)$, which we consider to be fixed from time 0 until the end of the planning horizon T, the future time t portfolio value $V(\varphi, t)$ is given by the random variable

$$V(\varphi, t) = \sum_{i=1}^{n} \varphi_i V_i(t)$$

with $V(\varphi, 0)$ denoting the portfolio value at time 0. The simulated portfolio value $V^k(\varphi, t)$ based on the values $V_i^k(t)$ is given by

$$V^k(\varphi, t) = \sum_{i=1}^{n} \varphi_i V_i^k(t), \; k = 1, ..., K.$$

To restrict the downside risk of the future portfolio value at time $t \in \mathcal{T}_\mathsf{B}$ we consider the discrete version of the lower partial moment of order $l \in \mathbb{N}$. Hereby, the investor specific benchmark $\mathsf{B}_t \in \mathbb{R}$ is defined by

$$LPM_l(\varphi, V, \mathsf{B}, t) = \sum_{\substack{k=1,...,K \\ V^k(\varphi,t) < \mathsf{B}_t}} p_k \cdot \left(\mathsf{B}_t - V^k(\varphi, t)\right)^l. \tag{6.120}$$

The lower partial moment only considers realizations of the future value of V below the investor specific benchmark measured to a power of l. For $l = 0$ this is the probability that the random future value falls below the given benchmark, which is referred to as shortfall probability. Setting the benchmark equal to 0 this is the probability of loss. For $l = 1$, the lower partial moment is the expected deviation of the future values below the benchmark, sometimes called (expected) regret. For $l = 2$, the lower partial moment is weighting the squared deviations below the benchmark and thus is the semi-variance if the benchmark is set equal to the expected future value. For the ease of exposition we assume that the simulated lower partial moment is equal to the real one. If this is not the case the equality has to be interpreted as an approximation where we suppose that the number and quality of the simulations is sufficiently high. For a more detailed discussion of lower partial moments see, e.g., Harlow (1991) or Zagst (2001).
A portfolio manager or trader may be restricted to specific trading limits. Therefore, we introduce absolute lower (s_i) and upper bounds (S_i) for the amount φ_i of security $i = 1, ..., n$ in the portfolio and claim that

$$s_i \leq \varphi_i \leq S_i, \; i = 1, ..., n.$$

For $A_\mathsf{B}^l(t), A_\mathsf{L}^l(t) \in \mathbb{R}$ and $l \in \{0, 1, 2\}$, let us consider the following optimization problem

$$(P)\begin{cases} \sum_{i=1}^{n} \varphi_i \cdot E^{\mathbf{P}}\left[V_i(T)\right] \rightarrow \max \\ LPM_l\left(\varphi, V, \mathsf{B}, t\right) \leq A_{\mathsf{B}}^l(t), \; l \in \{0,1,2\}, t \in T_{\mathsf{B}} \\ LPM_l\left(\varphi, C_0, 0, t\right) \leq A_{\mathsf{L}}^l(t), \; l \in \{0,1,2\}, t \in T_{\mathsf{L}} \\ s_i \leq \varphi_i \leq S_i, \; i = 0, ..., n \\ C_0\left(\varphi, T_l^{\mathsf{L}}\right) = C_0\left(\varphi, T_{l-1}^{\mathsf{L}}\right) \cdot e^{R_{ca} \cdot \left(T_l^{\mathsf{L}} - T_{l-1}^{\mathsf{L}}\right)} + \\ \qquad + C_0\left(\varphi, T_{l-1}^{\mathsf{L}}, T_l^{\mathsf{L}}\right) - L\left(T_l^{\mathsf{L}}\right), \; l = 1, ..., m_{\mathsf{L}} \\ C_0(\varphi, 0) = \varphi_0 = C_0 - \sum_{i=1}^{n} \varphi_i \cdot V_i(0). \end{cases}$$

Note, that we have set $R_{ca}\left(T_{l-1}^{\mathsf{L}}, T_l^{\mathsf{L}}\right) \equiv R_{ca}$ for the ease of exposition. To implement the optimization problem let us have a closer look at the general LPM constraint, i.e. the LPM restriction for the future value V. We choose the numbers $m_k < 0$ to be sufficiently small and the numbers $M_k > 0$, $m_k^l > 0$ and $M_k^l > 0$ to be sufficiently large and define the constraints

$$M_k y_k + V^k(\varphi, t) \geq \mathsf{B}_t \tag{A}$$

$$m_k(1 - y_k) + V^k(\varphi, t) < \mathsf{B}_t \tag{B}$$

$$0 \leq \left(V^k(\varphi, t) - \mathsf{B}_t\right)^l + (-1)^{l-1} w_k^l \leq M_k^l (1 - y_k) \tag{C}$$

$$0 \leq w_k^l \leq m_k^l y_k \tag{D}$$

with $w_k^l \in \mathbb{R}$, $y_k \in \{0,1\}$ for all $k = 1, ..., K$, where we consider $l \in \{0,1,2\}$ to be arbitrary but fixed as well as $t \in (0,T]$. A proof of the following lemma can be found in Zagst (2001).

Theorem 6.7.1.
Let $t \in (0,T]$, and $l \in \{0,1,2\}$ be arbitrary but fixed and $y_k \in \{0,1\}$ for all $k = 1, ..., K$.

a) Let condition (A) be satisfied. If $V^k(\varphi, t) < \mathsf{B}_t$, then $y_k = 1$ for all $k = 1, ..., K$.
b) Let condition (B) be satisfied. If $V^k(\varphi, t) \geq \mathsf{B}_t$, then $y_k = 0$ for all $k = 1, ..., K$.
c) Let conditions (A) and (B) be satisfied. Then, for all $k = 1, ..., K$:

$$V^k(\varphi, t) < \mathsf{B}_t \text{ if and only if } y_k = 1.$$

d) Let $l \in \{0,2\}$. Under conditions (A), (C), and (D), we have:

$$LPM_l(\varphi, V, \mathsf{B}, t) \leq \sum_{k=1}^{K} p_k w_k^l.$$

e) Under conditions (A), (B), (C), and (D), we have:

$$LPM_l(\varphi, V, \mathsf{B}, t) = \sum_{k=1}^{K} p_k w_k^l.$$

Note, that for the special case of $l = 0$, we can conclude that $w_k^l = 1$ if $y_k = 1$ and that $w_k^l = 0$ if $y_k = 0$ under conditions (C) and (D). Hence,

$$LPM_0(\varphi, V, \mathsf{B}, t) = \sum_{k=1}^{K} p_k y_k \tag{6.121}$$

under the additional conditions (A) and (B), with "$\leq$" instead of "$=$" if only condition (A) is satisfied in addition to (C) and (D). Using theorem 6.7.1 we can replace the lower partial moment constraint

$$LPM_l(\varphi, V, \mathsf{B}, t) \leq A^l \tag{LPM}$$

by the constraint

$$\sum_{k=1}^{K} p_k w_k^l \leq A^l \tag{E}$$

for $A^l \in \mathbb{R}$ and $l \in \{0, 1, 2\}$, if conditions (A), (B), (C), and (D) are satisfied. Hence, instead of using constraint (LPM), we can use inequality (E) if we add conditions (A), (B), (C), and (D) to the optimization problem. Furthermore, we can omit condition (B) if $l \in \{0, 2\}$. For $l = 0$ condition (LPM) is called shortfall constraint and the corresponding A^0, in this case chosen to be an element of $(0, 1)$, is called shortfall probability. Usually all commercial optimization tools use inequalities of the form $\leq$ or $\geq$ and a precision expressed by the smallest absolute number that can be recognized numerically within the tool and which will be denoted by $\varepsilon > 0$ here. That's why we rewrite equation (B) in the following form:

$$m_k(1 - y_k) + V^k(\varphi, t) \leq \mathsf{B}_t - \varepsilon. \tag{B'}$$

For $l \in \{0, 1, 2\}$ and $t \in (0, T]$ let us therefore denote the set of restrictions (A), (B'), (C), and (D) by

$$MIP_l\left(\varphi, y^k, w_k^l, V^k, \mathsf{B}, t\right), \; k = 1, ..., K.$$

This yields the following optimization problem (P_1) which is (approximately) equivalent to (P):

$$
(P_1)\begin{cases}
\sum_{k=1}^{K} p_k V^k(\varphi, T) \rightarrow \max \\
MIP_l\left(\varphi, y^k, w_k^l, V^k, \mathsf{B}, t\right), \; k = 1, ..., K, \; t \in T_{\mathsf{B}} \\
MIP_l\left(\varphi, y_0^k, w_{0k}^l, C_0, 0, t\right), \; k = 1, ..., K, \; t \in T_{\mathsf{L}} \\
y_k, y_0^k \in \{0, 1\}, \; k = 1, ..., K \\
w_k^l, w_{0k}^l \in \mathbb{R} \\
\sum_{k=1}^{K} p_k w_k^l \leq A_B^l, \; l \in \{0, 1, 2\} \\
\sum_{k=1}^{K} p_k w_{0k}^l \leq A_L^l, \; l \in \{0, 1, 2\} \\
s_i \leq \varphi_i \leq S_i, \; i = 1, ..., n \\
C_0\left(\varphi, T_l^{\mathsf{L}}\right) = C_0\left(\varphi, T_{l-1}^{\mathsf{L}}\right) e^{R_{ca}\left(T_l^{\mathsf{L}} - T_{l-1}^{\mathsf{L}}\right)} + \\
\qquad + C_0\left(\varphi, T_{l-1}^{\mathsf{L}}, T_l^{\mathsf{L}}\right) - L\left(T_l^{\mathsf{L}}\right), \; l = 1, ..., m_{\mathsf{L}} \\
C_0(\varphi, 0) = \varphi_0 = C_0 - \sum_{i=1}^{n} \varphi_i V_i(0).
\end{cases}
$$

The variables of (P_1) to be optimized are $\varphi_0, ..., \varphi_n$, $y_k, y_0^k \in \{0, 1\}$, and $w_k^l, w_{0k}^l \in \mathbb{R}$, $k = 1, ..., K$. Hence, we have to solve a mixed-integer linear program which can be rather easily done by commercial optimization tools[48].

6.7.3 Case Study: Optimizing a Sovereign Bond Portfolio

Data and Estimated Parameters. For the following case study we use process parameters estimated from observed data of German zero rates and credit spreads between AAA–rated German and AA–rated Italian as well as AAA–rated German and $(A-)$– rated Greek government bonds. As in

[48] If we define

$$
a_{ik} := \begin{cases} S_i, & \text{if } V_i^k(t) < 0 \\ s_i, & \text{if } V_i^k(t) \geq 0 \end{cases} \quad \text{and} \quad b_{ik} = \begin{cases} s_i, & \text{if } V_i^k(t) < 0 \\ S_i, & \text{if } V_i^k(t) \geq 0 \end{cases}
$$

for $i = 1, ..., n$ and $k = 1, ..., K$, we can choose

$$
M_k = \mathsf{B} - \sum_{i=1}^{n} a_{ik} V_i^k(t), \; m_k = \mathsf{B} - \varepsilon - \sum_{i=1}^{n} b_{ik} \cdot V_i^k(t),
$$

$$
M_k^l = \left[\sum_{i=1}^{n} b_{ik} V_i^k(t) - \mathsf{B}\right]^l, \text{ and } m_k^l = \left[\mathsf{B} - \sum_{i=1}^{n} a_{ik} \cdot V_i^k(t)\right]^l,
$$

$k = 1, ..., K$, $l \in \{0, 1, 2\}$, for a practical application.

Table 6.42. Estimated parameters for the German bond market.

Parameter	Estimation
θ_r	0.014413
a_r	0.238205
σ_r	0.015581
λ_r	−0.086076
r_0	0.042434

Table 6.43. Estimated parameters for the Italian bond market.

Parameter	Estimation	Parameter	Estimation
b_s^I	0.274800	θ_u^I	0.000031
a_s^I	0.047687	a_u^I	0.068696
σ_s^I	0.158324	σ_u^I	0.030482
λ_s^I	−1.898668	λ_u^I	−1.228143
s_0^I	0.001296	u_0^I	0.005112

section 6.6.3 we assume throughout that the German bonds are default risk free whereas the Italian and Greek bonds are defaultable. We consider a 12 months time series of daily bond prices from November $1st$, 1999, until October $23rd$, 2000, provided by Reuters Information Services. The data set is a subset of the data considered in section 6.6.3. Whereas in section 6.6.3 we wanted to demonstrate the in-sample and out-of-sample properties of our model, i.e. we used one part of the data for parameter estimation and the other part for out-of-sample model validation, we now estimate the parameters for the case study from the latest available data. In addition, this choice of the time frame ensures that we don't get too much noise in the data due to convergence effects of the yields of the members of the European Monetary Union. Therefore, we repeat the whole Kalman Filter estimation for this specific subset of data. The results of the Kalman filter estimation are summarized in tables 6.42-6.44. Note, that we make exactly the same theoretical assumptions on the model as in section 6.6.3, especially we consider the stochastic processes r, s^I, s^G, u^I, u^G, but for simplification of the case study simulation assume that the deterministic function $\theta_r(t)$ is constant, i.e. $\theta_r(t) = \theta_r$ for all $t \geq 0$. Hence, we have to estimate one additional parameter.

If the Kalman filter is set up correctly and the theoretical formulas based on the estimated parameters can explain the observed data well, the standardized residuals of the Kalman filter, i.e. the standardized difference between the new observations and the corresponding Kalman filter forecasts, should be standard normally distributed. One method that we have already applied in section 6.6.3 is to refer the test statistic

Table 6.44. Estimated parameters for the Greek bond market.

Parameter	Estimation	Parameter	Estimation
b_s^G	0.343225	θ_u^G	0.000392
a_s^G	0.167814	a_u^G	0.074952
σ_s^G	0.446885	σ_u^G	0.067231
λ_s^G	−0.215937	λ_u^G	−1.109091
s_0^G	0.002325	u_0^G	0.012704

Table 6.45. Statistics for testing the quality of the Kalman filter estimation for different maturities.

Maturity	L (Ger)	L (I)	L (G)
3 years	11.593	0.598	12.161
4 years	9.201	4.134	5.973
5 years	5.202	0.940	2.076
6 years	12.850	0.223	11.662
7 years	6.234	0.982	10.513
8 years	3.985	9.048	5.812
9 years	1.729	8.619	2.945

$$L = n \cdot \left(\frac{skewness^2}{6} + \frac{excess\ kurtosis^2}{24} \right)$$

to the chi-squared distribution with two degrees of freedom (see, e.g., Harvey (1989), page 260, for a similar test). The critical values for 5% and 1% are 5.99 and 9.21, respectively. The null hypothesis is rejected if L is greater than the corresponding critical value. According to table 6.45 we get the following results:

- Germany: For the maturities 5, 8, and 9 years the normal distribution hypothesis isn't rejected on a 5%-significance level, and for the maturities 4, 5, 7, 8, and 9 years it isn't rejected on a 1%−significance level. The normal distribution hypothesis is rejected on both significance levels for the maturities 3 and 6 years.
- Italy: The normal distribution hypothesis is not rejected on a 5% - significance level for the maturities 3, 4, 5, 6, and 7 years, and on a 1%− significance level for all maturities.
- Greece: The normal distribution hypothesis is not rejected on both, a 5%− and a 1%−significance level for the maturities 4, 5, 8, and 9 years. The normal distribution hypothesis is rejected on both significance levels for the maturities 3, 6, and 7 years.

One of the critical model assumptions is that there is no correlation between any of the standard Brownian motions driving the underlying processes. Be-

Table 6.46. Statistics for testing the correlation of the filtered residuals dW_r, dW_s^j, and dW_u^j, $j \in \{I, G\}$.

	$\mathbf{dW_s^I}$	$\mathbf{dW_s^G}$	$\mathbf{dW_u^I}$	$\mathbf{dW_u^G}$	$\mathbf{dW_r}$
$\mathbf{dW_s^I}$	1				
$\mathbf{dW_s^G}$	1.279	1			
$\mathbf{dW_u^I}$	−0.819	−1.122	1		
$\mathbf{dW_u^G}$	0.501	−1.684	1.974	1	
$\mathbf{dW_r}$	−0.945	−1.702	−0.114	0.472	1

lieving in the estimated parameters we now use the filtered values for r, s^j and u^j to test for this assumption. We therefore apply the test statistic

$$T_{n-2} = \frac{R \cdot \sqrt{n-2}}{\sqrt{1-R^2}}$$

with R denoting the well-known sample correlation coefficient of the corresponding filtered values for dW_r, dW_s^j and dW_u^j, $j \in \{I, G\}$. Larsen & Marx (2001), page 626, show that under the assumption (null hypothesis) that the residuals are uncorrelated, the test statistic follows a Student-t distribution with $n-2$ degrees of freedom. The critical values for 5% and 1% are 1.97 and 2.60, respectively. The null hypothesis is rejected if $|T_{n-2}|$ is greater than the corresponding critical value. According to table 6.46 the null hypothesis is only rejected on a 5%–significance level for dW_u^I and dW_u^G and is not rejected on a 1%–significance level for all residuals. In our simulations we therefore assume that dW_r, dW_s^j and dW_u^j, $j \in \{I, G\}$ are pairwise uncorrelated.

Monte Carlo Simulation and Optimization in Practice. For the case study we choose a planning horizon of $T = 2$ years and 5 sovereign bonds with annual coupon payments for each of the countries Germany, Italy, and Greece, i.e. $n = 15$. The specification and dirty prices of the bonds at time $t = 0$ with a notional of 100 *Euro* each are summarized in table 6.47.

Furthermore, we set $\mathcal{T}_\mathsf{B} = \mathcal{T}_\mathsf{L} = \{6M, 1Y, 1Y6M, 2Y\}$ with a liability stream of $(50{,}000; 100{,}000; 200{,}000; 400{,}000)$ *Euro*, $R_{ca} = R_{ca}^{d,I} = R_{ca}^{d,G} = 0$, and a budget of $C_0 = 1$ *Mio. Euro*. As mentioned above, $V_i(t)$ is considered to be the dirty price of bond $i \in \{1, ..., n\}$ plus accumulated cash flows over time accrued at an interest of R_{ca}. For the optimization we choose $K = 100$ simulations, absolute lower bounds of $s_i = 0$ and upper bounds of $S_i = +\infty$, $i \in \{1, ..., n\}$, a benchmark of $\mathsf{B}_t = C_0$ for all $t \in \mathcal{T}_\mathsf{L}$, $l = 0$, i.e. we are dealing with shortfall constraints, and $A_\mathsf{B}^0(t) = 1\% = A_\mathsf{L}^0(t)$. The probabilities of default and the resulting default boundaries $\xi^{d,I}$ and $\xi^{d,G}$ are given in table 6.48.

According to the rather low default probabilities, no default took place in our simulations. Therefore, the shortfall constraint for covering the liability

Table 6.47. Specification of the portfolio universe.

i	Country	Maturity	Coupon	Dirty Price
1	Germany	$1.0Y$	6%	101.40
2	Germany	$1.5Y$	6%	104.92
3	Germany	$2.0Y$	6%	102.42
4	Germany	$2.5Y$	6%	104.92
5	Germany	$3.0Y$	6%	103.17
6	Italy	$1.0Y$	7%	100.94
7	Italy	$1.5Y$	7%	104.88
8	Italy	$2.0Y$	7%	101.91
9	Italy	$2.5Y$	7%	104.80
10	Italy	$3.0Y$	7%	102.79
11	Greece	$1.0Y$	8%	101.96
12	Greece	$1.5Y$	8%	106.74
13	Greece	$2.0Y$	8%	103.54
14	Greece	$2.5Y$	8%	107.02
15	Greece	$3.0Y$	8%	104.79

Table 6.48. Default probabilities and default boundaries for Italy and Greece.

	Default probability	Default boundary
Italy	0.02%	0.014659
Greece	0.04%	0.047047

stream degenerates to an exact cash flow matching restriction as stated, e.g., in Elton & Gruber (1991), pages 565-566.

Example 6.7.1.
In the first example we do not consider any shortfall or liability constraints. Our aim is simply to maximize the expected final portfolio value, i.e. the dirty price of the portfolio plus the value of the cash account at time T, allowing for a budget of C_0. The optimal portfolio consists of 9,658 of the 2 year 8% Greek bonds, i.e. a cash value of 1 *Mio. Euro*. This portfolio has an expected final value (including previous coupon payments) of 1,120,337 *Euro* corresponding to an expected rate of return of 5.68%. The probabilities of falling below the benchmark C_0 at times $t \in \mathcal{T}_L$ are given by the vector $(5\%, 2\%, 1\%, 0\%)$. The liability stream is not covered by this portfolio. The value of the cash account compared to the liability stream is shown in figure 6.37.

Example 6.7.2.
For the second example we add shortfall constraints for the portfolio value at times $t \in \mathcal{T}_L$ to the budget constraint of example 6.7.1 and get an optimal portfolio consisting of

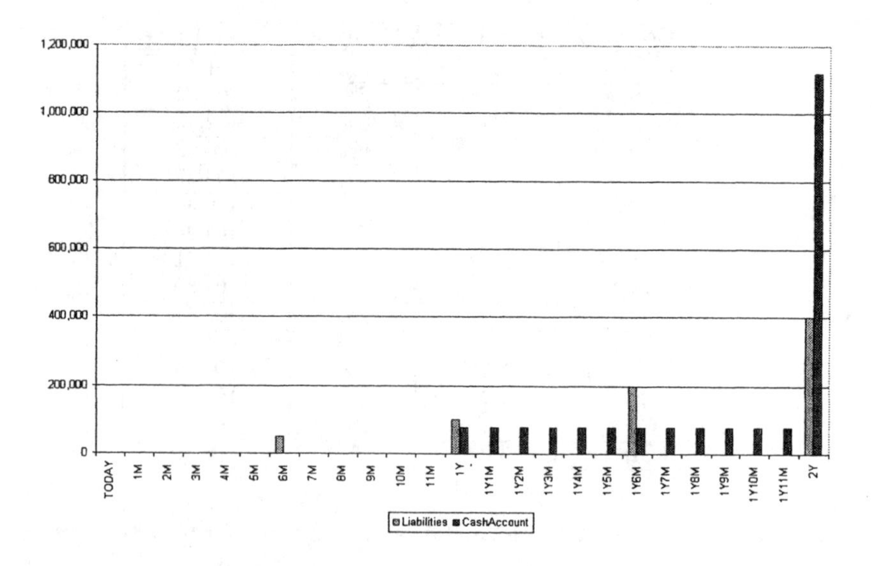

Fig. 6.37. Value of the cash account in example 6.7.1 over time compared to the liability stream.

3,008 (311,428 *Euro*)	of the 2 year 8% Greek bond
2,485 (253,287 *Euro*)	of the 2 year 7% Italian bond
3,604 (369,108 *Euro*)	of the 2 year 6% German bond
641 (66,177 *Euro*)	of the 3 year 6% German bond

This portfolio has an expected final value (including previous coupon payments) of 1,105,841 *Euro* corresponding to an expected rate of return of 5.03%. The probabilities of falling below the benchmark C_0 at times $t \in \mathcal{T}_L$ are given by the vector (1%, 1%, 0%, 1%). To satisfy the shortfall constraints, the investment into the Greek bond is reduced and shifted towards the less risky Italian and German bonds. Unfortunately, the liability stream is not covered by this portfolio. The optimal country allocation as well as the value of the cash account compared to the liability stream are shown in figures 6.38 and 6.39.

Example 6.7.3.
In this example we add the liability constraints (but not the shortfall constraints of example 6.7.2) at times $t \in \mathcal{T}_L$ to the budget constraint of example 6.7.1 and get an optimal portfolio which consists of

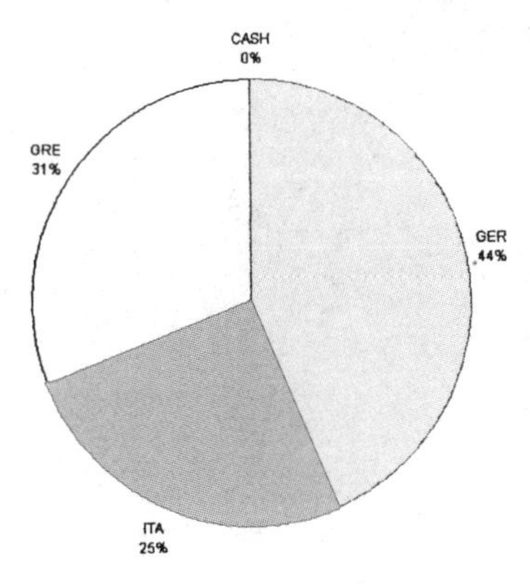

Fig. 6.38. Optimal country allocation in example 6.7.2.

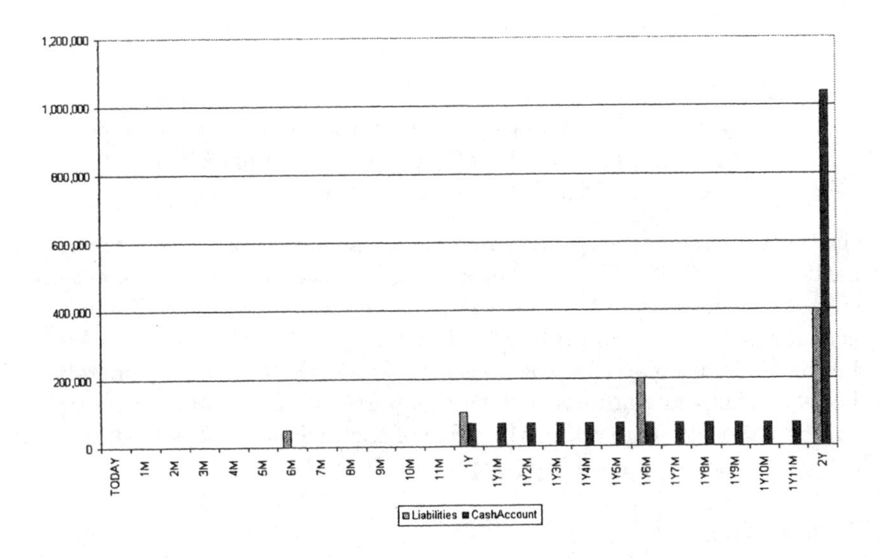

Fig. 6.39. Value of the cash account in example 6.7.2 over time compared to the liability stream.

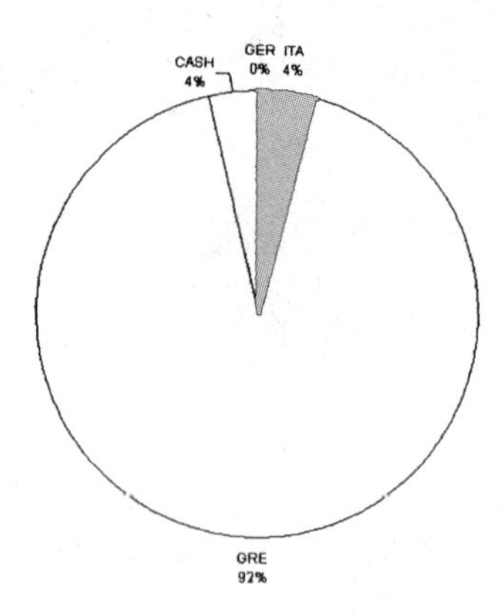

Fig. 6.40. Optimal country allocation in example 6.7.3.

7,009 (725,703 *Euro*)	of the 2 year 8% Greek bond
1,852 (197,672 *Euro*)	of the 1.5 year 8% Greek bond
411 (41,440 *Euro*)	of the 1 year 7% Italian bond
35,185 *Euro*	on the cash account.

This portfolio has an expected final value (including previous coupon payments) of 1,106,961 *Euro* corresponding to an expected rate of return of 5.08%. The maturities of the bonds are chosen according to the liability payment dates $t \in \mathcal{T}_L$. The probabilities of falling below the benchmark C_0 at times $t \in \mathcal{T}_L$ are given by the vector (5%, 2%, 0%, 0%). Unfortunately, the shortfall constraints are not met by this portfolio. The optimal country and maturity allocation as well as the value of the cash account compared to the liability stream are shown in figures 6.40, ??, and 6.41.

Example 6.7.4.
We add the shortfall and liability constraints at times $t \in \mathcal{T}_L$ to the budget constraint of example 6.7.1 and get an optimal portfolio which consists of

3,050 (315,815 *Euro*)	of the 2 year 8% Greek bond
2,133 (217,370 *Euro*)	of the 2 year 7% Italian bond
1,869 (196,040 *Euro*)	of the 1.5 year 7% Italian bond
465 (46,900 *Euro*)	of the 1 year 7% Italian bond
1,825 (186,959 *Euro*)	of the 2 year 6% German bond
36,916 *Euro*	on the cash account.

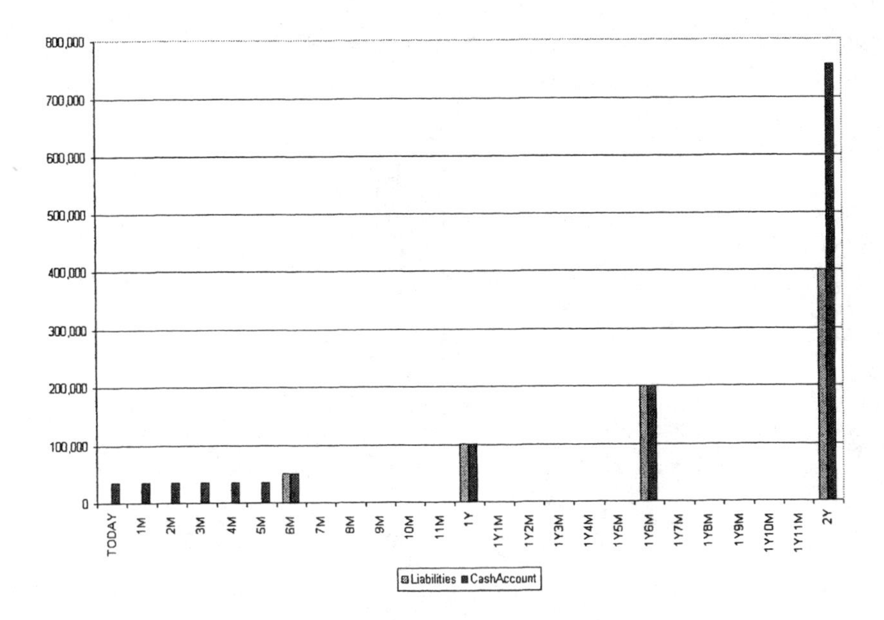

Fig. 6.41. Value of the cash account in example 6.7.3 over time compared to the liability stream.

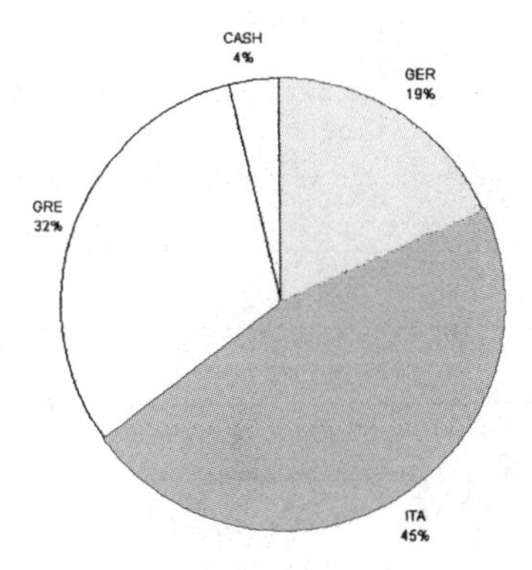

Fig. 6.42. Optimal country allocation in example 6.7.4.

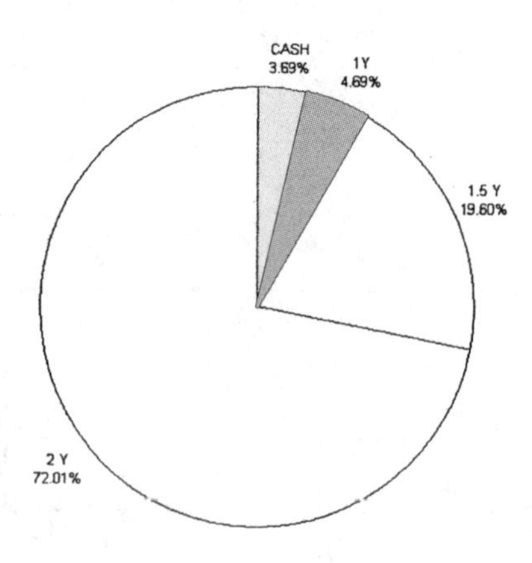

Fig. 6.43. Optimal maturity allocation in example 6.7.4.

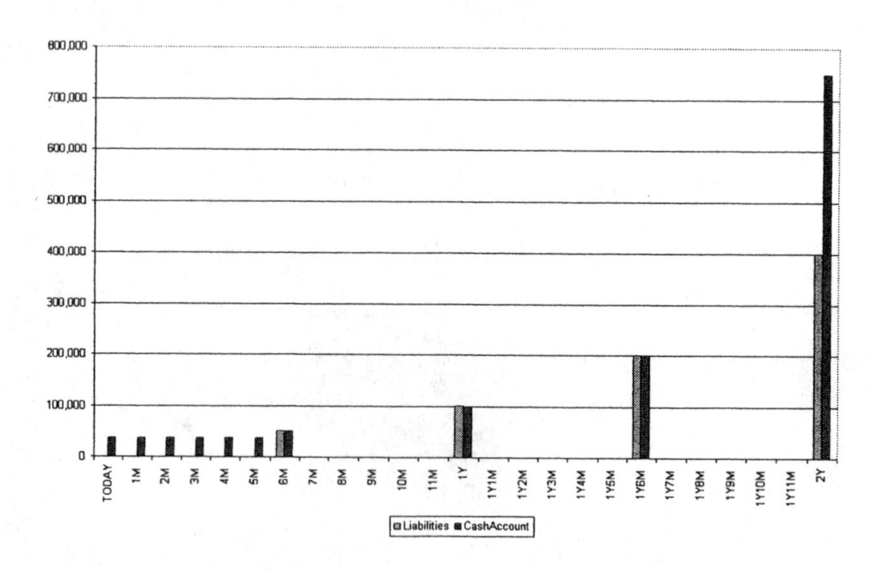

Fig. 6.44. Value of the cash account in example 6.7.4 over time compared to the liability stream.

This portfolio has an expected final value (including previous coupon payments) of $1,099,185$ *Euro* corresponding to an expected rate of return of 4.73%. We now have a portfolio with maturity times according to the liability payment dates $t \in \mathcal{T}_\mathsf{L}$ and a country diversification including Italian and German bonds to increase the safety of the portfolio, i.e. to reduce the probability of falling below the given benchmark. The probabilities of falling below the benchmark C_0 at times $t \in \mathcal{T}_\mathsf{L}$ are given by the vector $(1\%, 1\%, 0\%, 0\%)$. The optimal country and maturity allocation as well as the value of the cash account compared to the liability stream are shown in figures 6.42, 6.43, and 6.44.

A. Some Definitions of S&P

A.1 Definition of Credit Ratings

A.1.1 Issue Credit Ratings

"Standard & Poor's *Issue* Credit Rating is a current opinion of the creditworthiness of an obligor with respect to a specific financial obligation, a specific class of financial obligations, or a specific financial program. It takes into consideration the creditworthiness of guarantors, insurers, or other forms of credit enhancement on the obligation and takes into account the currency in which the obligation is denominated. [...] Issue Credit Ratings are based on current information furnished by the obligors or obtained by Standard & Poor's from other sources it considers reliable. Standard & Poor's does not perform an audit in connection with any credit rating and may, on occasion, rely on unaudited financial information. Credit ratings may be changed, suspended, or withdrawn as a result of changes in, or unavailability of, such information, or based on other circumstances. Issue Credit Ratings can be either long term or short-term. Short-term ratings are generally assigned to those obligations considered short-term in the relevant market. In the U.S., for example, that means obligations with an original maturity of no more than 365 days - including commercial paper. Short-term ratings are also used to indicate the creditworthiness of an obligor with respect to put features on long term obligations. The result is a dual rating, in which the short-term rating addresses the put feature, in addition to the usual long term rating. Medium-term notes are assigned long term ratings."[1]

A.1.2 Issuer Credit Ratings

"Standard & Poor's *Issuer* Credit Rating is a current opinion of an obligor's overall financial capacity (its creditworthiness) to pay its financial obligations. This opinion focuses on the obligor's capacity and willingness to meet its financial commitments as they come due. It does not apply to any specific financial obligation, as it does not take into account the nature of and provisions of the obligation, its standing in bankruptcy or liquidation, statutory

[1] Standard & Poor's, (*Ratings Performance 1996: Stability & Transition* 1997), page 56.

preferences, or the legality and enforceability of the obligation. In addition, it does not take into account the creditworthiness of the guarantors, insurers, or other forms of credit enhancement on the obligation. [...] Counterparty Credit Ratings, ratings assigned under the Corporate Credit Rating Service (formerly called the Credit Assessment Service) and Sovereign Credit Ratings are all forms of Issuer Credit Ratings. Issuer Credit Ratings are based on current information furnished by obligors or obtained by Standard & Poor's from other sources it considers reliable. Standard & Poor's does not perform an audit in connection with any credit rating and may, on occasion, rely on unaudited financial information. Credit ratings may be changed, suspended, or withdrawn as a result of changes in, or unavailability of, such information, or based on other circumstances. Issue Credit Ratings can be either long term or short-term. Short-Term Issuer Credit Ratings reflect the obligor's creditworthiness over a short-term time horizon"[2]

S&P's Long-Term Issuer Credit Ratings. Standard & Poor's gives the following long-term issuer credit rating definitions[3]:

- An obligor rated 'AAA' has an extremely strong capacity to meet its financial commitments. 'AAA' is the highest issuer credit rating assigned by Standard & Poor's.
- An obligor rated 'AA' has very strong capacity to meet its financial commitments. It differs from the highest rated obligors only in a small degree.
- An obligor rated 'A' has strong capacity to meet its financial commitments but is somewhat more susceptible to the adverse effects of changes in circumstances and economic conditions than obligors in higher-rated categories.
- An obligor rated 'BBB' has adequate capacity to meet its financial commitments. However, adverse economic conditions or changing circumstances are more likely to lead to a weakened capacity of the obligor to meet its financial commitments.
- Obligors rated 'BB', 'B', 'CCC', and 'CC' are regarded as having significant speculative characteristics. 'BB' indicates the least degree of speculation and 'CC' the highest. While such obligors will likely have some quality and protective characteristics, these may be outweighed by large uncertainties or major exposures to adverse conditions.
- An obligor rated 'BB' is less vulnerable in the near term than other lower-rated obligors. However, it faces major ongoing uncertainties and exposure to adverse business, financial, or economic conditions which could lead to the obligor's inadequate capacity to meet its financial commitments. An obligor rated 'B' is more vulnerable than the obligors rated 'BB', but the obligor currently has the capacity to meet its financial commitments.

[2] Standard & Poor's,(*Ratings Performance 1996: Stability & Transition* 1997), page 57.
[3] From www.standardandpoors.com.

Adverse business, financial, or economic conditions will likely impair the obligor's capacity or willingness to meet its financial commitments.

- An obligation rated 'B' is more vulnerable to nonpayment than obligations rated 'BB', but the obligor currently has the capacity to meet its financial commitment on the obligation. Adverse business, financial, or economic conditions will likely impair the obligor's capacity or willingness to meet its financial commitment on the obligation.
- An obligor rated 'CCC' is currently vulnerable, and is dependent upon favorable business, financial, and economic conditions to meet its financial commitments.
- An obligor rated 'CC' is currently highly vulnerable.
- The ratings from 'AA' to 'CCC' may be modified by the addition of a plus or minus sign to show relative standing within the major rating categories.
- A subordinated debt or preferred stock obligation rated 'C' is currently highly vulnerable to nonpayment. The 'C' rating may be used to cover a situation where a bankruptcy petition has been filed or similar action taken, but payments on this obligation are being continued. A 'C' also will be assigned to a preferred stock issue in arrears on dividends or sinking fund payments, but that is currently paying.
- An obligor rated 'R' is under regulatory supervision owing to its financial condition. During the pendency of the regulatory supervision the regulators may have the power to favor one class of obligations over others or pay some obligations and not others. Please see Standard & Poor's issue credit ratings for a more detailed description of the effects of regulatory supervision on specific issues or classes of obligations.
- An obligor rated 'SD' (Selective Default) or 'D' has failed to pay one or more of its financial obligations (rated or unrated) when it came due. A 'D' rating is assigned when Standard & Poor's believes that the default will be a general default and that the obligor will fail to pay all or substantially all of its obligations as they come due. An 'SD' rating is assigned when Standard & Poor's believes that the obligor has selectively defaulted on a specific issue or class of obligations but it will continue to meet its payment obligations on other issues or classes of obligations in a timely manner. Please see Standard & Poor's issue credit ratings for a more detailed description of the effects of a default on specific issues or classes of obligations.
- An issuer designated N.R. is not rated.
- Public Information Ratings: Ratings with a 'pi' subscript are based on an analysis of an issuer's published financial information, as well as additional information in the public domain. They do not, however, reflect in-depth meetings with an issuer's management and are therefore based on less comprehensive information than ratings without a 'pi' subscript. Ratings with a 'pi' subscript are reviewed annually based on a new year's financial statements, but may be reviewed on an interim basis if a major event occurs

that may affect the issuer's credit quality. Outlooks are not provided for ratings with a 'pi' subscript, nor are they subject to potential CreditWatch listings. Ratings with a 'pi' subscript generally are not modified with '+' or '-' designations. However, such designations may be assigned when the issuer's credit rating is constrained by sovereign risk or the credit quality of a parent company or affiliated group.

S&P's Short-Term Issuer Credit Ratings. Standard & Poor's gives the following short-term issuer credit rating definitions[4]:

- An obligor rated 'A-1' has strong capacity to meet its financial commitments. It is rated in the highest category by Standard & Poor's. Within this category, certain obligors are designated with a plus sign (+). This indicates that the obligor's capacity to meet its financial commitments is extremely strong.
- An obligor rated 'A-2' has satisfactory capacity to meet its financial commitments. However, it is somewhat more susceptible to the adverse effects of changes in circumstances and economic conditions than obligors in the highest rating category.
- An obligor rated 'A-3' has adequate capacity to meet its financial obligations. However, adverse economic conditions or changing circumstances are more likely to lead to a weakened capacity of the obligor to meet its financial commitments.
- An obligation rated 'B' is more vulnerable to nonpayment than obligations rated 'BB', but the obligor currently has the capacity to meet its financial commitment on the obligation. Adverse business, financial, or economic conditions will likely impair the obligor's capacity or willingness to meet its financial commitment on the obligation.
- A subordinated debt or preferred stock obligation rated 'C' is currently highly vulnerable to nonpayment. The 'C' rating may be used to cover a situation where a bankruptcy petition has been filed or similar action taken, but payments on this obligation are being continued. A 'C' also will be assigned to a preferred stock issue in arrears on dividends or sinking fund payments, but that is currently paying.
- An obligor rated 'R' is under regulatory supervision owing to its financial condition. During the pendency of the regulatory supervision the regulators may have the power to favor one class of obligations over others or pay some obligations and not others. Please see Standard & Poor's issue credit ratings for a more detailed description of the effects of regulatory supervision on specific issues or classes of obligations.
- An obligor rated 'SD' (Selective Default) or 'D' has failed to pay one or more of its financial obligations (rated or unrated) when it came due. A 'D' rating is assigned when Standard & Poor's believes that the default

[4] www.standardandpoors.com

will be a general default and that the obligor will fail to pay all or substantially all of its obligations as they come due. An 'SD' rating is assigned when Standard & Poor's believes that the obligor has selectively defaulted on a specific issue or class of obligations but it will continue to meet its payment obligations on other issues or classes of obligations in a timely manner. Please see Standard & Poor's issue credit ratings for a more detailed description of the effects of a default on specific issues or classes of obligations.
- An issuer designated N.R. is not rated.

A.2 Definition of Default

A.2.1 S&P's definition of corporate default

"A default is recorded upon the first occurrence of a payment default on any financial obligation, rated or unrated, other than a financial obligation subject to a bona fide commercial dispute; an exception occurs when an interest payment missed on the due date is made within the grace period. Preferred stock is not considered a financial obligation; thus, a missed preferred stock dividend is not normally equated to a default. Distressed exchanges, on the other hand, are considered defaults whenever the debtholders are offered substitute instruments with lower coupons, longer maturities, or any other diminished financial terms."
- *Ratings Performance 1999: Stability & Transition* (2000)

A.2.2 S&P's definition of sovereign default

"Sovereign debt is considered in default in any of the following circumstances:

- For local and foreign currency bonds, notes, and bills, when either scheduled debt service is not paid on the due date, or an exchange offer of new debt contains terms less favorable than the original issue.
- For central bank currency, when notes are converted into new currency of less than equivalent face value.
- For bank loans, when either scheduled debt service is not paid on the due date, or a rescheduling of principal and/or interest is agreed to by creditors at less favorable terms than the original loan. Such rescheduling agreements covering short- and long-term bank debt are considered defaults even when, for legal or regulatory reasons, creditors deem forced rollover of principal to be voluntary.

In addition, many rescheduled sovereign bank loans are ultimatively extinguished at a discount from their original face value. Typical deals have included exchange offers (such as for those linked to the issuance of Brady

bonds), debt/equity swaps related to government privatization programs, and/or buybacks for cash. Standard & Poor's considers such transactions as defaults because they contain terms less favorable to creditors than the original obligation."
- *Ratings Performance 2000: Default, Transition, Recovery, and Spreads* (2001)

B. Technical Proofs

B.1 Proof of Lemma 6.2.1

Proof.
We want to find solutions of the SDE (6.4). First we show, that there exists a weak solution:
$\theta_r(t)$ is bounded on $[0,T^*]$, i.e. there exists a $\theta_r \geq 0$ such that $|\theta_r(t)| \leq \theta_r$ $\forall\, t \in [0,T^*]$. Because $b(.)$ and $\sigma(.)$ are continuous functions, we know that for every given initial distribution of (r_0, u_0, s_0) with compact support there exists a solution $X = (X_t)_{0\leq t\leq T^*}$ which possibly blows up or explodes in finite time[1]. But since there exists a positive constant K, e.g.,

$$K = \max\left(a_r^2 + \sigma_r^2 + 2\theta_r a_r, a_u^2 + b_s^2 + \sigma_u^2 + 2\left(\theta_u a_u + b_s a_s\right),\right.$$
$$\left. a_s^2 + \sigma_s^2 + 2b_s a_s, \theta_r^2 + \theta_u^2 + \sigma_r^2 + \sigma_u^2 + \sigma_s^2 + 2\left(\theta_u a_u + \theta_r a_r\right)\right)$$

such that ($\|\cdot\|$ denotes the Euclidean norm)

$$\begin{aligned}
&\|\sigma(r_t, u_t, s_t)\|^2 + \|b(r_t, u_t, s_t)\|^2 \\
&= \sum_{j=1}^{3}\sum_{i=1}^{3} |\sigma_{ij}(r_t, u_t, s_t)|^2 + \sum_{i=1}^{3} |b_i(r_t, u_t, s_t)|^2 \\
&= \sigma_r^2 |r_t|^{2\beta} + \sigma_u^2 |u_t| + \sigma_s^2 |s_t| + |\theta_r(t) - a_r r_t|^2 + |\theta_u - a_u u_t|^2 + |b_s u_t - a_s s_t|^2 \\
&\leq \left(a_r^2 + \sigma_r^2 + 2\theta_r a_r\right) r_t^2 + \left(a_u^2 + b_s^2 + \sigma_u^2 + 2\left(\theta_u a_u + b_s a_s\right)\right) u_t^2 \\
&\quad + \left(a_s^2 + \sigma_s^2 + 2b_s a_s\right) s_t^2 + \left(\theta_r^2 + \theta_u^2 + \sigma_r^2 + \sigma_u^2 + \sigma_s^2 + 2\left(\theta_u a_u + \theta_r a_r\right)\right) \\
&\leq K\left(1 + r_t^2 + u_t^2 + s_t^2\right) = K(1 + \|(r_t, s_t, u_t)\|^2),
\end{aligned}$$

[1] Use theorem 2.3, p. 173, in Ikeda & Watanabe (1989b): Given continuous functions
$\sigma : \mathbf{R}^d \to \mathbf{R}^d \times \mathbf{R}^r$ and $b : \mathbf{R}^d \to \mathbf{R}^d$. Consider the stochastic differential equation

$$dX(t) = \sigma(X(t))\, dW(t) + b(X(t))\, dt.$$

Then for any probability μ on $\left(\mathbf{R}^d, \mathcal{B}\left(\mathbf{R}^d\right)\right)$ with compact support, there exists a solution
$X = (X(t))$ such that the law of $X(0)$ coincides with μ, i.e., a solution does exist locally but, in general, blows up (or explodes) in finite time. As always, $\mathcal{B}$ denotes the Borel $\sigma-$algebra.

we can conclude that the solution doesn't have any blow ups or explosions and therefore must coincide with an ordinary weak solution[2].
In the second part of the proof we show that a weak solution is pathwise unique[3]:
Let us suppose that there are two weak solutions X and $\widetilde{X}$ with $X_0 = \widetilde{X}_0$ a.s. We define their difference by

$$\begin{aligned}\Delta X_t &= \left(\Delta X_t^1, \Delta X_t^2, \Delta X_t^3\right)' = X_t - \widetilde{X}_t \\ &= \int_0^t \left\{b\left(X_l\right) - b\left(\widetilde{X}_l\right)\right\} dl + \int_0^t \left\{\sigma\left(X_l\right) - \sigma\left(\widetilde{X}_l\right)\right\} dW_l.\end{aligned}$$

For x_j, $y_j \in \mathbb{R}$, $j = 1, 2, 3$,

$$|\sigma_{ii}(x_1, x_2, x_3) - \sigma_{ii}(y_1, y_2, y_3)| \leq \max(\sigma_r, \sigma_u, \sigma_s)\sqrt{|x_i - y_i|} \tag{B.1}$$

and

$$|b_i\left(x_1, x_2, x_3\right) - b_i\left(y_1, y_2, y_3\right)| \leq \max\left(a_r, a_u, a_s, b_s\right) \sum_{i=1}^{3} |x_i - y_i| \tag{B.2}$$

for $i = 1, 2, 3$. Let $h : [0, \infty) \to [0, \infty)$ with $h\left(x\right) = \max(\sigma_r, \sigma_u, \sigma_s)\sqrt{x}$. Then h has the following properties:

1. $h\left(0\right) = 0$.

[2] Use theorem 2.4, p. 177, in Ikeda & Watanabe (1989b): Given the continuous functions $\sigma : \mathbb{R}^{\mathrm{d}} \to \mathbb{R}^{\mathrm{d}} \times \mathbb{R}^{\mathrm{r}}$ and $b : \mathbb{R}^{\mathrm{d}} \to \mathbb{R}^{\mathrm{d}}$ satisfying the condition

$$\|\sigma\left(X\right)\|^2 + \|b\left(X\right)\|^2 \leq K\left(1 + \|X\|^2\right)$$

for some positive constant K, then for any solution of equation (6.4) such that $E\left(\|X\left(0\right)\|^2\right) < \infty$, we have $E\left(\|X\left(t\right)\|^2\right) < \infty$ for all $t > 0$.

[3] There are two concepts of uniqueness that can be associated with weak solutions:

1. Suppose that whenever $(X, W), (\Omega, \mathcal{F}, \mathbf{P}), (\mathcal{F}_t)$ and $\left(\widetilde{X}, W\right), (\Omega, \mathcal{F}, \mathbf{P}), \left(\widetilde{\mathcal{F}}_t\right)$ are weak solutions to (6.4) with common Brownian motion W (relative to possibly different filtrations) on a common probability space $(\Omega, \mathcal{F}, \mathbf{P})$ and with common initial value, i.e $\mathbf{P}\left[X_0 = \widetilde{X}_0\right] = 1$, the two processes X and $\widetilde{X}$ are indistinguishable: $\mathbf{P}\left[X_t = \widetilde{X}_t; \forall\, 0 \leq t < \infty\right] = 1$. We say then that pathwise uniqueness holds for equation (6.4).
2. We say that uniqueness in the sense of probability law holds for equation (6.4) if, for any two weak solutions (X, W), $(\Omega, \mathcal{F}, \mathbf{P})$, $(\mathcal{F}_t)$ and $\left(\widetilde{X}, \widetilde{W}\right)$, $\left(\widetilde{\Omega}, \widetilde{\mathcal{F}}, \widetilde{\mathbf{P}}\right)$, $\left(\widetilde{\mathcal{F}}_t\right)$, with the same initial distribution, i.e., $\mathbf{P}\left[X_0 \in \Gamma\right] = \widetilde{\mathbf{P}}\left[\widetilde{X}_0 \in \Gamma\right]\ \forall\, \Gamma \in \mathcal{B}\left(\mathbb{R}^d\right)$, the two processes X and $\widetilde{X}$ have the same law.

Uniqueness in the sense of probability law does not imply pathwise uniqueness, but pathwise uniqueness does imply uniqueness in the sense of probability law.

2. h is strictly increasing.
3. $\int_{(0,\varepsilon)} h^{-2}(x)\,dx = \infty \;\forall\; \varepsilon > 0.$

Define a sequence $(a_n)_{n=0}^{\infty}$ such that

$$a_0 = 1,\; a_n = a_{n-1} \cdot \exp\left[-n \max^2(\sigma_r, \sigma_u, \sigma_s)\right],\; n \geq 1.$$

Then the following statements hold:

1. $(a_n)_{n=0}^{\infty} \subseteq (0,1]$.
2. $(a_n)_{n=0}^{\infty}$ is strictly decreasing with $\lim_{n\to\infty} a_n = 0$.
3. $\int_{a_n}^{a_{n-1}} h^{-2}(x)\,dx = \int_{a_n}^{a_{n-1}} \frac{1}{x \max^2(\sigma_r,\sigma_u,\sigma_s)}dx = \frac{1}{\max^2(\sigma_r,\sigma_u,\sigma_s)} \ln \frac{a_{n-1}}{a_n} = n.$

Define a function $\rho_n : \mathbb{R} \times [0,1] \times \left(0, \frac{a_{n-1}-a_n}{2}\right] \to \mathbb{R}$ such that

$$\rho_n(x,\delta,\varepsilon) = \begin{cases} \frac{1+\delta}{nh^2(a_n+\varepsilon)\varepsilon}(x-a_n) & \text{, if } x \in (a_n, a_n+\varepsilon] \\ \frac{1+\delta}{nh^2(x)} & \text{, if } x \in (a_n+\varepsilon, a_{n-1}-\varepsilon) \\ \frac{1+\delta}{nh^2(a_{n-1}-\varepsilon)\varepsilon}(a_{n-1}-x) & \text{, if } x \in [a_{n-1}-\varepsilon, a_{n-1}) \\ 0 & \text{, if } x \notin (a_n, a_{n-1}) \end{cases}.$$

Then the following statements hold:

1. ρ_n is a continuous function on each three-dimensional compact subinterval of $\mathbb{R} \times [0,1] \times \left(0, \frac{a_{n-1}-a_n}{2}\right]$.
2. $$0 \leq \rho_n(x,\delta,\varepsilon) \leq \frac{2}{nh^2(x)} \tag{B.3}$$
   on $\mathbb{R} \times [0,1] \times \left(0, \frac{a_{n-1}-a_n}{2}\right]$.
3. $\int_{a_n}^{a_{n-1}} \rho_n(x,\delta,\varepsilon)\,dx$ is continuous on each two-dimensional compact subinterval of $[0,1] \times \left(0, \frac{a_{n-1}-a_n}{2}\right]$.
4. There exists a pair $(\delta^*, \varepsilon^*) \in [0,1] \times \left(0, \frac{a_{n-1}-a_n}{2}\right]$ such that $\int_{a_n}^{a_{n-1}} \rho_n(x,\delta^*,\varepsilon^*)\,dx = 1$:
   $\lim_{\varepsilon\to 0,\varepsilon>0} \int_{a_n}^{a_{n-1}} \rho_n(x,1,\varepsilon)\,dx = \int_{a_n}^{a_{n-1}} \frac{2}{nh^2(x)}dx = 2$. Hence, there exists an $\varepsilon^* \in \left(0, \frac{a_{n-1}-a_n}{2}\right]$ such that $\int_{a_n}^{a_{n-1}} \rho_n(x,1,\varepsilon^*)\,dx > 1$. On the other hand, $\int_{a_n}^{a_{n-1}} \rho_n(x,0,\varepsilon^*)\,dx < 1$. Because of the continuity of $\int_{a_n}^{a_{n-1}} \rho_n(x,\delta,\varepsilon^*)\,dx$ in δ, there exists a $\delta^* \in [0,1]$ such that $\int_{a_n}^{a_{n-1}} \rho_n(x,\delta^*,\varepsilon^*)\,dx = 1$. We define $\rho_n^*(x) = \rho_n(x,\delta^*,\varepsilon^*)$.

For $k = 1,2,3$ we define approximation sequences $\Psi_n^{(k)} : \mathbb{R}^3 \to \mathbb{R}$ such that

$$\Psi_n^{(k)}(x_1,x_2,x_3) = \int_0^{|x_k|} \int_0^y \rho_n^*(u)\,du dy.$$

Then the following hold:

1. $\Psi_n^{(k)}$ is even and twice continuously differentiable.
2. $\frac{\partial \Psi_n^{(k)}}{\partial x_j} = 0 \; \forall \; j \neq k, \forall \; n.$
3. $\left|\frac{\partial \Psi_n^{(k)}}{\partial x_j}\right| \leq 1 \; \forall \; j, \forall \; n:$
   $\left|\frac{\partial \Psi_n^{(k)}}{\partial x_k}\right| = \left|\int_0^{|x_k|} \rho_n^* (u) \, du\right| \leq \int_{a_n}^{a_{n-1}} \rho_n^* (u) \, du = 1.$
4. For all k:
$$\Psi_n^{(k)}(x_1, x_2, x_3) \longrightarrow |x_k| \tag{B.4}$$
   for $n \to \infty$ and for all $x \in \mathbb{R}^3$:
   $\lim_{n\to\infty} \Psi_n^{(k)}(x_1, x_2, x_3) = \lim_{n\to\infty} \int_0^{|x_k|} \int_0^y \rho_n^* (u) \, dudy$
   $= \int_0^{|x_k|} \lim_{n\to\infty} \int_{a_n}^{a_{n-1}} \rho_n^* (u) \, dudy = |x_k| \, .$
5. $\left|\Psi_n^{(k)} (x_1, x_2, x_3)\right| \leq \int_0^{|x_k|} \int_0^y |\rho_n^* (u)| \, dudy \leq \int_0^{|x_k|} \int_{a_n}^{a_{n-1}} \rho_n^* (u) \, dudy$
   $\leq |x_k| \, .$ Then,
$$E\left[\left|\Psi_n^{(k)} (x_1, x_2, x_3)\right|\right] \leq E\left[|x_k|\right] < \infty, \tag{B.5}$$
   as $E\left[|x_k|^2\right] < \infty$ by footnote 2.

If we apply Itô's formula (see footnote 6.110) to $\Psi_n^{(k)}$ and ΔX_t, we get

$$\begin{aligned}
&\Psi_n^{(k)} (\Delta X_t) \\
&= \sum_{i=1}^{3} \int_0^t \frac{\partial \Psi_n^{(k)}}{\partial x_i} (\Delta X_l) \, d\Delta X_i \, (l) + \frac{1}{2} \sum_{1 \leq i,j \leq 3} \int_0^t \frac{\partial \Psi_n^{(k)}}{\partial x_i \partial x_j} (\Delta X_l) \, d \left[\Delta X_i, \Delta X_j\right]_l \\
&= \int_0^t \frac{\partial \Psi_n^{(k)}}{\partial x_k} (\Delta X_l) \left\{b_k (X_l) - b_k \left(\widetilde{X}_l\right)\right\} dl \\
&\quad + \frac{1}{2} \int_0^t \frac{\partial \Psi_n^{(k)}}{\partial^2 x_k} (\Delta X_l) \left\{\sigma_{kk} (X_l) - \sigma_{kk} \left(\widetilde{X}_l\right)\right\}^2 dl \\
&\quad + \int_0^t \frac{\partial \Psi_n^{(k)}}{\partial x_k} (\Delta X_l) \left\{\sigma_{kk} (X_l) - \sigma_{kk} \left(\widetilde{X}_l\right)\right\} dW_l.
\end{aligned}$$

Because $\left|\frac{\partial \Psi_n^{(k)}}{\partial x_k} (x)\right| \leq 1$ for $x \in \mathbb{R}^3$ and equation (B.1) holds,

$$\begin{aligned}
&E\left[\int_0^t \left|\frac{\partial \Psi_n^{(k)}}{\partial x_k} (\Delta X_l) \left\{\sigma_{kk} (X_l) - \sigma_{kk} \left(\widetilde{X}_l\right)\right\}\right|^2 dl\right] \\
&\leq E\left[\int_0^t \left|\sigma_{kk} (X_l) - \sigma_{kk} \left(\widetilde{X}_l\right)\right|^2 dl\right] \\
&\leq \max(\sigma_r, \sigma_u, \sigma_s)^2 E\left[\int_0^t \left|\Delta X_l^k\right| dl\right] \\
&\leq t \cdot \max(\sigma_r, \sigma_u, \sigma_s)^2 \max_{0 \leq l \leq t} E\left|\Delta X_l^k\right| \\
&< \infty \; \forall \; 0 \leq t < \infty, \; \forall \; k.
\end{aligned}$$

The last inequality holds by theorem 2.4, page 177, in Ikeda & Watanabe (1989b)[4]. Hence, we can apply theorem 3.2.1 (iii) in Oksendal (1998)[5], page 30, and get

$$E\left[\int_0^t \frac{\partial \Psi_n^{(k)}}{\partial x_k}(\Delta X_l)\left\{\sigma_{kk}(X_l)-\sigma_{kk}\left(\widetilde{X}_l\right)\right\}dW_l\right]=0.$$

Using inequalities (B.1) and (B.3) we get

$$E\left[\frac{1}{2}\int_0^t \frac{\partial \Psi_n^{(k)}}{\partial^2 x_k}(\Delta X_l)\left\{\sigma_{kk}(X_l)-\sigma_{kk}\left(\widetilde{X}_l\right)\right\}^2 dl\right]\le \frac{t}{n}.$$

Applying inequality (B.2) yields

$$E\left[\int_0^t \frac{\partial \Psi_n^{(k)}}{\partial x_k}(\Delta X_l)\left\{b_k(X_l)-b_k\left(\widetilde{X}_l\right)\right\}dl\right]$$
$$\le \max(a_r,a_u,a_s,b_s)\sum_{k=1}^{3}\int_0^t E\left|\Delta X_l^k\right|dl.$$

Hence, we find that

$$E\Psi_n^{(i)}(\Delta X_t)\le \frac{t}{n}+\max(a_r,a_u,a_s,b_s)\sum_{k=1}^{3}\int_0^t E\left|\Delta X_l^k\right|dl,\ \forall\, 0\le t,\forall\, k.$$

For

$$\Psi_n=\sum_{k=1}^{3}\Psi_n^{(k)}$$

we get

$$E\Psi_n(\Delta X_t)\le \frac{3t}{n}+3\cdot\max(a_r,a_u,a_s,b_s)\sum_{k=1}^{3}\int_0^t E\left|\Delta X_l^k\right|dl.$$

Hence by using properties (B.4), (B.5) and the dominated convergence theorem, as $n\to\infty$, we finally find that

$$E\sum_{k=1}^{3}\left|\Delta X_t^k\right|\le 3\cdot\max(a_r,a_u,a_s,b_s)\int_0^t E\sum_{k=1}^{3}\left|\Delta X_l^k\right|dl,\ \forall\, 0\le t,\forall\, k.$$

By applying Gronwall's lemma[6] and sample path continuity, we get

$$E\sum_{k=1}^{3}\left|\Delta X_t^k\right|=0,\ \forall\, 0\le t,\ \text{and therefore}\ \left|\Delta X_t^k\right|=0\ \mathbf{P}\text{-}a.s.\ \forall\, k.$$

[4] See footnote 2.

[5] This is a basic property of Itó integrals.

[6] See, e.g., Karatzas & Shreve (1988), pp. 287-288: Suppose that the continuous function $g(t)$ satisfies

B.2 Proof of Theorem 6.3.1 for $\beta = \frac{1}{2}$

Proof.
If $\beta = \frac{1}{2}$, plugging in the partial derivatives of P^d yields the following system of ODEs

$$\frac{1}{2}\sigma_r^2 \left(B^d\right)^2 + \hat{a}_r B^d - B_t^d - 1 = 0 \tag{B.6}$$

$$\frac{1}{2}\sigma_s^2 \left(C^d\right)^2 + \hat{a}_s C^d - C_t^d - 1 = 0$$

$$\frac{1}{2}\sigma_u^2 \left(D^d\right)^2 + \hat{a}_u D^d - D_t^d - b_s C^d = 0$$

$$A^d \left(\theta_r B^d + \theta_u D^d\right) - A_t^d = 0 \tag{B.7}$$

The solutions for C^d and D^d are the same as in the case $\beta = 0$. As in equation (6.27), the solution to equation (B.6) that satisfies the boundary condition $B^d(T,T) = 0$ is

$$B^d(t,T) = \frac{1 - e^{-\delta_r(T-t)}}{\kappa_1^{(r)} - \kappa_2^{(r)} e^{-\delta_r(T-t)}},$$

where $\kappa_{1/2}^{(r)}$ is defined as in equation (6.25) and δ_r is given by equation (6.28). By direct substitution, the solution to equation (B.7) for A^d that satisfies the boundary condition $A^d(T,T) = 1$ is

$$A^d(t,T) = A^d(0,T) \exp\left[\int_0^t \theta_r(\tau) B^d(\tau,T) + \theta_u D^d(\tau,T)\, d\tau\right].$$

B.3 Proofs of Lemma 6.3.1 and Lemma 6.4.2

Proof.
As usual let $y = (y_r, y_s, y_u)$, $z = (r, s, u)$, and $x = (x_1, x_2, x_3)$. Let $\alpha > 0$ be a constant and $G^{(\alpha)}(y, \check{t}, z, t)$ the solution of the PDE

$$\begin{aligned} 0 = &\frac{1}{2}\sigma_r^2 r^{2\beta} G_{rr}^{(\alpha)} + \frac{1}{2}\sigma_s^2 s G_{ss}^{(\alpha)} + \frac{1}{2}\sigma_u^2 u G_{uu}^{(\alpha)} \\ &+ \left[\theta_r(t) - \hat{a}_r r\right] G_r^{(\alpha)} + \left[b_s u - \hat{a}_s s\right] G_s^{(\alpha)} \\ &+ \left[\theta_u - \hat{a}_u u\right] G_u^{(\alpha)} + G_t^{(\alpha)} - (r + \alpha \cdot s)\, G^{(\alpha)}, \end{aligned}$$

$$0 \le g(t) \le \alpha(t) + \beta \int_0^t g(s)\, ds, \; 0 \le t \le T,$$

with $\beta \ge 0$ and $\alpha : [0,T] \to \mathbf{R}$ integrable. Then

$$g(t) \le \alpha(t) + \beta \int_0^t \alpha(s)\, e^{\beta(t-s)} ds, 0 \le t \le T.$$

with boundary condition

$$G^{(\alpha)}\left(y,\check{t},z,\check{t}\right)=\delta\left(r-y_r\right)\delta\left(s-y_s\right)\delta\left(u-y_u\right).$$

The Fourier transformation of $G^{(\alpha)}\left(y,\check{t},z,t\right)$,

$$\widetilde{G}^{(\alpha)}\left(x,\check{t},z,t\right)=\frac{1}{(2\pi)^{3/2}}\iiint\limits_{\mathbf{R}^3}e^{ixy'}\widetilde{G}^{(\alpha)}\left(y,\check{t},z,t\right)dy,$$

is the solution of the PDE,

$$\begin{aligned}0=&\frac{1}{2}\sigma_r^2r^{2\beta}\widetilde{G}_{rr}^{(\alpha)}+\frac{1}{2}\sigma_s^2s\widetilde{G}_{ss}^{(\alpha)}+\frac{1}{2}\sigma_u^2u\widetilde{G}_{uu}^{(\alpha)}\\&+\left[\theta_r\left(t\right)-\hat{a}_rr\right]\widetilde{G}_r^{(\alpha)}+\left[b_su-\hat{a}_ss\right]\widetilde{G}_s^{(\alpha)}\\&+\left[\theta_u-\hat{a}_uu\right]\widetilde{G}_u^{(\alpha)}+\widetilde{G}_t^{(\alpha)}-\left(r+\alpha\cdot s\right)\widetilde{G}^{(\alpha)},\end{aligned}\tag{B.8}$$

with boundary condition

$$\widetilde{G}^{(\alpha)}\left(x,\check{t},z,\check{t}\right)=e^{izx'}.$$

Assume that $\widetilde{G}^{(\alpha)}\left(x,\check{t},z,t\right)$ is given by

$$\widetilde{G}^{(\alpha)}\left(x,\check{t},z,t\right)=A^{\widetilde{G}^{(\alpha)}}\left(x,\check{t},t\right)e^{-B^{\widetilde{G}^{(\alpha)}}(x,\check{t},t)r-C^{\widetilde{G}^{(\alpha)}}(x,\check{t},t)s-D^{\widetilde{G}^{(\alpha)}}(x,\check{t},t)u}.$$

Then solving the PDE (B.8) is equivalent to solving the following system of ODE's:
If $\beta=0$,

$$\hat{a}_rB^{\widetilde{G}^{(\alpha)}}-B_t^{\widetilde{G}^{(\alpha)}}-1=0,\tag{B.9}$$

$$\frac{1}{2}\sigma^2\left(C^{\widetilde{G}^{(\alpha)}}\right)^2+\hat{a}_sC^{\widetilde{G}^{(\alpha)}}-C_t^{\widetilde{G}^{(\alpha)}}-\alpha=0,\tag{B.10}$$

$$\frac{1}{2}\sigma_u^2\left(D^{\widetilde{G}^{(\alpha)}}\right)^2+\hat{a}_uD^{\widetilde{G}^{(\alpha)}}-D_t^{\widetilde{G}^{(\alpha)}}-b_sC^{\widetilde{G}^{(\alpha)}}=0,\tag{B.11}$$

$$A^{\widetilde{G}^{(\alpha)}}\left(\theta_rB^{\widetilde{G}^{(\alpha)}}+\theta_uD^{\widetilde{G}^{(\alpha)}}-\frac{1}{2}\sigma_r^2\left(B^{\widetilde{G}^{(\alpha)}}\right)^2\right)-A_t^{\widetilde{G}^{(\alpha)}}=0,\tag{B.12}$$

and if $\beta=\frac{1}{2}$,

$$\begin{aligned}\frac{1}{2}\sigma_r^2\left(B^{\widetilde{G}^{(\alpha)}}\right)^2+\hat{a}_rB^{\widetilde{G}^{(\alpha)}}-B_t^{\widetilde{G}^{(\alpha)}}-1&=0,\qquad\text{(B.13)}\\\frac{1}{2}\sigma_s^2\left(C^{\widetilde{G}^{(\alpha)}}\right)^2+\hat{a}_sC^{\widetilde{G}^{(\alpha)}}-C_t^{\widetilde{G}^{(\alpha)}}-\alpha&=0,\\\frac{1}{2}\sigma_u^2\left(D^{\widetilde{G}^{(\alpha)}}\right)^2+\hat{a}_uD^{\widetilde{G}^{(\alpha)}}-D_t^{\widetilde{G}^{(\alpha)}}-b_sC^{\widetilde{G}^{(\alpha)}}&=0,\\A^{\widetilde{G}^{(\alpha)}}\left(\theta_rB^{\widetilde{G}^{(\alpha)}}+\theta_uD^{\widetilde{G}^{(\alpha)}}\right)-A_t^{\widetilde{G}^{(\alpha)}}&=0,\qquad\text{(B.14)}\end{aligned}$$

in both cases with boundary conditions

$$A^{\widetilde{G}^{(\alpha)}}\left(x,\check{t},\check{t}\right)=1, \tag{B.15}$$

$$B^{\widetilde{G}^{(\alpha)}}\left(x,\check{t},\check{t}\right)=-ix_1, \tag{B.16}$$

$$C^{\widetilde{G}^{(\alpha)}}\left(x,\check{t},\check{t}\right)=-ix_2, \tag{B.17}$$

$$D^{\widetilde{G}^{(\alpha)}}\left(x,\check{t},\check{t}\right)=-ix_3. \tag{B.18}$$

Consider the case $\beta=0$:
The solution of equation (B.9) that satisfies the boundary condition (B.16) is

$$B^{\widetilde{G}^{(\alpha)}}\left(x,t,\check{t}\right)=-ix_1e^{-\hat{a}_r\left(\check{t}-t\right)}+\frac{1}{\hat{a}_r}\left(1-e^{-\hat{a}_r\left(\check{t}-t\right)}\right).$$

Consider the case $\beta=\frac{1}{2}$:
Equation (B.13) is the same as equation (B.6) in the bond pricing section and can be solved in the same way. All solutions are given by

$$B^{\widetilde{G}^{(\alpha)}}\left(x,t,\check{t}\right)=-\frac{2}{\sigma_r^2}\frac{\alpha_1\kappa_1^{(r)}e^{\kappa_1^{(r)}t}+\alpha_2\kappa_2^{(r)}e^{\kappa_2^{(r)}t}}{\alpha_1e^{\kappa_1^{(r)}t}+\alpha_2e^{\kappa_2^{(r)}t}}, \tag{B.19}$$

where α_1 and α_2 are some constants and $\kappa_{1/2}^{(r)}$ is defined as in equation (6.25). In order to satisfy the boundary condition (B.16),

$$\alpha_1=\alpha_2e^{\left(\kappa_2^{(r)}-\kappa_1^{(r)}\right)\check{t}}\varpi\left(x_1\right), \tag{B.20}$$

with

$$\varpi\left(j\right)=\frac{ij\sigma_r^2-2\kappa_2^{(r)}}{2\kappa_1^{(r)}-ij\sigma_r^2}. \tag{B.21}$$

Then, $B^{\widetilde{G}^{(\alpha)}}\left(x,t,\check{t}\right)$ can be determined from equations (B.19) and (B.20). By applying the same methods as in the proof of theorem 6.3.1, the solution to equation (B.10) with boundary condition (B.17) is given by

$$C^{\widetilde{G}^{(\alpha)}}\left(x,t,\check{t}\right)=\frac{\kappa_3^{(s,\alpha)}\left(x\right)-e^{-\delta_s^{(\alpha)}\left(\check{t}-t\right)}}{\kappa_4^{(s,\alpha)}\left(x\right)-\kappa_5^{(s,\alpha)}\left(x\right)e^{-\delta_s^{(\alpha)}\left(\check{t}-t\right)}},$$

where

$$\delta_x^{(\alpha)}=\sqrt{\hat{a}_x^2+2\alpha\sigma_x^2}, \tag{B.22}$$

$$\kappa_3^{(s,\alpha)}\left(x\right)=\frac{1+ix_2\kappa_2^{(s,\alpha)}}{1+ix_2\kappa_1^{(s,\alpha)}}, \tag{B.23}$$

$$\kappa_4^{(s,\alpha)}\left(x\right)=\frac{\kappa_1^{(s,\alpha)}-\frac{1}{2}ix_2\sigma_s^2}{1+ix_2\kappa_1^{(s,\alpha)}}, \tag{B.24}$$

$$\kappa_5^{(s,\alpha)}\left(x\right)=\frac{\kappa_2^{(s,\alpha)}-\frac{1}{2}ix_2\sigma_s^2}{1+ix_2\kappa_1^{(s,\alpha)}}, \tag{B.25}$$

with

$$\kappa_{1/2}^{(x,\alpha)} = \frac{\hat{a}_x}{2} \pm \frac{1}{2}\sqrt{\hat{a}_x^2 + 2\alpha\sigma_x^2}.$$

The solution to equation (B.11) with boundary condition (B.18) is given by

$$D^{\widetilde{G}^{(\alpha)}}\left(x,t,\check{t}\right) = \frac{-2\left(v^{\widetilde{G}^{(\alpha)}}\right)'\left(x,t,\check{t}\right)}{\sigma_u^2 v^{\widetilde{G}^{(\alpha)}}\left(x,t,\check{t}\right)},$$

where

$$\begin{aligned} & v^{\widetilde{G}^{(\alpha)}}\left(x,t,\check{t}\right) \\ &= \vartheta_1\left(\sigma_u^2 e^{-\delta_s^{(\alpha)}(\check{t}-t)}\right)^{\frac{\hat{a}_u}{2\delta_s^{(\alpha)}}-\phi\left(\kappa_4^{(s,\alpha)}(x),\kappa_3^{(s,\alpha)}(x)\right)} F_1^{\widetilde{G}^{(\alpha)}}\left(x,t,\check{t}\right) \\ &\quad +\vartheta_2\left(\sigma_u^2 e^{-\delta_s^{(\alpha)}(\check{t}-t)}\right)^{\frac{\hat{a}_u}{2\delta_s^{(\alpha)}}+\phi\left(\kappa_4^{(s,\alpha)}(x),\kappa_3^{(s,\alpha)}(x)\right)} F_3^{\widetilde{G}^{(\alpha)}}\left(x,t,\check{t}\right), \end{aligned} \tag{B.26}$$

where ϑ_1 and ϑ_2 are some constants and

$$\begin{aligned} F_1^{\widetilde{G}^{(\alpha)}}\left(x,t,\check{t}\right) = F\Big(&-\phi^{(\alpha)}\left(\kappa_5^{(s,\alpha)}(x)\right) - \phi^{(\alpha)}\left(\kappa_4^{(s,\alpha)}(x),\kappa_3^{(s,\alpha)}(x)\right), \\ &\phi^{(\alpha)}\left(\kappa_5^{(s,\alpha)}(x)\right) - \phi^{(\alpha)}\left(\kappa_4^{(s,\alpha)}(x),\kappa_3^{(s,\alpha)}(x)\right), \\ &1-2\phi^{(\alpha)}\left(\kappa_4^{(s,\alpha)}(x),\kappa_3^{(s,\alpha)}(x)\right), \\ &\kappa_5^{(s,\alpha)}(x)/\kappa_4^{(s,\alpha)}(x)\, e^{-\delta_s^{(\alpha)}(\check{t}-t)}\Big), \end{aligned}$$

$$\begin{aligned} F_3^{\widetilde{G}^{(\alpha)}}\left(x,t,\check{t}\right) = F\Big(&-\phi^{(\alpha)}\left(\kappa_5^{(s,\alpha)}(x)\right) + \phi^{(\alpha)}\left(\kappa_4^{(s,\alpha)}(x),\kappa_3^{(s,\alpha)}(x)\right), \\ &\phi^{(\alpha)}\left(\kappa_5^{(s,\alpha)}(x)\right) + \phi^{(\alpha)}\left(\kappa_4^{(s,\alpha)}(x),\kappa_3^{(s,\alpha)}(x)\right), \\ &1+2\phi^{(\alpha)}\left(\kappa_4^{(s,\alpha)}(x),\kappa_3^{(s,\alpha)}(x)\right), \\ &\kappa_5^{(s,\alpha)}(x)/\kappa_4^{(s,\alpha)}(x)\, e^{-\delta_s^{(\alpha)}(\check{t}-t)}\Big), \end{aligned}$$

with

$$\phi^{(\alpha)}(g,h) = \frac{1}{2\delta_s^{(\alpha)}}\sqrt{\frac{\hat{a}_u^2 g + 2b_s\sigma_u^2 h}{g}} \text{ and } \phi^{(\alpha)}(g) = \phi^{(\alpha)}(g,1),$$

and - as usual - $F(a,b,c,z)$ is the hypergeometric function. Differentiation of $v^{\widetilde{G}^{(\alpha)}}\left(x,t,\check{t}\right)$ yields

$$\left(v^{\widetilde{G}^{(\alpha)}}\right)'\left(x,t,\check{t}\right) = y_1^{\widetilde{G}^{(\alpha)}}\left(x,t,\check{t}\right) + y_2^{\widetilde{G}^{(\alpha)}}\left(x,t,\check{t}\right), \tag{B.27}$$

where

$$y_1^{\widetilde{G}^{(\alpha)}}\left(x,t,\check{t}\right) = \vartheta_1 \left(\sigma_u^2 e^{-\delta_s^{(\alpha)}\left(\check{t}-t\right)}\right)^{\frac{\hat{a}_u}{2\delta_s^{(\alpha)}} - \phi^{(\alpha)}\left(\kappa_4^{(s,\alpha)}(x),\kappa_3^{(s,\alpha)}(x)\right)} \varphi_2^{(\alpha)}\left(x,t,\check{t}\right),$$
$$y_2^{\widetilde{G}^{(\alpha)}}\left(x,t,\check{t}\right) = -\vartheta_2 \left(\sigma_u^2 e^{-\delta_s^{(\alpha)}\left(\check{t}-t\right)}\right)^{\frac{\hat{a}_u}{2\delta_s^{(\alpha)}} + \phi^{(\alpha)}\left(\kappa_4^{(s,\alpha)}(x),\kappa_3^{(s)}(x)\right)} \varphi_1^{(\alpha)}\left(x,t,\check{t}\right),$$

with

$$\varphi_1^{(\alpha)}\left(x,t,\check{t}\right) = \zeta_2^{(\alpha)}\left(x\right) e^{-\delta_s^{(\alpha)}\left(\check{t}-t\right)} F_4^{\widetilde{G}^{(\alpha)}}\left(x,t,\check{t}\right) - \xi_1^{(\alpha)}\left(x\right) F_3^{\widetilde{G}^{(\alpha)}}\left(x,t,\check{t}\right),$$
$$\varphi_2^{(\alpha)}\left(x,t,\check{t}\right) = \xi_2^{(\alpha)}\left(x\right) F_1^{\widetilde{G}^{(\alpha)}}\left(x,t,\check{t}\right) - \zeta_1^{(\alpha)}\left(x\right) e^{-\delta_s^{(\alpha)}\left(\check{t}-t\right)} F_2^{\widetilde{G}^{(\alpha)}}\left(x,t,\check{t}\right),$$

where

$$\begin{aligned} F_2^{\widetilde{G}^{(\alpha)}}\left(x,t,\check{t}\right) = F\Big(& 1-\phi^{(\alpha)}\left(\kappa_5^{(s,\alpha)}\left(x\right)\right) - \phi^{(\alpha)}\left(\kappa_4^{(s,\alpha)}\left(x\right),\kappa_3^{(s,\alpha)}\left(x\right)\right), \\ & 1+\phi^{(\alpha)}\left(\kappa_5^{(s,\alpha)}\left(x\right)\right) - \phi^{(\alpha)}\left(\kappa_4^{(s,\alpha)}\left(x\right),\kappa_3^{(s,\alpha)}\left(x\right)\right), \\ & 2-2\phi^{(\alpha)}\left(\kappa_4^{(s,\alpha)}\left(x\right),\kappa_3^{(s,\alpha)}\left(x\right)\right), \\ & \kappa_5^{(s,\alpha)}\left(x\right)/\kappa_4^{(s,\alpha)}\left(x\right) e^{-\delta_s^{(\alpha)}\left(\check{t}-t\right)}\Big), \end{aligned}$$

$$\begin{aligned} F_4^{\widetilde{G}^{(\alpha)}}\left(x,t,\check{t}\right) = F\Big(& 1-\phi^{(\alpha)}\left(\kappa_5^{(s,\alpha)}\left(x\right)\right) + \phi^{(\alpha)}\left(\kappa_4^{(s,\alpha)}\left(x\right),\kappa_3^{(s,\alpha)}\left(x\right)\right), \\ & 1+\phi^{(\alpha)}\left(\kappa_5^{(s,\alpha)}\left(x\right)\right) + \phi^{(\alpha)}\left(\kappa_4^{(s,\alpha)}\left(x\right),\kappa_3^{(s,\alpha)}\left(x\right)\right), \\ & 2+2\phi^{(\alpha)}\left(\kappa_4^{(s,\alpha)}\left(x\right),\kappa_3^{(s,\alpha)}\left(x\right)\right), \\ & \kappa_5^{(s,\alpha)}\left(x\right)/\kappa_4^{(s,\alpha)}\left(x\right) e^{-\delta_s^{(\alpha)}\left(\check{t}-t\right)}\Big), \end{aligned}$$

and

$$\xi_{1/2}^{(\alpha)}\left(x\right) = \frac{\hat{a}_u}{2} \pm \delta_s^{(\alpha)} \phi^{(\alpha)}\left(\kappa_4^{(s,\alpha)}\left(x\right),\kappa_3^{(s,\alpha)}\left(x\right)\right),$$
$$\zeta_{1/2}^{(\alpha)}\left(x\right) = \delta_s^{(\alpha)} \frac{\kappa_5^{(s,\alpha)}\left(x\right)}{\kappa_4^{(s,\alpha)}\left(x\right)} \frac{\left(\phi^{(\alpha)}\right)^2\left(\kappa_5^{(s,\alpha)}\left(x\right)\right) - \left(\phi^{(\alpha)}\right)^2\left(\kappa_4^{(s,\alpha)}\left(x\right),\kappa_3^{(s,\alpha)}\left(x\right)\right)}{1 \mp 2\phi^{(\alpha)}\left(\kappa_4^{(s,\alpha)}\left(x\right),\kappa_3^{(s,\alpha)}\left(x\right)\right)}.$$

In order to satisfy the boundary condition $D^{\widetilde{G}^{(\alpha)}}\left(x,\check{t},\check{t}\right) = -ix_3$, ϑ_1 equals

$$\vartheta_1 = \vartheta_2 \left(\sigma_u^2\right)^{2\phi^{(\alpha)}\left(\kappa_4^{(s,\alpha)}(x),\kappa_3^{(s,\alpha)}(x)\right)} \frac{\pi_1^{(\alpha)}\left(x\right)}{\pi_2^{(\alpha)}\left(x\right)}, \tag{B.28}$$

where

$$\pi_1^{(\alpha)}\left(x\right) = 2\varphi_1^{(\alpha)}\left(x,\check{t},\check{t}\right) + ix_3\sigma_u^2 F_3^{\widetilde{G}^{(\alpha)}}\left(x,\check{t},\check{t}\right),$$

and

$$\pi_2^{(\alpha)}(x) = 2\varphi_2^{(\alpha)}\left(x, \check{t}, \check{t}\right) - ix_3\sigma_u^2 F_1^{\widetilde{G}^{(\alpha)}}\left(x, \check{t}, \check{t}\right).$$

Then $D^{\widetilde{G}^{(\alpha)}}\left(x, t, \check{t}\right)$ can be easily determined as

$$\begin{aligned}
& D^{\widetilde{G}^{(\alpha)}}\left(x, t, \check{t}\right) \\
&= \frac{-2}{\sigma_u^2}\left(\pi_1^{(\alpha)}(x)\,\varphi_2^{(\alpha)}\left(x, t, \check{t}\right)\right. \\
&\left. \quad -e^{-2\delta_s^{(\alpha)}\phi^{(\alpha)}\left(\kappa_4^{(s,\alpha)}(x),\kappa_3^{(s,\alpha)}(x)\right)\left(\check{t}-t\right)}\pi_2^{(\alpha)}(x)\,\varphi_1^{(\alpha)}\left(x, t, \check{t}\right)\right) / \\
&\quad \left(\pi_1^{(\alpha)}(x)\,F_1^{\widetilde{G}^{(\alpha)}}\left(x, t, \check{t}\right)\right. \\
&\left. \quad +e^{-2\delta_s^{(\alpha)}\phi^{(\alpha)}\left(\kappa_4^{(s,\alpha)}(x),\kappa_3^{(s,\alpha)}(x)\right)\left(\check{t}-t\right)}\pi_2^{(\alpha)}(x)\,F_3^{\widetilde{G}^{(\alpha)}}\left(x, t, \check{t}\right)\right).
\end{aligned}$$

The equations for $A^{\widetilde{G}^{(\alpha)}}\left(x, \check{t}, t\right)$ follow directly from equations (B.12), (B.14), and (B.15):
If $\beta = 0$, by direct substitution, the solution of equation (B.12) for $A^{\widetilde{G}^{(\alpha)}}$ that satisfies the boundary condition (B.15) is

$$\begin{aligned}
& A^{\widetilde{G}^{(\alpha)}}\left(x, t, \check{t}\right) \\
&= e^{-\int_{\check{t}}^{t}\left(\frac{1}{2}\sigma_r^2\left(B^{\widetilde{G}^{(\alpha)}}\left(x,\tau,\check{t}\right)\right)^2 - \theta_r(\tau)B^{\widetilde{G}^{(\alpha)}}\left(x,\tau,\check{t}\right) - \theta_u D^{\widetilde{G}^{(\alpha)}}\left(x,\tau,\check{t}\right)\right)d\tau}.
\end{aligned}$$

If $\beta = \frac{1}{2}$, by direct substitution, the solution to equation (B.14) for $A^{\widetilde{G}^{(\alpha)}}$ that satisfies the boundary condition (B.15) is

$$A^{\widetilde{G}^{(\alpha)}}\left(x, t, \check{t}\right) = \exp\left[\int_{\check{t}}^{t}\left(\theta_r(\tau)\,B^{\widetilde{G}^{(\alpha)}}\left(x, \tau, \check{t}\right) + \theta_u D^{\widetilde{G}^{(\alpha)}}\left(x, \tau, \check{t}\right)\right)d\tau\right].$$

B.4 Proof of Lemma 6.4.3

Proof.
We substitute

$$\widetilde{y}_r = \frac{c_1 y_r}{c_0},\ \widetilde{y}_s = \frac{c_2 y_s}{c_0},\ \text{and } \widetilde{y}_u = \frac{c_1 y_u}{c_0},$$

and

$$\widetilde{a}_i = \frac{c_0 a_i}{c_i} \text{ for } i = 1, 2, 3.$$

Then, we get

$$\Lambda^C(a_1,a_2,a_3,c_0,c_1,c_2,c_3)$$
$$= \frac{c_0^3}{c_1c_2c_3} \iiint\limits_{\mathbf{R}^3} e^{-i\widetilde{a}_1\widetilde{y}_r - i\widetilde{a}_2\widetilde{y}_s - i\widetilde{a}_3\widetilde{y}_u} 1_{\{\widetilde{y}_r+\widetilde{y}_s+\widetilde{y}_u\le 1\}} d\widetilde{y}_r d\widetilde{y}_s d\widetilde{y}_u$$
$$= \frac{c_0^3}{c_1c_2c_3} \int_0^1 \int_0^{1-\widetilde{y}_u} \int_0^{1-\widetilde{y}_s-\widetilde{y}_u} e^{-i\widetilde{a}_1\widetilde{y}_r - i\widetilde{a}_2\widetilde{y}_s - i\widetilde{a}_3\widetilde{y}_u} d\widetilde{y}_r d\widetilde{y}_s d\widetilde{y}_u$$
$$= \frac{ic_0^3}{c_1c_2c_3\widetilde{a}_1} \left(e^{-i\widetilde{a}_1} \int_0^1 e^{-i(\widetilde{a}_3-\widetilde{a}_1)\widetilde{y}_u} \int_0^{1-\widetilde{y}_u} e^{-i(\widetilde{a}_2-\widetilde{a}_1)\widetilde{y}_s} d\widetilde{y}_s d\widetilde{y}_u \right.$$
$$\left. - \int_0^1 e^{-i\widetilde{a}_3\widetilde{y}_u} \int_0^{1-\widetilde{y}_u} e^{-i\widetilde{a}_2\widetilde{y}_s} d\widetilde{y}_s d\widetilde{y}_u \right)$$
$$= -\frac{c_0^3}{c_1c_2c_3\widetilde{a}_1} \cdot$$
$$\left(\frac{e^{\;i\widetilde{a}_1}}{\widetilde{a}_2-\widetilde{a}_1} \left(e^{-i(\widetilde{a}_2-\widetilde{a}_1)} \int_0^1 e^{-i(\widetilde{a}_3-\widetilde{a}_2)\widetilde{y}_u} d\widetilde{y}_u - \int_0^1 e^{-i(\widetilde{a}_3-\widetilde{a}_1)\widetilde{y}_u} d\widetilde{y}_u \right) \right.$$
$$\left. -\frac{1}{\widetilde{a}_2} \left(e^{-i\widetilde{a}_2} \int_0^1 e^{-i(\widetilde{a}_3-\widetilde{a}_2)\widetilde{y}_u} d\widetilde{y}_u - \int_0^1 e^{-i\widetilde{a}_3\widetilde{y}_u} d\widetilde{y}_u \right) \right)$$
$$= -\frac{ic_0^3}{c_1c_2c_3\widetilde{a}_1} \cdot$$
$$\left(\frac{e^{-i\widetilde{a}_3} - e^{-i\widetilde{a}_2}}{(\widetilde{a}_3-\widetilde{a}_2)(\widetilde{a}_2-\widetilde{a}_1)} - \frac{e^{-i\widetilde{a}_3} - e^{-i\widetilde{a}_1}}{(\widetilde{a}_3-\widetilde{a}_1)(\widetilde{a}_2-\widetilde{a}_1)} + \frac{e^{-i\widetilde{a}_3} - 1}{\widetilde{a}_2\widetilde{a}_3} - \frac{e^{-i\widetilde{a}_3} - e^{-i\widetilde{a}_2}}{\widetilde{a}_2(\widetilde{a}_3-\widetilde{a}_2)} \right).$$

The proof for Λ^P works similarly.

B.5 Tools for Pricing Non-Defaultable Contingent Claims

Lemma B.5.1.
The solution to the PDE

$$\begin{aligned} 0 = &\frac{1}{2}\sigma_r^2 r^{2\beta} \overline{G}_{rr} + \frac{1}{2}\sigma_s^2 s \overline{G}_{ss} + \frac{1}{2}\sigma_u^2 u \overline{G}_{uu} \\ &+ [\theta_r(t) - \hat{a}_r r]\,\overline{G}_r + [b_s u - \hat{a}_s s]\,\overline{G}_s \\ &+ [\theta_u - \hat{a}_u u]\,\overline{G}_u + \overline{G}_t - r\overline{G}, \end{aligned}$$

with boundary condition

$$\overline{G}(y,\check{t},z,\check{t}) = \delta(r-y_r)\,\delta(s-y_s)\,\delta(u-y_u),$$

is given by

$$\overline{G}(y,\check{t},z,t) = \frac{1}{(2\pi)^{3/2}} \iiint\limits_{\mathbf{R}^3} e^{-ixy'} \underline{G}(x,\check{t},z,t)\,dx \qquad \text{(B.29)}$$

where

$$\underline{G}\left(x,\check{t},z,t\right)=A^{\underline{G}}\left(x,t,\check{t}\right)e^{-B^{\tilde{G}}\left(x,t,\check{t}\right)r-C^{\underline{G}}\left(x,t,\check{t}\right)s-D^{\underline{G}}\left(x,t,\check{t}\right)u}$$

with

$$C^{\underline{G}}\left(x,t,\check{t}\right)=\frac{2\hat{a}_s x_2}{\left(\sigma_s^2x_2+2i\hat{a}_s\right)e^{-\hat{a}_s\left(t-\check{t}\right)}-\sigma_s^2x_2},$$

$$D^{\underline{G}}\left(x,t,\check{t}\right)=\frac{-2\left(v^{\underline{G}}\right)'\left(x,t,\check{t}\right)}{\sigma_u^2v^{\underline{G}}\left(x,t,\check{t}\right)},$$

and if $\beta=0:$

$$A^{\underline{G}}\left(x,t,\check{t}\right)=e^{-\int_t^{\check{t}}\left(-\frac{1}{2}\sigma_r^2\left(B^{\tilde{G}}\right)^2\left(x,\tau,\check{t}\right)+\theta_r(\tau)B^{\tilde{G}}\left(x,\tau,\check{t}\right)+\theta_u D^{\underline{G}}\left(x,\tau,\check{t}\right)\right)d\tau},$$

if $\beta=\frac{1}{2}:$

$$A^{\underline{G}}\left(x,t,\check{t}\right)=e^{-\int_t^{\check{t}}\left(\theta_r(\tau)B^{\tilde{G}}\left(x,\tau,\check{t}\right)+\theta_u D^{\underline{G}}\left(x,\tau,\check{t}\right)\right)d\tau}.$$

$v^{\underline{G}}\left(x,\check{t},t\right)$ *and* $\left(v^{\underline{G}}\right)'\left(x,\check{t},t\right)$ *are defined as in equations (B.44), (B.45) and (B.46) below.* $B^{\tilde{G}}\left(x,\check{t},t\right)$ *is given by equation (6.52).*

Proof.
The Fourier transformation of $\overline{G}\left(y,\check{t},z,t\right)$,

$$\underline{G}\left(x,\check{t},z,t\right)=\frac{1}{\left(2\pi\right)^{3/2}}\iiint_{\mathbf{R}^3}e^{ixy'}\overline{G}\left(y,\check{t},z,t\right)dy,$$

is the solution of the PDE

$$\begin{aligned}0=&\frac{1}{2}\sigma_r^2r^{2\beta}_{rr}\underline{G}_{rr}+\frac{1}{2}\sigma_s^2s\underline{G}_{ss}+\frac{1}{2}\sigma_u^2u\underline{G}_{uu}\\&+\left[\theta_r\left(t\right)-\hat{a}_rr\right]\underline{G}_r+\left[b_su-\hat{a}_ss\right]\underline{G}_s\\&+\left[\theta_u-\hat{a}_uu\right]\underline{G}_u+\underline{G}_t-r\underline{G},\end{aligned}\tag{B.30}$$

with boundary condition

$$\underline{G}\left(x,\check{t},z,\check{t}\right)=e^{izx'}.$$

Assume that $\underline{G}\left(x,\check{t},z,t\right)$ is given by

$$\underline{G}\left(x,\check{t},z,t\right)=A^{\underline{G}}\left(x,t,\check{t}\right)e^{-B^{\underline{G}}\left(x,t,\check{t}\right)r-C^{\underline{G}}\left(x,t,\check{t}\right)s-D^{\underline{G}}\left(x,t,\check{t}\right)u}.$$

Then solving the PDE B.30 is equivalent to solving the following system of ODE's:

If $\beta = 0$,

$$\hat{a}_r B^{\underline{G}} - B_t^{\underline{G}} - 1 = 0, \tag{B.31}$$

$$\frac{1}{2}\sigma^2 \left(C^{\underline{G}}\right)^2 + \hat{a}_s C^{\underline{G}} - C_t^{\underline{G}} = 0, \tag{B.32}$$

$$\frac{1}{2}\sigma_u^2 \left(D^{\underline{G}}\right)^2 + \hat{a}_u D^{\underline{G}} - D_t^{\underline{G}} - b_s C^{\underline{G}} = 0, \tag{B.33}$$

$$A^{\underline{G}} \left(\theta_r B^{\underline{G}} + \theta_u D^{\underline{G}} - \frac{1}{2}\sigma_r^2 \left(B^{\underline{G}}\right)^2\right) - A_t^{\underline{G}} = 0, \tag{B.34}$$

and if $\beta = \frac{1}{2}$,

$$\frac{1}{2}\sigma_r^2 \left(B^{\underline{G}}\right)^2 + \hat{a}_r B^{\underline{G}} - B_t^{\underline{G}} - 1 = 0, \tag{B.35}$$

$$\frac{1}{2}\sigma_s^2 \left(C^{\underline{G}}\right)^2 + \hat{a}_s C^{\underline{G}} - C_t^{\underline{G}} = 0,$$

$$\frac{1}{2}\sigma_u^2 \left(D^{\underline{G}}\right)^2 + \hat{a}_u D^{\underline{G}} - D_t^{\underline{G}} - b_s C^{\underline{G}} = 0,$$

$$A^{\underline{G}} \left(\theta_r B^{\underline{G}} + \theta_u D^{\underline{G}}\right) - A_t^{\underline{G}} = 0, \tag{B.36}$$

in both cases with boundary conditions

$$A^{\underline{G}} \left(x, \check{t}, \check{t}\right) = 1, \tag{B.37}$$

$$B^{\underline{G}} \left(x, \check{t}, \check{t}\right) = -ix_1, \tag{B.38}$$

$$C^{\underline{G}} \left(x, \check{t}, \check{t}\right) = -ix_2, \tag{B.39}$$

$$D^{\underline{G}} \left(x, \check{t}, \check{t}\right) = -ix_3. \tag{B.40}$$

Equations (B.31), (B.35), and (B.38) are the same as equations (B.9), (B.13), and (B.16). Therefore,

$$B^{\underline{G}} \left(x, t, \check{t}\right) = B^{\widetilde{G}} \left(x, t, \check{t}\right).$$

where $B^{\widetilde{G}}$ is defined by equation (6.52). Equation (B.32) has Bernoulli form

$$C_t^{\underline{G}} = \hat{a}_s C^{\underline{G}} + \frac{1}{2}\sigma^2 \left(C^{\underline{G}}\right)^2.$$

The solution is

$$C^{\underline{G}} \left(x, t, \check{t}\right) = \frac{1}{w^{\underline{G}} \left(x, t, \check{t}\right)},$$

where $w^{\underline{G}} \left(\check{t}, t\right)$ satisfies

$$\left(w^{\underline{G}}\right)' = -\hat{a}_s w^{\underline{G}} - \frac{1}{2}\sigma_s^2. \tag{B.41}$$

All solutions of ODE (B.41) are given by

$$w^{\underline{G}} = -\frac{\sigma_s^2}{2\hat{a}_s} + \alpha e^{-\hat{a}_s t},$$

where α is some constant. Then all solutions of equation (B.32) are given by

$$C^{\underline{G}}\left(x,t,\check{t}\right) = \frac{2\hat{a}_s}{2\hat{a}_s \alpha e^{-\hat{a}_s t} - \sigma_s^2}.$$

In order to satisfy the boundary condition (B.39),

$$\alpha = \frac{x_2 \sigma_s^2 + 2i\hat{a}_s}{2x_2\hat{a}_s} e^{\hat{a}_s \check{t}},$$

and therefore

$$C^{\underline{G}}\left(x,t,\check{t}\right) = \frac{2\hat{a}_s x_2}{\left(\sigma_s^2 x_2 + 2i\hat{a}_s\right) e^{-\hat{a}_s\left(t-\check{t}\right)} - \sigma_s^2 x_2}.$$

Equation (B.33) has Ricatti form

$$D_t^{\underline{G}} = \frac{1}{2}\sigma_u^2 \left(D^{\underline{G}}\right)^2 + \hat{a}_u D^{\underline{G}} - b_s C^{\underline{G}}.$$

The solution is

$$D^{\underline{G}}\left(x,t,\check{t}\right) = \frac{-2\left(v^{\underline{G}}\right)'\left(x,t,\check{t}\right)}{\sigma_u^2 v^{\underline{G}}\left(x,t,\check{t}\right)}, \tag{B.42}$$

where $v^{\underline{G}}\left(\check{t},t\right)$ satisfies

$$\left(v^{\underline{G}}\right)'' - \hat{a}_u\left(v^{\underline{G}}\right)' - \frac{1}{2} C^{\underline{G}} b_s \sigma_u^2 v^{\underline{G}} = 0. \tag{B.43}$$

The solutions of equation (B.43) are given by

$$v^{\underline{G}}\left(x,t,\check{t}\right) = \vartheta_1 F_1^{\underline{G}}\left(x,t,\check{t}\right) + \vartheta_2\left(e^{\hat{a}_s t} b_s \sigma_u^2 x_2\right)^{\frac{\hat{a}_u}{\hat{a}_s}} F_3^{\underline{G}}\left(x,t,\check{t}\right), \tag{B.44}$$

where ϑ_1 and ϑ_2 are some constants and

$$F_1^{\underline{G}}\left(x,t,\check{t}\right) = F\left(-\varrho_1, -\varrho_2, 1 - \frac{\hat{a}_u}{\hat{a}_s}, e^{-\hat{a}_s\left(\check{t}-t\right)}\eta\left(x\right)\right),$$

$$F_3^{\underline{G}}\left(x,t,\check{t}\right) = F\left(\varrho_2, \varrho_1, 1 + \frac{\hat{a}_u}{\hat{a}_s}, e^{-\hat{a}_s\left(\check{t}-t\right)}\eta\left(x\right)\right),$$

with

$$\varrho_{1/2} = \frac{\hat{a}_u\sigma_s \pm \sqrt{-4b_s\hat{a}_s\sigma_u^2 + \hat{a}_u^2\sigma_s^2}}{2\hat{a}_s\sigma_s} \text{ and } \eta\left(x\right) = \frac{x_2\sigma_s^2}{2i\hat{a}_s + x_2\sigma_s^2}.$$

Differentiation of $v^{\underline{G}}\left(x,t,\check{t}\right)$ yields

$$\left(v^{\underline{G}}\right)'\left(x,t,\check{t}\right)=y_1^{\underline{G}}\left(x,t,\check{t}\right)+y_2^{\underline{G}}\left(x,t,\check{t}\right), \tag{B.45}$$

where

$$y_1^{\underline{G}}\left(x,t,\check{t}\right)=\vartheta_1 e^{-\hat{a}_s\left(\check{t}-t\right)}\eta\left(x\right)\iota_2 F_2^{\underline{G}}\left(x,t,\check{t}\right),$$
$$y_2^{\underline{G}}\left(x,t,\check{t}\right)=\vartheta_2\left(e^{\hat{a}_s t}b_s\sigma_u^2 x_2\right)^{\frac{\hat{a}_u}{\hat{a}_s}}$$
$$\left(\hat{a}_u F_3^{\underline{G}}\left(x,t,\check{t}\right)-e^{-\hat{a}_s\left(\check{t}-t\right)}\eta\left(x\right)\iota_1 F_4^{\underline{G}}\left(x,t,\check{t}\right)\right),$$

with

$$\iota_{1/2}=\frac{b_s\hat{a}_s\sigma_u^2}{\sigma_s^2\left(\hat{a}_s\pm\hat{a}_u\right)}$$

and

$$F_2^{\underline{G}}\left(x,t,\check{t}\right)=F\left(1-\varrho_1,1-\varrho_2,2-\frac{\hat{a}_u}{\hat{a}_s},e^{-\hat{a}_s\left(\check{t}-t\right)}\eta\left(x\right)\right),$$
$$F_4^{\underline{G}}\left(x,t,\check{t}\right)=F\left(1+\varrho_2,1+\varrho_1,2+\frac{\hat{a}_u}{\hat{a}_s},e^{-\hat{a}_s\left(\check{t}-t\right)}\eta\left(x\right)\right).$$

In order to satisfy the boundary condition (B.40),

$$\vartheta_1=\vartheta_2\left(e^{\hat{a}_s\check{t}}b_s\sigma_u^2 x_2\right)^{\frac{\hat{a}_u}{\hat{a}_s}}\mu\left(x\right), \tag{B.46}$$

with

$$\mu\left(x\right)=\frac{\left(2\hat{a}_u-ix_3\sigma_u^2\right)F_3^{\underline{G}}\left(x,t,\check{t}\right)-2e^{-\hat{a}_s\left(\check{t}-t\right)}\eta\left(x\right)\iota_1 F_4^{\underline{G}}\left(x,t,\check{t}\right)}{ix_3\sigma_u^2 F_1^{\underline{G}}\left(x,t,\check{t}\right)-2e^{-\hat{a}_s\left(\check{t}-t\right)}\eta\left(x\right)\iota_2 F_2^{\underline{G}}\left(x,t,\check{t}\right)}.$$

Then $D^{\underline{G}}\left(x,\check{t},t\right)$ can be determined from equations (B.42), (B.44), (B.45) and (B.46).
Consider the case $\beta=0$:
The solution of equation (B.34) for $A^{\underline{G}}$ that satisfies the boundary condition (B.37) is

$$A^{\underline{G}}\left(x,t,\check{t}\right)$$
$$=e^{-\int_t^{\check{t}}\left(-\frac{1}{2}\sigma_r^2\left(B^{\tilde{G}}\right)^2\left(x,\tau,\check{t}\right)+\theta_r(\tau)B^{\tilde{G}}\left(x,\tau,\check{t}\right)+\theta_u D^{\underline{G}}\left(x,\tau,\check{t}\right)\right)d\tau}.$$

Consider the case $\beta=\frac{1}{2}$:
The solution of equation (B.36) for $A^{\underline{G}}$ that satisfies the boundary condition (B.37) is

$$A^{\underline{G}}\left(x,t,\check{t}\right)=e^{-\int_t^{\check{t}}\left(\theta_r(\tau)B^{\tilde{G}}\left(x,\tau,\check{t}\right)+\theta_u D^{\underline{G}}\left(x,\tau,\check{t}\right)\right)d\tau}.$$

C. Pricing of Credit Derivatives: Extensions

If we let $\beta = 0$, we must take into account that r may become negative. Therefore, instead of considering the sets $\mathcal{B}_1$, $\mathcal{B}_2$, $\mathcal{B}_3$, and $\mathcal{B}_4$, we have to work with

$$\widetilde{\mathcal{B}}_1 = \left\{ r \in \mathbb{R},\ (s,u) \in \mathbb{R}^2_+ | B(T_O,T)\, r + C^d(T_O,T)\, s + D^d(T_O,T)\, u \le K^* \right\},$$
$$\widetilde{\mathcal{B}}_2 = \left\{ r \in \mathbb{R},\ (s,u) \in \mathbb{R}^2_+ | K^* \le B(T_O,T)\, r + C^d(T_O,T)\, s + D^d(T_O,T)\, u \right\},$$

and

$$\widetilde{\mathcal{B}}_3 = \left\{ r \in \mathbb{R},\ (s,u) \in \mathbb{R}^2_+ | \right.$$
$$\left. C^d(T_O,T)\, s + D^d(T_O,T)\, u \le (T - T_O)\, K + \ln\left(A^d(T_O,T) / A(T_O,T)\right) \right\},$$
$$\widetilde{\mathcal{B}}_4 = \left\{ r \in \mathbb{R},\ (s,u) \in \mathbb{R}^2_+ | \right.$$
$$\left. (T - T_O)\, K + \ln\left(A^d(T_O,T) / A(T_O,T)\right) \le C^d(T_O,T)\, s + D^d(T_O,T)\, u \right\}.$$

The only consequences are more complicated calculations of Λ^C and Λ^P. Everything else works like in the case $\beta = \frac{1}{2}$ which we presented in section 6.4. In the following we show the result for Λ^C, the calculation of Λ^P works identically:

$$\begin{aligned} &\Lambda^C(a_1,a_2,a_3,c_0,c_1,c_2,c_3) \\ &= \frac{c_0^3}{c_1 c_2 c_3} \iiint\limits_{\mathbf{R}^3} e^{-i\widetilde{a}_1\widetilde{y}_r - i\widetilde{a}_2\widetilde{y}_s - i\widetilde{a}_3\widetilde{y}_u} 1_{\{\widetilde{y}_r+\widetilde{y}_s+\widetilde{y}_u \le 1\}} d\widetilde{y}_r d\widetilde{y}_s d\widetilde{y}_u \\ &= \frac{c_0^3}{c_1 c_2 c_3} \int_0^1 e^{-i\widetilde{a}_3\widetilde{y}_u} \int_0^{1-\widetilde{y}_u} e^{-i\widetilde{a}_2\widetilde{y}_s} \int_{-\infty}^{1-\widetilde{y}_s-\widetilde{y}_u} e^{-i\widetilde{a}_1\widetilde{y}_r} d\widetilde{y}_r d\widetilde{y}_s d\widetilde{y}_u, \end{aligned}$$

where $\widetilde{y}_r$, $\widetilde{y}_s$, $\widetilde{y}_u$, and $\widetilde{a}_i$, $i = 1,2,3$, are defined as in section B.4. Because

$$\int_{-\infty}^{1-\widetilde{y}_s-\widetilde{y}_u} e^{-i\widetilde{a}_1\widetilde{y}_r} d\widetilde{y}_r = \int_{-\infty}^{0} e^{-i\widetilde{a}_1\widetilde{y}_r} d\widetilde{y}_r + \int_0^{1-\widetilde{y}_s-\widetilde{y}_u} e^{-i\widetilde{a}_1\widetilde{y}_r} d\widetilde{y}_r,$$

we can concentrate on the calculation of

$$\begin{aligned} \int_{-\infty}^{0} e^{-i\widetilde{a}_1\widetilde{y}_r} d\widetilde{y}_r &= \int_0^{\infty} e^{i\widetilde{a}_1\widetilde{y}_r} d\widetilde{y}_r \\ &= \int_{-\infty}^{\infty} H(\widetilde{y}_r)\, e^{i\widetilde{a}_1\widetilde{y}_r} d\widetilde{y}_r, \end{aligned}$$

where $H(\widetilde{y}_r)$ is the Heaviside step function[1]. According to Evans, Blackledge & Yardley (2000), page 43, the Fourier transform of $H(\widetilde{y}_r)$ is given by

$$\frac{1}{\sqrt{2\pi}}\left(\pi\delta(\widetilde{a}_1)-\frac{1}{i\widetilde{a}_1}\right),$$

where δ is the Dirac delta function[2]. Hence,

$$\int_{-\infty}^{0} e^{-i\widetilde{a}_1\widetilde{y}_r}\,d\widetilde{y}_r=\pi\delta(\widetilde{a}_1)-\frac{1}{i\widetilde{a}_1}.$$

Therefore,

$$\begin{aligned}
&\Lambda^C(a_1,a_2,a_3,c_0,c_1,c_2,c_3)\\
&=-\frac{ic_0^3}{c_1c_2c_3a_1}\cdot\\
&\left(\frac{e^{-i\widetilde{a}_3}-e^{-i\widetilde{a}_2}}{(\widetilde{a}_3-\widetilde{a}_2)(\widetilde{a}_2-\widetilde{a}_1)}-\frac{e^{-i\widetilde{a}_3}-e^{-i\widetilde{a}_1}}{(\widetilde{a}_3-\widetilde{a}_1)(\widetilde{a}_2-\widetilde{a}_1)}+\frac{e^{-i\widetilde{a}_3}-1}{\widetilde{a}_2\widetilde{a}_3}-\frac{e^{-i\widetilde{a}_3}}{\widetilde{a}_2(\widetilde{a}_3-\widetilde{a}_2)}\right)\\
&+\frac{c_0^3}{c_1c_2c_3}\left(\pi\delta(\widetilde{a}_1)+\frac{i}{\widetilde{a}_1}\right)\int_0^1 e^{-i\widetilde{a}_3\widetilde{y}_u}\int_0^{1-\widetilde{y}_u} e^{-i\widetilde{a}_2\widetilde{y}_s}\,d\widetilde{y}_s\,d\widetilde{y}_u\\
&=-\frac{ic_0^3}{c_1c_2c_3}\left(\frac{e^{-i\widetilde{a}_3}-e^{-i\widetilde{a}_2}}{\widetilde{a}_1(\widetilde{a}_3-\widetilde{a}_2)(\widetilde{a}_2-\widetilde{a}_1)}-\frac{e^{-i\widetilde{a}_3}-e^{-i\widetilde{a}_1}}{\widetilde{a}_1(\widetilde{a}_3-\widetilde{a}_1)(\widetilde{a}_2-\widetilde{a}_1)}+\frac{e^{-i\widetilde{a}_3}-1}{\widetilde{a}_1\widetilde{a}_2\widetilde{a}_3}\right.\\
&\left.-\frac{e^{-i\widetilde{a}_3}-e^{-i\widetilde{a}_2}}{\widetilde{a}_1\widetilde{a}_2(\widetilde{a}_3-\widetilde{a}_2)}+\frac{e^{-i\widetilde{a}_3}-e^{-i\widetilde{a}_2}}{\widetilde{a}_1\widetilde{a}_2(\widetilde{a}_3-\widetilde{a}_2)}-\frac{e^{-i\widetilde{a}_3}-1}{\widetilde{a}_1\widetilde{a}_2\widetilde{a}_3}\right)\\
&-\frac{c_0^3}{c_1c_2c_3}\pi\delta(\widetilde{a}_1)\left(\frac{e^{-i\widetilde{a}_3}-e^{-i\widetilde{a}_2}}{\widetilde{a}_2(\widetilde{a}_3-\widetilde{a}_2)}-\frac{e^{-i\widetilde{a}_3}-1}{\widetilde{a}_2\widetilde{a}_3}\right)\\
&=\frac{c_0^3}{c_1c_2c_3}\left(\frac{e^{-i\widetilde{a}_3}-e^{-i\widetilde{a}_2}}{i\widetilde{a}_1(\widetilde{a}_3-\widetilde{a}_2)(\widetilde{a}_2-\widetilde{a}_1)}-\frac{e^{-i\widetilde{a}_3}-e^{-i\widetilde{a}_1}}{i\widetilde{a}_1(\widetilde{a}_3-\widetilde{a}_1)(\widetilde{a}_2-\widetilde{a}_1)}\right.\\
&\left.-\pi\delta(\widetilde{a}_1)\left(\frac{e^{-i\widetilde{a}_3}-e^{-i\widetilde{a}_2}}{\widetilde{a}_2(\widetilde{a}_3-\widetilde{a}_2)}-\frac{e^{-i\widetilde{a}_3}-1}{\widetilde{a}_2\widetilde{a}_3}\right)\right).
\end{aligned}$$

[1] For a definition of the Heaviside step function, see, e.g., section 6.3.5.
[2] For a definition of the Dirac delta function, see, e.g., section 6.3.5.

List of Figures

List of Tables

References

Aalen, O. O. & Johansen, S. (1978). An empirical transition matrix for nonhomogeneous markov chains based on censored observations, *Scandinavian Journal of Statistics* **5**: 141–150.

Abrahams, J. (1986). *A Survey of Recent Progress on Level Crossing Problems*, Springer.

Altman, E. (2001). Defaults, recoveries, and returns in the US high yield bond market.

Altman, E. I. (1997). Rating migration of corporate bonds - comparative results and investor/lender implications. Mimeo, Salomon Brothers.

Altman, E. I. (1998). The importance and subtlety of credit rating migration, *Journal of Banking & Finance* **22**: 1231–1247.

Altman, E. I., Brady, B., Resti, A. & Sironi, A. (2003). The link between default and recovery rates: Theory, empirical evidence and implications. Working Paper.

Altman, E. I. & Eberhart, A. C. (1994). Do seniority provisions protect bondholders' investments, *Journal of Portfolio Management* **20**: 67–75.

Altman, E. I. & Kao, D. L. (1992a). The implications of corporate bond ratings drift, *Financial Analysts Journal* pp. 64– 75.

Altman, E. I. & Kao, D. L. (1992b). The implications of corporate bond ratings drift, *Financial Analysts Journal* pp. 64–75.

Altman, E. I. & Kao, D. L. (1992c). Rating drift in high-yield bonds, *The Journal of Fixed Income* pp. 15–20.

Altman, E. I. & Kishore, V. M. (1996). Almost everything you wanted to know about recoveries on defaulted bonds, *Financial Analysts Journal* pp. 57–63.

Altman, E. & Kishore, V. M. (1997). Defaults and returns on high yield bonds: Analysis through 1996. Special Report, New York University Salomon Center, New York.

Amin, K. (1995). Option pricing trees, *The Journal of Derivatives* pp. 34–46.

A New Capital Adequacy Framework (1999). Consultative paper issued by the Basel Committee on Banking Supervision, Basel.

Artzner, P. & Delbaen, F. (1992). Credit risk and prepayment option, *ASTIN Bulletin* **22**: 81–96.

Arvanitis, A., Gregory, J. & Laurent, J.-P. (1999). Building models for credit spreads, *Journal of Derivatives* pp. 27–43.

Asarnow, E. & Edwards, D. (1995). Measuring loss on defaulted bank loans: A 24-year study, *The Journal of Commercial Lending* pp. 11–23.

Babbs, S. H. (1993). Generalised vasicek models of the term structure, *in* J. Janssen & C. H. Skiadas (eds), *Applied Stochastic Models of Data Analysis*, Vol. 1, pp. 49–62. 6th Annual Symposium Proceedings.

Babbs, S. H. & Nowman, K. B. (1999). Kalman filtering of generalized Vasicek term structure models, *Journal of Financial and Quantitative Analysis* **34**(1): 115–130.

Bakshi, G., Madan, D. & Zhang, F. (2001). Understanding the role of recovery in default risk models: Empirical comparisons and implied recovery rates.

Balduzzi, P., Das, S. R., Foresi, S. & Sundaram, R. (1996). A simple approach to three factor affine term structure models, *Journal of Fixed Income* **6**: 43–53.

Baldwin, D. (1999). Business is booming, *RISK: Credit Risk Special Report* p. 8.

Bangia, A., Diebold, F., Kronimus, A., Schagen, C. & Schuermann, T. (26). Ratings migration and the business cycle, with applications to credit portfolio stress testing, *Journal of Banking and Finance* **26**(2/3): 445–474.

Barrett, B., Gosnell, T. & Heuson, A. (1995). Yield curve shifts and the selection of immunization strategies, *The Journal of Fixed Income* pp. 53–64.

Basin, V. (1996). On the credit risk of OTC derivative users. Discussion Paper, Board of Governors of the Federal Reserve System.

Bawa, V. (1975). Optimal rules for ordering uncertain prospects, *Journal of Financial Economics* **2**: 95–121.

BBA Credit Derivatives Report 1999/2000 (2000). British Bankers' Association.

Beaglehole, D. R. & Tenney, M. S. (1991). General solutions of some interest rate - contingent claim pricing equations, *Journal of Fixed Income* pp. 69–83.

Behar, R. & Nagpal, K. (1999). Dynamics of rating transition. Working Paper, Standard & Poor's.

Berhad, R. A. M. (2003). RAM's rating performance and default study 2002 - a statistical review (1992 - 2002).

Best Practices for Credit Risk Disclosure (2000). Basel Committee on Banking Supervision, Basel.

Bicksler, J. & Chen, A. (1986). An economic analysis of interest rate swaps, *Journal of Finance* **41**: 645–656.

Bielecki, T. & Rutkowski, M. (2002). *Credit Risk: Modeling, Valuation and Hedging*, Springer Finance, Springer, Berlin.

Bingham, N. H. & Kiesel, R. (1998). *Risk-Neutral Valuation: Pricing and Hedging of Financial Derivatives*, first edn, Springer.

Bishop, Y., Fienberg, S. & Holland, P. (1975). *Discrete Multivariate Analysis: Theory and Practice*, MIT Press.

Black, F. & Cox, J. (1976). Valuing corporate securities: Some effects of bond indenture provisions, *Journal of Finance* **31**: 351–367.

Bluhm, C., Overbeck, L. & Wagner, C. (2003). *An Introduction to Credit Risk Modeling*, Financial Mathematics Series, Chapman & Hall, Boca Raton.

Bouye, E., Durrelmann, V., Nikeghbali, A., Riboulet, G. & Roncalli, T. (2000). Copulas for finance: A reading guide and some applications. Working Paper, Credit Lyonnais, Paris.

Boyle, P. P. (1988). A lattice framework for option pricing with two state variables, *Journal of Financial and Quantitative Analysis* **23**(1): 1–12.

Brémaud, P. (1981). *Point Processes and Queues*, Springer Series in Statistics, Springer, New York.

Brennan, M. & Schwartz, E. (1980). Analysing convertible bonds, *Journal of Financial and Quantitative Analysis* **15**: 907–929.

Brennan, W., McGirt, D., Roche, J. & Verde, M. (1998). *Bank Loans: Secondary Market and Portfolio Management*, Frank J. Fabozzi Associates, New Hope, PA, chapter Bank Loan Ratings, pp. 57–69.

Brigo, D. & Mercurio, F. (2001). *Interest Rate Models - Theory and Practice*, Springer Finance, Springer, Berlin.

Briys, E., Bellalah, M., Mai, H. & de Varenne, F. (1998). *Options, Futures and Exotic Derivatives*, Wiley Frontiers in Finance, John Wiley and Sons.

Briys, E. & de Varenne, F. (1997). Valuing risky fixed rate debt: An extension, *Journal of Financial and Quantitative Analysis* **32**(2): 239–248.

Bronstein, I. N. & Semendjajew, K. A. (1991). *Taschenbuch der Mathematik*, 25th edn, Verlag Nauka, Moskau.

Brooks, R. & Yan, D. Y. (1999). London inter-bank offer rate (LIBOR) versus treasury rate: Evidence from the parsimonious term structure model, *The Journal of Fixed Income* pp. 71–83.

Brown, R. H. & Schaefer, S. M. (1994). Interest rate volatility and the shape of the term structure, *Philosophical Transactions of the Royal Society of London* **347**: 563–576.

Burton, F. & Inoue, H. (1985). The influence of country factors on the interest differentials on international lendings to sovereign borrowers, *Applied Economics* **17**: 491–507.

Cantor, R. & Packer, F. (1996). Determinants and inputs of sovereign credit ratings, *FRBNY Economic Policy Review* **2**(2): 37–53.

Caouette, D. J., Altman, F. & Narayanan, V. (1998). *Managing Credit Risk: The Next Great Financial Challenge*, John Wiley & Sons, Inc., New York.

Carthy, L. & Fons, J. (1994). Measuring changes in corporate credit quality, *The Journal of Fixed Income* **4**: 27–41.

Carty, L. & Lieberman, D. (1996). Defaulted bank loan recoveries. Moody's Special Report.

Carty, L. V. (1997). Moody's rating migration and credit quality correlation, 1920 - 1996. Special Comment, Moody's Investors Service.

Carty, L. V. (1998). *Moody's Rating Migration and Credit Quality Correlation, 1920-1996*, John Wiley & Sons (Asia) Pte Ltd, Singapore, chapter 10, pp. 349–384.

Carty, L. V. & Fons, J. (1993). Measuring changes in credit quality. Moody's Special Report, Moody's Investors Service.

Carty, L. V. & Lieberman, D. (1998). *Historical Default Rates of Corporate Bond Issuers, 1920 - 1996*, John Wiley & Sons (Asia) Pte Ltd, Singapore, chapter 9, pp. 317–348.

Cathcart, L. & El-Jahel, L. (1998). Valuation of defaultable bonds, *The Journal of Fixed Income* pp. 65–78.

Chen, L. (1996). *Interest Rate Dynamics, Derivatives Pricing, and Risk Management*, number 435 in *Lecture Notes in Economics and Mathematical Systems*, first edn, Springer.

Chen, R.-R. & Yang, T. T. (1996). An integrated model for the term and volatility structures of interest rates.

Chen, R. & Scott, L. (1992). Pricing interest rate options in a two-factor cox-ingersoll-ross model of the term structure, *Review of Financial Studies* **5**: 613–636.

Chen, R. & Scott, L. (1995). Multi-factor Cox-Ingersoll-Ross models of the term structure: Estimates and tests from a Kalman filter model. Unpublished Paper, Rutgers University.

Christensen, J. & Lando, D. (2002). Confidence sets for continuous-time rating transition probabilities.

Cifuentes, A., Efrat, I., Gluck, J. & Murphy, E. (1998). *Buying and Selling Credit Risk: A Perspective on Credit Linked Obligations*, Risk Books.

Cooper, I. & Mello, A. (1991). The default risk of swaps, *Journal of Finance* **46**(2): 597–620.

Cox, D. & Miller, H. (1972). *The Theory of Stochastic Processes*, Chapman and Hall, London and New York.

Cox, J., Ingersoll, J. & Ross, S. (1980). An analysis of variable rate loan contracts, *Journal of Finance* **15**: 389–403.

Cox, J., Ingersoll, J. & Ross, S. (1985). A theory of the term structure of interest rates, *Econometrica* **36**(4): 385–407.

CreditMetrics - Technical Document (1997). J.P. Morgan, New York.

CreditRisk+ A Credit Risk Management Framework (1997). Credit Suisse Financial Products.

Crouhy, M. & Mark, R. (1998). A comparative analysis of current credit risk models. Working Paper, Canadian Imperial Bank of Commerce.

Dai, Q. & Singleton, K. J. (1998). Specification analysis of affine term structure models. Working Paper, Graduate School of Business, Stanford University.

Das, S. R. (1995). Credit risk derivatives, *The Journal of Derivatives* pp. 7–23.

Das, S. R. (1997). Pricing credit derivatives. Working Paper, Harvard Business School & NBER.

Das, S. R. & Tufano, P. (1996). Pricing credit-sensitive debt when interest rates, credit ratings and credit spreads are stochastic, *The Journal of Financial Engineering* **5**(2): 161–198.

Davis, M. & Lo, V. (1999). Infectious defaults. Working Paper, Imperial College, London.

Demchak, B. (2000). Modelling credit migration, *RISK* pp. 99–103.

Dickey, R. & Fuller, W. (1981). Likelihood ratio tests for autoregressive time series with a unit root, *Econometrica* **49**: 1057–1072.

Driessen, J. (2002). Is default event risk priced in corporate bonds ? Working Paper, University of Amsterdam.

Duffee, G. (1999). Estimating the price of default risk, *Review of Financial Studies* **12**: 197–226.

Duffee, G. R. (1996a). Estimating the price of default risk. Working Paper, Federal Reserve Board, Washington.

Duffee, G. R. (1996b). The relation between treasury yields and corporate bond yield spreads. Working Paper, Federal Reserve Board, Washington.

Duffie, D. (1992). *Dynamic Asset Pricing Theory*, Princeton University Press, Princeton.

Duffie, D. (1998a). Credit swap valuation. Working Paper, Graduate School of Business, Stanford University, Forthcoming: Financial Analyst's Journal.

Duffie, D. (1998b). Defaultable term structure models with fractional recovery of par. Working Paper, Graduate School of Business, Stanford University.

Duffie, D. (1998c). First-to-default valuation. Working Paper, Graduate School of Business, Stanford University.

Duffie, D. & Huang, M. (1996). Swap rates and credit quality, *The Journal of Finance* **51**(3): 921–949.

Duffie, D. & Kan, R. (1994). Multi-factor term structure models, *Philosophical Transactions of the Royal Society of London* **347**: 577–586.

Duffie, D. & Kan, R. (1996). A yield-factor model of interest rates, *Mathematical Finance* **6**(4): 379–406.

Duffie, D. & Lando, D. (1997). Term structures of credit spreads with incomplete accounting information. Working Paper, Graduate School of Business, Stanford University.

Duffie, D. & Pan, J. (1997). An overview of value at risk, *Journal of Derivatives* **4**(3): 7–49.

Duffie, D., Pan, J. & Singleton, K. (1998). Transform analysis and option pricing for affine jump-diffusions. Working Paper, Graduate School of Business, Stanford University.

Duffie, D., Schroder, M. & Skiadas, C. (1996). Recursive valuation of defaultable securities and the timing of resolution of uncertainty, *The Annals of Applied Probability* **6**(4): 1075–1090.

Duffie, D. & Singleton, K. (1998a). Simulating correlated defaults. Working Paper, Stanford University.

Duffie, D. & Singleton, K. J. (1997). Modeling term structures of defaultable bonds. Working Paper, Graduate School of Business, Stanford University.

Duffie, D. & Singleton, K. J. (1998b). Credit risk for financial institutions: Management and pricing.

Duffie, D. & Singleton, K. J. (2003). *Credit Risk*, Princeton Series in Finance, Princeton University Press.

Duffy, D. & Santner, T. (1989). *The Statistical Analysis of Discrete Data*, Springer.

Düllmann, K. & Windfuhr, M. (2000). Credit spreads between German and Italian Sovereign bonds - do affine models work ?

Eberhart, A., Altman, E. & Aggarwal, R. (1998). The equity performance of firms emerging from bankruptcy, *Journal of Finance* **54**(5): 1855–1868.

Eberhart, A. & Sweeney, R. (1992). Does the bond market predict bankruptcy settlements, *Journal of Finance* **47**(3): 943–980.

Eberhart, A. & Weiss, L. (1998). The importance of deviations from the absolute priority rule in chapter 11 bankruptcy proceedings, *Financial Management* **27**(4): 106–110.

Edwards, S. (1984). LDC foreign borrowing and default risk: An empirical default risk, 1976-1980, *American Economic Review* **74**: 726–734.

Edwards, S. (1986). The pricing of bonds and bank loans in international markets, *European Economic Review* **30**: 565–589.

Eichengreen, B. & Mody, A. (1998). What explains changing spreads on emerging market debt? fundamentals or market sentiment? NBER Working Paper, No. 6408.

Elsas, R., Ewert, R., Krahnen, J., Rudolph, B. & Weber, M. (1999). Risikoorientiertes Kreditmanagement Deutscher Banken, *Die Bank* **3**: 190–199.

Elton, E. J. & Gruber, M. J. (1991). *Modern Portfolio Theory and Investment Analysis*, John Wiley & Sons.

Embrechts, P., Lindskog, F. & McNeil, A. (2001). Modeling dependence with copulas and applications to risk management. Working Paper, ETH Zuerich.

Estimating the Term Structure of Interest Rates (1997). *Monthly report*, Bundesbank.

Evans, G., Blackledge, J. & Yardley, P. (2000). *Analytic Methods for Partial Differential Equations*, Springer Undergraduate Mathematics Series, Springer, London.

Fabozzi, F. & Goodman, L. (2001). *Investing in Collateralized Debt Obligations*, Frank J. Fabozzi Associates.

Fons, J. S. (1994). Using default rates to model the term structure of credit risk, *Financial Analysts Journal* **50**(5): 25–32.

Franks, J. & Torous, W. (1994). A comparison of financial recontracting in distressed exchanges and chapter 11 reorganizations, *Journal of Financial Economics* **35**: 349–370.

Frees, E. & Valdez, E. A. (1997). Understanding relationships using copulas. Working Paper.

Frey, R., McNeil, A. & Nyfeler, M. (2001). Copulas and credit models, *RISK* **14**(10): 111–114.

Frye, J. (2000a). Collateral damage, *RISK* pp. 91–94.

Frye, J. (2000b). Collateral damage detected. Working Paper, Federal Reserve Bank of Chicago, Emerging Issues Series.

Garbade, K. (2001). *Pricing Corporate Securities as Contingent Claims*, MIT Press, Cambridge, MA.

Gastineau, G. (1996). *Dicitionary of Financial Risk Management*, Frank J. Fabozzi Associates, New York.

Geske, R. (1977). The valuation of corporate liabilities as compound options, *Journal of Financial and Quantitative Analysis* pp. 541–552.

Geweke, J., Marshall, R. C. & Zarkin, G. A. (1986). Mobility indices in continuous time markov chains, *Econometrica* **54**(6): 1407–1423.

Geyer, A. L. J. & Pichler, S. (1996). A state-space approach to estimate and test multi-factor Cox-Ingersoll-Ross models of the term structure.

Gihman, I. & Skorohod, A. (1980). *Introduction à la Théorie des Processus Aléatoires*, Mir.

Gluck, J. & Remeza, H. (2000). Moody's approach to rating multisector CDOs. Moody's Investor Service.

Gordy, M. B. (1998). A comparative anatomy of credit risk models. Working Paper, Board of Governors of the Federal Reserve System, Washington.

Green, J., Locke, J. & Paul-Choudhury, S. (1998). Strength through adversity, *CreditRisk - A Risk Special Report* pp. 6–9.

Greene, W. H. (2000). *Econometric Analysis*, fourth edn, Prentice Hall International, Inc., New Jersey.

Grossman, R., O'Shea, S. & Bonelli, S. (2001). Bank loan and bond recovery study: 1997-2000. Fitch Loan Products Special Report.

Gupton, G., Gates, D. & Carty, L. (Moody's Special Comment). Bank loan loss given default.

Gupton, G. M. & Stein, R. M. (2002). LossCalc: Moody's model for predicting LGD. Moody's Investors Service, Global Credit Research, Special Comment.

Hamilton, D. T., Cantor, R. & Ou, S. (2002). Default and recovery rates of corporate bond issuers. Moody's Investors Service, Global Credit Research, Special Comment.

Hamilton, D. T., Cantor, R., West, M. & Fowlie, K. (2002). Default and recovery rates of european corporate bond issuers 1985-2001. Moody's Investors Service, Global Credit Research, Special Comment.

Haque, N., Kumar, M. & Mathieson, D. (1996). The economic contents of indicators of developing country creditworthiness, *IMF Staff Papers* **43**(4): 688–724.

Hargreaves, T. (2000). Default swaps drive growth, *CreditRisk - A Risk Special Report* pp. 2–3.

Harlow, W. (1991). Asset allocation in a downside-risk framework, *Financial Analysts Journal* **47**(5): 28–40.

Harlow, W. & Rao, K. (1989). Asset pricing in a generalized mean-lower partial moment framework: Theory and evidence, *Journal of Financial and Quantitative Analysis* **24**: 285–311.

Harrison, J. M. (1990). *Brownian Motion and Stochastic Flow Systems*, Krieger Publishing Company, Florida.

Harrison, J. & Pliska, S. (1981). Martingales and stochastic integrals in the theory of continuous trading, *Stochastic Processes and Their Applications* **11**: 215–260.

Harvey, A. C. (1981). *The Econometric Analysis of Time Series*, Philip Allan Publishers Limited.

Harvey, A. C. (1989). *Forecasting, Structural Time Series Models and the Kalman Filter*, Cambridge University Press, Cambridge.

Heath, D., Jarrow, R. & Morton, A. (1992). Bond pricing and the term structure of interest rates: A new methodology for contingent claims valuation, *Econometrica* **60**: 77–105.

Helwege, J. (1999). How long do junk bonds spend in default, *Journal of Finance* **54**(1): 341–357.

Helwege, J. & Kleiman, P. (1996). Understanding aggregate default rates of high yield bonds, *Federal Reserve Bank of New York Current Issues in Economics and Finance* **2**(6): 1–6.

Historical Default Rates of Corporate Bond Issuers, 1920-1999 (2000a). Special Comment, Moody's Investors Services, Global Credit Research.

Historical Default Rates of Corporate Bond Issuers, 1920-1999 (2000b). Special Comment, Moody's Investors Service, Global Credit Research.

Ho, T. & Lee, S.-B. (1986). Term structure movements and the pricing of interest rate contingent claims, *The Journal of Finance* **41**: 1011–1029.

Ho, T. & Singer, R. (1982). Bond indenture provisions and the risk of corporate debt, *Journal of Financial Economics* (10): 375–406.

Hu, Y.-T., Kiesel, R. & Perraudin, W. (2001). The estimation of transition matrices for sovereign credit ratings.

Hu, Y.-T. & Perraudin, W. (2002). The dependence of recovery rates and defaults. Working Paper, Birkbeck College.

Hull, J. (1997). *Options, Futures, and Other Derivatives*, Prentice-Hall International, Inc., London.

Hull, J. C. & White, A. (2000). Valuing credit default swaps I: No counterparty default risk, *The Journal of Derivatives* pp. 29–40.

Hull, J. C. & White, A. D. (1993). One-factor interest-rate models and the valuation of interest-rate derivative securities, *Journal of Financial and Quantitative Analysis* **28**: 235–254.

Hull, J. & White, A. (1990). Pricing interest-rate-derivative securities, *The Review of Financial Studies* **3**(4): 573–592.

Hull, J. & White, A. (1994). Numerical procedures for implementing term structure models II: Two-factor models, *The Journal of Derivatives* pp. 37–48.

Hull, J. & White, A. (2001). Valuing credit default swaps II: Modeling default correlations, *Journal of Derivatives* **8**(3): 12–22.

Ikeda, N. & Watanabe, S. (1989a). *Stochastic Differential Equations and Diffusion Processes*, 1st edn, North-Holland.

Ikeda, N. & Watanabe, S. (1989b). *Stochastic Differential Equations and Diffusion Processes*, 2nd edn, North-Holland.

Ingersoll, J. E. (1987). *Theory of Financial Decision Making*, Studies in Financial Economics, Rowman & Littlefield Publishers.

Israel, R. B., Rosenthal, J. S. & Wei, J. Z. (2001). Finding generators for markov chains via empirical transition matrices, with applications to credit ratings, *Mathematical Finance* **11**(2): 245–265.

Jacod, J. & Shiryaev, A. (1987). *Limit Theorems for Stochastic Processes*, A Series of Comprehensive Studies in Mathematics, Springer, Berlin.

Jafry, Y. & Schuermann, T. (2003). Metrics for comparing credit migration matrices.

James, J. (1999). How much should they cost, *CreditRisk - A Risk Special Report* pp. 8–10.

James, J. & Webber, N. (2000). *Interest Rate Modelling*, Wiley Series in Financial Engineering, John Wiley & Sons, LTD.

Jamshidian, F. (1989). An exact bond option formula, *Journal of Finance* **44**: 205–209.

Jamshidian, F. (1995). A simple class of square root interest rate models, *Applied Mathematical Finance* **2**: 61–72.

Jamshidian, F. (1996). Bond, futures and option valuation in the quadratic interest rate model, *Applied Mathematical Finance* **3**: 93–115.

Jarrow, R. (2001). Default parameter estimation using market prices, *Financial Analysts Journal* **57**(5): 75–92.

Jarrow, R. A., Lando, D. & Turnbull, S. M. (1997). A markov model for the term structure of credit risk spreads, *The Review of Financial Studies* **10**(2): 481–523.

Jarrow, R. & Turnbull, S. (1995). Pricing options on financial securities subject to default risk, *Journal of Finance* **50**: 53–86.

Jarrow, R. & Turnbull, S. (1998). The intersection of market and credit risk. Working Paper.

Jarrow, R. & Yu, F. (2000). Counterparty risk and the pricing of defaultable securities. Working Paper, Johnson GSM, Cornell University.

Jokivuolle, E. & Peura, S. (2000). A model for estimating recovery rates and collateral haircuts for bank loans. Bank of Finland Discussion Papers.

Jones, E., Mason, S. & Rosenfeld, E. (1984). Contingent claim analysis of corporate capital structures: An empirical investigation, *Journal of Finance* **39**: 611–625.

Jones, F. (1991). Yield curve strategies, *Journal of Fixed Income* pp. 43–51.

Jonsson, J. & Fridson, M. (1996). Forecasting default rates on high-yield bonds, *The Journal of Fixed Income* pp. 69–77.

Jordan, J. & Mansi, S. (2000). Estimation of the term structure from on-the-run treasuries. Working Paper, National Economic Research Associates, and George Washington University.

Juttner, J. & McCarthy, J. (1998). Modelling a rating crisis.

Kalbfleisch, J. & Prentice, R. (1980). *The Statistical Analysis of Failure Time Data*, John Wiley and Sons, New York.

Kamin, S. & Kleist, K. (1999). The evolution and determinants of emerging market credit spreads in the 1990s. BIS Working Paper.

Karatzas, I. (1988). On the pricing of american options, *Applied Mathematics and Optimization* **17**: 37–60.

Karatzas, I. & Shreve, S. (1988). *Brownian Motion and Stochastic Calculus*, Springer.

Kavvathas, D. (2000). Estimating credit rating transition probabilities for corporate bonds. Working Paper, Department of Economics, University of Chicago.

Kealhofer, S. (1998). Portfolio management of default risk, *Net Exposure* **1**(2).

Keenan, S. & Sobehart, J. (2000). A credit risk catwalk, *RISK* pp. 84–88.

Kiesel, R. & Schmid, B. (2000). Aspekte der stochastischen Modellierung von Ausfallwahrscheinlichkeiten in Kreditportfoliomodellen, *in* A. Oehler (ed.), *Kreditrisikomanagement - Portfoliomodelle und Derivate*, Schäffer Poeschel, Stuttgart.

Kijima, M. (2000). Valuation of a credit swap of the basket type, *Review of Derivatives Research* **4**: 81–97.

Kim, J., Ramaswamy, K. & Sundaresan, S. (1992). The valuation of corporate fixed income securities.

Kim, J., Ramaswamy, K. & Sundaresan, S. (Autumn 1993). Does default risk in coupons affect the valuation of corporate bonds ? A contingent claims model, *Financial Management* pp. 117–131.

Klein, J. & Moeschberger, M. (1997). *Survival Analysis: Techniques for Censored and Truncated Data*, Springer, New York.

Korn, R. & Korn, E. (1999). *Optionsbewertung und Portfolio-Optimierung*, first edn, Vieweg, Braunschweig/Wiesbaden.

Koyluoglu, H. U. & Hickman, A. (1998). A generalized framework for credit risk portfolio models. Working Paper, Oliver Wyman and Company and CSFP Capital, Inc., New York.

Lamberton, D. & Lapeyre, B. (1996). *Introduction to Stochastic Calculus Applied to Finance*, Chapman & Hall.

Lancaster, T. (1990). *The Econometric Analysis of Econometric Data*, Cambridge University Press, UK.

Lando, D. (1994). *Three Essays on Contingent Claims Pricing*, PhD thesis, Cornell University.

Lando, D. (1995). On jump-diffusion option pricing from the viewpoint of semimartingale characteristics, *Surveys in Applied and Industrial Mathematics* **2**(4): 605–625.

Lando, D. (1996). Modelling bonds and derivatives with default risk. Working Paper, Department of Operations Research, University of Copenhagen.

Lando, D. (1998). On Cox processes and credit risky securities. Working Paper, Department of Operations Research, University of Copenhagen.

Lando, D. (1999). Some elements of rating-based credit risk modelling.

Lando, D. & Skodeberg, T. (2002). Analyzing rating transitions and rating drift with continuous observations, *Journal of Banking and Finance* **26**(2/3): 423–444.

Larsen, R. J. & Marx, M. L. (2001). *An Introduction to Mathematical Statistics and Its Applications*, third edn, Prentice Hall, New Jersey.

Leland, H. E. (1994). Corporate debt value, bond covenants, and optimal capital structure, *The Journal of Finance* **49**(4): 1213–1252.

Leland, H. E. & Toft, K. (1996). Optimal capital structure, endogenous bankruptcy, and the term structure of credit spreads., *Journal of Finance* **51**: 987–1019.

Li, D. (2000). On default correlation: A copula function approach, *Journal of Fixed Income* pp. 115–118.

Litterman, R. & Scheinkman, J. (1991). Common factors affecting bond returns, *The Journal of Fixed Income* pp. 54–61.

Longstaff, F. A. & Schwartz, E. S. (1992). Interest rate volatility and the term structure: A two-factor general equilibrium model, *Journal of Finance* **47**: 1259–1282.

Longstaff, F. A. & Schwartz, E. S. (1995a). Valuing credit derivatives, *The Journal of Fixed Income* pp. 6–12.

Longstaff, F. & Schwartz, E. (1995b). A simple approach to valuing risky fixed and floating rate debt, *The Journal of Finance* **50**(3): 789–819.

Lopez, J. A. & Saidenberg, M. R. (1998). Evaluating credit risk models. Working Paper, Federal Reserve Bank of San Francisco and Federal Reserve Bank of New York.

Lucas, D. J. & Lonski, J. G. (1992). Changes in corporate credit quality 1970 - 1990, *Journal of Fixed Income* pp. 7–14.

Madan, D. B. & Unal, H. (1994). Pricing the risks of default. The Wharton School, University of Pennsylvania.

Madan, D. & Unal, H. (1998). Pricing the risks of default, *Review of Derivatives Research* **2**(2/3): 121–160.

Mansi, J. V. J. A. S. A. (2000). How well do constant-maturity treasuries approximate the on-the-run term structure, *The Journal of Fixed Income* **10**(2): 35–62.

Markowitz, H. (1952). Portfolio selection, *Journal of Finance* **7**: 77–91.

Markowitz, H. (1991). *Portfolio Selection*, Blackwell.

Martin, J., Cox, S. & MacMinn, R. (1988). *The Theory of Finance, Evidence and Applications*, Dryden.

Mason, S. & Bhattacharya, S. (1981). Risky debt, jump processes, and safety covenants, *Journal of Financial Economics* **9**: 281–307.

Masters, B. (1998). Credit derivatives and the management of credit risk, *Net Exposure* **1**(2).

McLeish, N. (2000). European credit markets: The case for growth, *CreditRisk: A Risk Special Report* pp. 4–6.

Merton, R. (1974). On the pricing of corporate debt: The risk structure of interest rates, *Journal of Finance* **29**: 449–470.

Min, H. (1998). Determinants of emerging market bond spreads: Do economic fundamentals matter ? World Bank Working Paper.

Monfort, B. & Mulder, C. (2000). Using credit ratings for capital requirement on lending to emerging market economies: Possible impact of basel accord.

Musiela, M. & Rutkowski, M. (1997). *Martingale Methods in Financial Modelling*, Springer.

Nelder, J. A. & Mead, R. (1965). A simplex method for function minimization, *The Computer Journal* **7**: 308–313.

Nelsen, R. (1999). *An Introduction to Copulas*, Springer, New York.

Nelson, C. & Siegel, A. (1987). Parsimonious modeling of yield curves, *Journal of Business* **60**: 473–489.

Nickell, P., Perraudin, W. & Varotto, S. (1998). Stability of rating transitions. Working Paper, Bank of England.

Nickell, P., Perraudin, W. & Varotto, S. (2000). Stability of rating transitions, *Journal of Banking and Finance* **24**: 203–227.

Nielsen, L., Saà-Requejo, J. & Santa-Clara, P. (1993). Default risk and interest rate risk: The term structure of default spreads. Working Paper, INSEAD.

Nielsen, L. T. (1999). *Pricing and Hedging of Derivative Securities*, Oxford University Press.

Nunes, J. (1998). Interest rate derivatives in a duffie and kan model with stochastic volatility: Application of green's functions.

O'Kane, D. (2000). Introduction to asset swaps, *Technical report*, Lehman Brothers.

O'Kane, D. (2001). Credit derivatives explained. Lehman Brothers, Credit Derivatives Explained, Structured Credit Research.

Oksendal, B. (1998). *Stochastic Differential Equations - An Introduction with Applications*, Universitext, 5th edn, Springer.

Ong, M. K. (1999). *Internal Credit Risk Models: Capital Allocation and Performance Measurement*, Risk Books, London.

Overview of the New Basel Capital Accord (2001). Consultative Document, Basel Committee on Banking Supervision, Basel.

Overview of The New Basel Capital Accord (2003). Consultative Document, Basel Commitee on Banking Supervision, Basel.

Partnoy, F. (2002). The paradox of credit ratings. Law and Economics Research Working Paper No. 20.

Patel, N. (2001). Credit derivatives: Vanilla volumes challenged, *RISK* pp. 32–34.

Pedrosa, M. & Roll, R. (1998). Systematic risk in corporate bond credit spreads, *The Journal of Fixed Income* **8**(3): 7–26.

Pelsser, A. (1996). Efficient methods for valuing and managing interest rate and other derivative securities. Pdh Thesis, Erasmus University, Rotterdam.

Perraudin, W. (2001). *Credit Explorer - Documentation for Analysts*, 1 edn, Risk Control Limited, London.

Protter, P. (1992). *Stochastic Integration and Differential Equations*, Springer, Berlin.

Ratings Performance 1996: Stability & Transition (1997). New York.

Ratings Performance 1997: Stability & Transition (1998). New York.

Ratings Performance 1998: Stability & Transition (1999). New York.

Ratings Performance 1999: Stability & Transition (2000). New York.

Ratings Performance 2000: Default, Transition, Recovery, and Spreads (2001). New York.

Rendleman, R. J. (1992). How risks are shared in interest rate swaps, *Journal of Financial Services Research* pp. 5–34.

Rhee, J. (1999). *Interest Rate Models*, PhD thesis, University of Warwick.

Richard, S. (1978). An arbitrage model of the term structure of interest rates, *Journal of Financial Economics* **6**: 33–57.

Ross, S. M. (1996). *Stochastic Processes*, Wiley Series in Probability and Mathematical Statistics, second edn, John Wiley & Sons, Inc., New York.

Saá-Requejo, J. & Santa-Clara, P. (1997). Bond pricing with default risk.

Sandmann, K. & Sondermann, D. (1997). A note on the stability of lognormal interest rate models and pricing of eurodollar futures, *Mathematical Finance* **7**(2): 119–126.

Sarig, O. & Warga, A. (1989). Some empirical estimates of the term structure of interest rates, *Journal of Finance* **44**(5): 1351–1360.

Schmid, B. (1997). CreditMetrics - setting a benchmark, *Solutions* **1**(3-4): 35–53.

Schmid, B. (1998a). Credit Risk - Verschiedene Methoden zur Berechnung von Kreditrisiken. RiskLab Research Paper No. 9807.

Schmid, B. (1998b). Quantifizierung von Ausfallrisiken - Alternativen zu CreditMetrics, *Solutions* **2**(1).

Schmid, B. (2000). Das RiskLab Kreditrisiko Modell, *Solutions* **4**(3/4): 39–54.

Schmid, B. (2001). The pricing of defaultable fixed and floating rate debt, *The International Journal of Finance* **13**(2): 1871–1894.

Schmid, B. (2002). *Pricing Credit Linked Instruments - Theory and Empirical Evidence*, Vol. 516 of *Lecture Notes in Economics and Mathematical Systems*, Springer, Berlin.

Schmid, B. & Kalemanova, A. (2002a). Applying a three-factor defaultable term strucutre model to the pricing of credit default options, *International Review of Financial Analysis* **11**(2): 139–158.

Schmid, B. & Kalemanova, A. (2002b). Credit spreads between german and italian zero coupon government bonds, *Research in Interational Business and Finance* **16**: 497–533.

Schmid, B. & Zagst, R. (2000). A three-factor defaultable term structure model, *The Journal of Fixed Income* **10**(2): 63–79.

Schönbucher, P. J. (1996). The term structure of defaultable bond prices. Discussion Paper B-384, University of Bonn, Department of Statistics.

Schönbucher, P. J. (2000). *Credit Risk Modelling and Credit Derivatives*, PhD thesis, Rheinische Friedrich-Wilhelms-Universität Bonn.

Schönbucher, P. J. & Schubert, D. (2001). Copula-dependent default risk intensity models. Working Paper, Bonn University.

Schuermann, T. (2003). What do we know about loss-given-default. Working Paper, Federal Reserve Bank of New York.

Scott, L. (1995). The valuation of interest rate derivatives in a multifactor term structure model with deterministic components. Working Paper, University of Georgia.

Sharpe, W. (1964). Capital asset prices: A theory of market equilibrium under conditions of risk, *Journal of Finance* **29**: 425–442.

Shimko, D., Tejima, N. & Deventer, D. V. (1993). The pricing of risky debt when interest rates are stochastic, *The Journal of Fixed Income* pp. 58–65.

Shorrocks, A. F. (1978). The measurement of mobility, *Econometrica* **46**: 1013–1024.

Sklar, A. (1959). Fonction de repartition à n dimension et leur marges, *Publications de l'Institute Statistique l'Université de Paris* **8**: 229–231.

Skodeberg, T. (1998). Statistical analysis of rating transitions - a survival analytic approach. Master's thesis, University of Copenhagen.

Sorensen, E. & Bollier, T. (1994). Pricing swap default risk, *Financial Analysts Journal* **50**(3): 23–33.

S&P CDO Surveillance (2002). S&P Structured Finance Seminar, Florida.

Special Report: Ratings Performance 2001 (2002).

Special Report: Ratings Performance 2002 (2003).

Steeley, J. M. (1991). Estimating the gilt-edged term structure: Basis splines and confidence intervals, *Journal of Business Finance and Accounting* **18**: 513–530.

Structured Products/ ABS Market Monthly (1999). J. P. Morgan Securities Inc., SP/ABS Credit Research, New York.

Sundaresan, S. (1991). *Valuation of Swaps*, Elsevier (North-Holland), chapter 12.

Szatzschneider, W. (2000). CIR model in financial markets. Working Paper, Anahuac University.

Taurén, M. (1999). A comparison of bond pricing models in the pricing of credit risk. Working Paper, Indiana University, Bloomington.

The Internal Ratings-Based Approach (2001). Consultative Document, Supporting Document to the New Basel Capital Accord, Basel Committee on Banking Supervision, Basel.

The New Basel Capital Accord (2001). Consultative Document, Basel Committee on Banking Supervision, Basel.

The Standardised Approach to Credit Risk (2001). Consultative Document, Supporting Document to the New Basel Capital Accord, Basel Committee on Banking Supervision, Basel.

Thornburn, K. (2000). Bankruptcy auctions: Costs, debt recovery and firm survival, *Journal of Financial Economics* **58**: 337–368.

Titman, S. & Torous, W. (1989). Valuing commercial mortgages: An empirical investigation of the contingent-claim approach to pricing risky debt, *Journal of Finance* **44**: 345–373.

Treacy, W. & Carey, M. (2000). Credit risk rating systems at large u.s. banks, *Journal of Banking and Finance* **24**: 167–201.

Update on Work on a New Capital Adequacy Framework (1999). Basel Committee on Banking Supervision, Basel.

Van de Castle, K., Keisman, D. & Yang, R. (2000). Suddenly structure mattered: Insights into recoveries of defaulted debt.

Vasicek, O. (1977). An equilibrium characterization of the term structure, *Journal of Financial Economics* **7**: 117–161.

Vasicek, O. A. (1997). Credit valuation, *Net Exposure* **1**: 1–12.

Wagner, H. (1996). The pricing of bonds in bankruptcy and financial restructuring, *The Journal of Fixed Income* pp. 40–47.

Wang, D. (1999). Pricing defaultable debt: Some exact results, *International Journal of Theoretical and Applied Finance* **2**: 95–99.

Wang, S. S. (2000). Aggregation of correlated risk portfolios: Models and algorithms. Working Paper, CAS.

Wei, D. G. & Guo, D. (1997). Pricing risky debt: An empirical comparison of the Longstaff and Schwartz and Merton models, *The Journal of Fixed Income* pp. 8–28.

Willner, R. (1996). A new tool for portfolio managers: Level, slope, and curvature durations, *The Journal of Fixed Income* pp. 48–59.

Wilmott, P., Dewynne, J. & Howison, S. (1993). *Option Pricing - Mathematical Models and Computation*, first edn, Oxford Financial Press.

Wilson, T. (1997a). Measuring and managing credit portfolio risk: Part I: Modelling systematic default risk, *The Journal of Lending and Credit Risk Management* pp. 61–72.

Wilson, T. (1997b). Measuring and managing credit portfolio risk: Part II: Portfolio loss distributions, *The Journal of Lending and Credit Risk Management* pp. 67–78.

Wilson, T. (1997c). Portfolio credit risk (I), *RISK* **10**(9).

Wilson, T. (1997d). Portfolio credit risk (II), *RISK* **10**(10).

Wilson, T. (1998). Trends in credit risk management, *The Journal of Financial Engineering* **7**(3/4): 217–240.

Young, G. & Bhagat, C. (2000). Credit risk's softer side, *Asia Risk* pp. 30–32.

Yu, F. (2003). Default correlation in reduced-form models. Working Paper, University of California, Irvine.

Zagst, R. (2001). *Interest Rate Management*, Springer.

Zagst, R., Kehrbaum, J. & Schmid, B. (2003). Portfolio optimization under credit risk, *Computational Statistics* **18**(3): 317–338. RiskLab Research Paper No. 0102.

Zhou, C. (1997). A jump-diffusion approach to modeling credit risk and valuing defaultable securities. Working Paper, Federal Reserve Board, Washington.

Zhou, C. (2001). An analysis of default correlations and multiple defaults, *Review of Financial Studies* **14**: 555–576.

Index